Food Bioactive Ingredients

Series Editor

Seid Mahdi Jafari, Department of Food Materials and Process Design Engineering
Gorgan University of Agricultural Sciences and Natural Resources
Gorgan, Iran

The Food Bioactive Ingredients Series covers recent advances and research on the science, properties, functions, technology, engineering and applications of food bioactive ingredients and their relevant products. The series also covers health-related aspects of these bioactive components, which have been shown to play a critical role in preventing or delaying different diseases and to have many health-improving properties. The books in this series target professional scientists, academics, researchers, students, industry professionals, governmental organizations, producing industries and all experts performing research on functional food development, pharmaceuticals, cosmetics and agricultural crops.

Seid Mahdi Jafari • Saurabh Kumar Jha
Vineet Kumar Goswami
Editors

Nutraceuticals: A New Millennium Approach for Preventive Medicine

Springer

Editors
Seid Mahdi Jafari
Department of Food Materials and Process Design Engineering
Gorgan University of Agricultural Sciences and Natural Resources
Gorgan, Iran

Saurabh Kumar Jha
Zoology
Kalindi College, University of Delhi
New Delhi, Delhi, India

Vineet Kumar Goswami
Biotechnology
Delhi Technological University
New Delhi, Delhi, India

ISSN 2661-8958 ISSN 2661-8966 (electronic)
Food Bioactive Ingredients
ISBN 978-3-031-76483-7 ISBN 978-3-031-76481-3 (eBook)
https://doi.org/10.1007/978-3-031-76481-3

This Springer imprint is published by the registered company Springer Nature Switzerland AG
The registered company address is: Gewerbestrasse 11, 6330 Cham, Switzerland

Preface and Acknowledgments

The field of nutraceuticals and bioactive compounds is expanding quickly, as are related subfields such as functional foods and dietary supplements. These compounds, which provide potential health advantages beyond basic nutrition, lie in between food and drugs. Vitamins, minerals, herbal extracts, probiotics, omega-3 fatty acids, and other bioactive components are all included in the broader category of nutraceuticals. Capsules, pills, powders, liquids, and functional meals are just some of the forms they come in that may be taken orally. The main goal of nutraceuticals is to promote general health and well-being by delivering beneficial nutrients or bioactive substances. The chance of developing chronic illnesses like diabetes, obesity, and cardiovascular problems may be decreased, while the immune system can be strengthened, mental performance can be improved, heart health can be promoted, joint health can be supported, energy levels can be increased, and so on.

The present book entitled "Nutraceuticals: A New Millennium Approach for Preventive Medicine" is focusing on very important aspects of health science and disease management, through nutraceuticals. In fact, the preventive use of nutraceuticals is a promising radical new trend for a variety of diseases and disorders. In the first part of the book, an overview of nutraceuticals is presented (Chap. 1) as well as foundation of nutraceuticals in preventive medicine (Chap. 2). Chapter 3 addresses the nutrients in metabolism and Xenobiotics. Chapter 4 dedicates to role of nutraceuticals in the prevention of cancer. Chapters 5 and 6 explain the importance of nutraceuticals in metabolic syndrome and diabetes, and prevention of chronic renal disease, respectively. The influence of nutraceuticals on the treatment and/or prevention of some other diseases/disorders has been discussed in Chaps. 7, 8, 9, and 10, including chronic pancreatitis (Chap. 7), chronic obstructive pulmonary disease (Chap. 8), inflammatory autoimmune diseases (Chap. 9), and infertility (Chap. 10). The final part of the book is dedicated to the relationship between nutraceuticals and suboptimal health (Chap. 11), and effect of excess intake of nutrients on human health (Chap. 12).

We are conscious of other diseases/disorders and their treatment/prevention with nutraceuticals which could have been addressed. We look forward that this edition will be well received and hopefully in the close future, other health issues can be addressed in this emerging area. We really appreciate the great cooperation of all authors of the chapters for taking time from their busy schedules to contribute to this project. Also, it is necessary to express our sincere thanks to all the editorial staff at Springer for their help and support throughout the project. Finally, special acknowledgment is to our family for their understanding and encouragement during the editing of this great project.

Gorgan, Iran — Seid Mahdi Jafari
New Delhi, Delhi, India — Saurabh Kumar Jha
New Delhi, Delhi, India — Vineet Kumar Gowasmi

Contents

Chapter 1
A Brief Overview of Nutraceuticals

Ritika Parashar, Abhimanyu Kumar Jha, Saurabh Kumar Jha, and Seid Mahdi Jafari

1 Introduction

Medicines derived from plant and animal extracts have been utilized by humans for a wide variety of reasons going all the way back to antiquity. Dr. Stephen first used the word "Nutraceuticals"in 1989 to describe a product that combines nutritional supplements with pharmaceuticals. Since the advent of modern pet care in the 1990s, nutraceuticals have seen significant development, mirroring developments in the human health industry at the same time. The definition of nutraceuticals was broadened by the passing of the Dietary Supplement Health and Education Act of 1994 to include vitamins, minerals, herbs and other botanicals, amino acids, and dietary substances for human use as dietary supplements. When compared to the more well-known dietary supplements, the word "nutraceutical" is not widely recognised on worldwide regulatory systems. Today, consumers may choose from more than 470 nutraceutical and functional food items that have been shown to improve health.

Traditional treatments work because someone figured out how to combine foods in such a way that their effects complement one another. Medical professionals now

R. Parashar · A. K. Jha
Department of Biotechnology, Sharda University, Greater Noida, Uttar Pradesh, India

S. K. Jha
Department of Zoology, Kalindi College, University of Delhi, New Delhi, India

S. M. Jafari (✉)
Department of Food Materials and Process Design Engineering, Gorgan University of Agricultural Sciences and Natural Resources, Gorgan, Iran

Halal Research Center of IRI, Iran Food and Drug Administration, Ministry of Health and Medical Education, Tehran, Iran
e-mail: smjafari@gau.ac.ir

S. M. Jafari et al. (eds.), *Nutraceuticals: A New Millennium Approach for Preventive Medicine*, Food Bioactive Ingredients,
https://doi.org/10.1007/978-3-031-76481-3_1

have a mythical status as saviours who can perform miracles and heal any ailment. Seventy percent of patients, according to a recent survey, see a doctor before or during conventional treatment, suggesting that these individuals are not fond of alternative medicine. Patients, on the other hand, have a good idea of the potential dangers posed by chemical agents in both short and long-term treatment. As a result, there has been a rise in the popularity of preventative medicine, with the subsequent push for further investigation into alternative treatments, particularly those that make use of nutritional techniques. This article summarizes key information on the therapeutic use of nutraceuticals as both modern and traditional medicines (Padmavathi 2018).

The field of nutraceuticals is expanding quickly, as are related subfields such as functional foods and dietary supplements. These items, which provide potential health advantages beyond basic nutrition, lie in between food and drugs. Vitamins, minerals, herbal extracts, probiotics, omega-3 fatty acids, and other bioactive components are all included in the broader category of nutraceuticals. Capsules, pills, powders, liquids, and functional meals are just some of the forms they come in that may be taken orally. The main goal of nutraceuticals is to promote general health and well-being by delivering beneficial nutrients or bioactive substances. The chance of developing chronic illnesses like diabetes, obesity, and cardiovascular problems may be decreased, while the immune system can be strengthened, mental performance can be improved, heart health can be promoted, joint health can be supported, energy levels can be increased, and so on.

Nutraceuticals have shown promise for a variety of uses, including disease prevention. People who want more natural and holistic approaches to their health problems often turn to these methods as a supplement to conventional care. Nutraceuticals are thought to be effective because they modulate biological processes, increase antioxidant capacity, and reduce inflammation. They are also useful for treating vitamin deficits brought on by improper eating habits or medical problems. Rising consumer awareness and interest in preventative healthcare have been major factors in the development of the global nutraceuticals market in recent years. More and more people are taking responsibility for their health and looking for methods to make positive changes, such as adopting healthier eating habits and using nutritional supplements, to achieve better overall health. More effective and specific nutraceutical products have been developed thanks to developments in scientific research and manufacturing technology (Pandey et al. 2010).

Different nations have various standards for the safety, effectiveness, labelling, and marketing claims that may be made for nutraceuticals. The manufacturing, labelling, and marketing of nutraceuticals are subject to strict rules in several parts of the world, including the United States and the European Union. These rules exist to protect consumers from harm and to stop false or misleading statements from being made about the items' purported health advantages. While it's true that nutraceuticals show great promise, each person's experience with them may be unique due to variables including genetics, lifestyle, and pre-existing illnesses. Also, people should be careful and talk to their doctors before adding nutraceuticals to their daily routine, particularly if they are already on medicine or have pre-existing health issues. For the same reasons, it's important to only buy nutraceuticals from dependable brands (Verma et al. 2010).

2 History

The idea of employing food as medicine goes back to ancient cultures, demonstrating the longevity of the nutraceutical industry. Nutritional supplements and functional foods have been used for their health advantages for centuries, even before the word "nutraceutical" was coined in the late twentieth century. The Egyptians, Greeks, and Chinese, among others, were well aware of the healing potential of certain foods and plants. Honey, garlic, turmeric, and ginger were among the compounds employed for their therapeutic effects. Indian traditional medicine, known as Ayurveda, places a strong emphasis on the curative and preventative powers of herbs, spices, and carefully planned meal combinations. Hippocrates, the ancient Greek physician often credited as the "father of modern medicine," understood the crucial role diet had in his patients' health. Among his most well-known quotes is "Let food be thy medicine and medicine be thy food." Hippocrates promoted the use of certain meals for illness prevention and treatment, with an emphasis on the medicinal qualities of plants and herbs (Verma et al. 2010).

There was a revival of curiosity in the curative powers of foods and plants throughout the Renaissance. Explorers of this period brought back exotic plants and spices from all over the globe, many of which were highly sought for their purported medicinal properties. Citrus fruits' high vitamin C content, for instance, proved useful to explorers in warding off scurvy. Some of the most important nutrients for human health were first singled out and isolated by scientists in the late 19th and early 20th centuries. The finding of vitamin C and, subsequently, the B vitamins, transformed our knowledge of nutrition and its effect on health. To fight iodine shortage and prevent goitre, essential foods like salt were fortified during this time period by being iodized.

Dr. Stephen DeFelice, who founded and chairs the Foundation for Innovation in Medicine (FIM), is credited with coining the word "nutraceutical" in 1989. Dr. DeFelice used the word to refer to compounds that could have positive effects on health beyond those of regular eating. He saw nutraceuticals as a link between healthy eating and medications, emphasizing both fields' abilities to combat illness. Advances in scientific research, rising public interest in preventive healthcare, and the invention of dietary supplements all contributed to the late twentieth-century explosion in the area of nutraceuticals. Nutritional supplements including vitamins, minerals, and botanical extracts have helped propel the nutraceutical business into the mainstream (Andlauer and Fürst 2002).

Increased consumer knowledge and desire for natural health products drove explosive expansion in the dietary supplement business in the 1990s. In the United States, the Dietary Supplement Health and Education Act (DSHEA) of 1994 established dietary supplements as a distinct sector of the food and medicine industries. Functional meals, drinks, and customized supplements are just some of the nutraceutical goods that have seen rapid growth in the twenty-first century. As people looked for easy methods to integrate healthful components into their diets, functional foods, which are fortified or supplemented with certain nutrients or bioactive chemicals, rose in popularity.

During this time period, scientists made strides in their knowledge of nutraceuticals as they uncovered the mechanisms of action and possible health advantages of diverse substances. The advantages of probiotics for gut health and immunological function, for instance, and the importance of omega-3 fatty acids in cardiovascular health, both of which received a lot of press recently. Increasing consumer awareness, aging demographics, and a focus on preventative healthcare have all contributed to the expansion of the worldwide nutraceutical business. It covers a broad spectrum of items, from nutritional supplements to functional meals and drinks aimed at improving one's health in a particular way (Andlauer and Fürst 2002).

The nutraceutical business, however, still faces obstacles. Nutraceutical product safety, effectiveness, and quality remain a concern despite the fact that regulatory structures differ from country to country. The efficacy of various nutraceuticals and the best way to employ them in disease management and prevention is also a topic of continuing discussion and study.

3 Nutraceuticals

As people are living longer and experiencing a greater variety of lifestyle-related illnesses, nutraceuticals have become an absolute must-have for consumers. It falls much below the definition of "Food for special dietary use" in India. Nutraceuticals are a vast and varied class of substances that are found naturally in plants. The term is used to describe substances derived from food that are said to have therapeutic effects on humans. Dr. Stephen DeFelice combined the terms "nutrition" and "pharmaceutical"to create the term "Nutraceutical" in 1989. DeFelice defines a nutraceutical as a food or dietary component that has health benefits beyond basic nutrition, such as the ability to treat or prevent disease. Despite being classified as non-nutritive, several of these chemicals have beneficial properties such as antioxidant, anti-mutagenic, anti-estrogenic, anti-carcinogenic, and anti-inflammatory effects that might aid in disease prevention and the maintenance of genomic stability. From genetically modified designer meals and natural ingredients to processed foods like cereals, soups, and beverages, the items cover a wide variety of needs. The term "nutraceutical" is widely used in advertising without a corresponding regulatory definition in the United States. The demand for chemical-free, all-natural food additives is a major emerging trend in the food industry (Lunin 1881).

The food, pharmaceutical, and nutraceutical sectors all saw major shifts as a result of the rising interest in health and wellness. The demand for nutraceuticals has skyrocketed, both in developed economies and those on the rise. More and more people are turning to nutraceutical products in an effort to maintain a healthy weight, lower their blood pressure, and reduce their risk of developing diabetes and cardiovascular disease (Sharma and Majumdar 2009) (Fig. 1.1).

They have a long track record of success, a relatively modest price tag, a high level of patient acceptance, and a solid reputation as a result of these factors. The most crucial reason is that herbal products give a sensible approach to the treatment

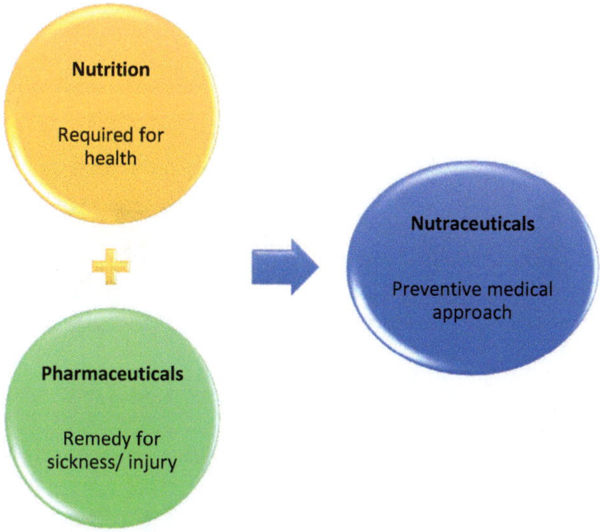

Fig. 1.1 The term Nutraceutical was derived from two words nutrition and pharmaceuticals

of many illnesses that may be tenacious and incurable. Extracts and refined bioactive components are often the foundation from which nutraceutical products are constructed. The capacity of the costly plant *Psoralea corylifolia* to carry a nutraceutical with energetic biocomponents is being investigated. Hippocrates, the widely acknowledged father of modern medicine (460–377 B.C), popularized the adage "Let food be thy remedy, and medicine be thy meals" to stress the importance of a healthy diet to one's overall well-being (Sharma and Majumdar 2009).

3.1 Types of Nutraceuticals

In addition to their nutritional value, nutraceuticals also have additional positive effects on human health. They're meant to help in general, help with certain health issues, and fill a void between food and medications. Vitamins, minerals, herbal extracts, probiotics, omega-3 fatty acids, and many more bioactive chemicals are only some of the topics covered in this article on nutraceuticals (Ahmad et al. 2011) (Fig. 1.2).

Vitamins Vitamins are chemical molecules that the body needs to function properly. Vitamin C and the B vitamins are examples of water-soluble vitamins, whereas vitamins A, D, E, and K are examples of fat-soluble vitamins. Vitamins play a crucial role in ensuring proper metabolic function, bolstering the immune system, fostering healthy development, and warding against vitamin shortages. You may get them in pill, tablet, and multivitamin forms, all of which are used as dietary supplements.

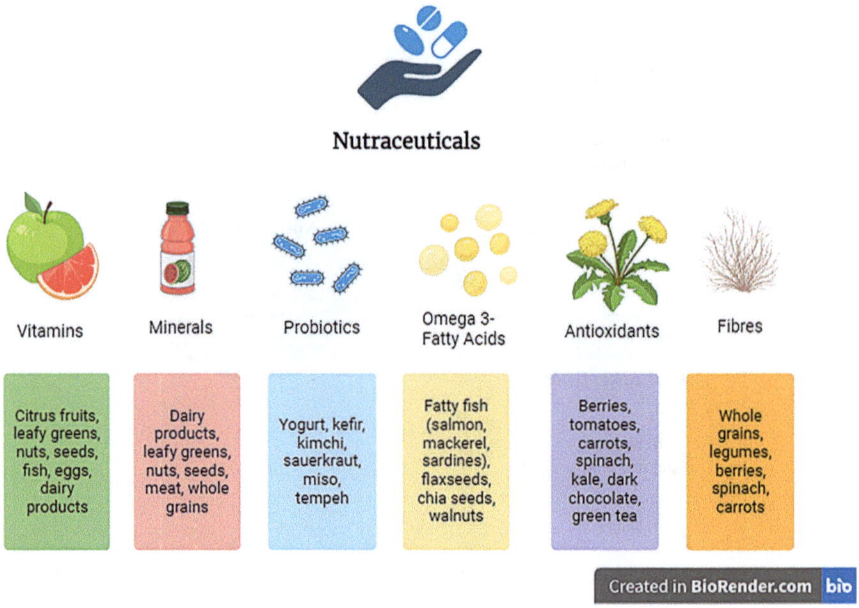

Fig. 1.2 Types of Nutraceuticals. (Created in BioRender.com)

Minerals The body can't operate properly without minerals, which are inorganic compounds. Calcium, magnesium, zinc, iron, and even selenium are all in there, and they're all really important. Bone and muscle health, energy generation, enzyme functioning, and many other physiological activities all rely on minerals. Mineral-specific dietary supplements are popular for their purported ability to treat deficiency symptoms, alleviate the symptoms of certain diseases, and improve general health.

Herbal Extracts Herbal extracts, made from a wide variety of plants, have been utilized for millennia in alternative medicine. They have health-promoting bioactive substances. Ginkgo biloba, turmeric (curcumin), green tea, and garlic all have extracts that are widely used. Antioxidant, anti-inflammatory, antibacterial, and other therapeutic activities may be present in these extracts. Capsules, powders, and teas are only some of the delivery methods for herbal extracts.

Probiotics Probiotics are beneficial bacteria that may be eaten orally. They are most often linked to helping maintain a good microbiome balance in the stomach. Beneficial bacteria like Lactobacillus and Bifidobacterium are often used in probiotics. These microbes have been shown to aid in the digestive process, increase nutritional absorption, strengthen the immune system, and keep the gut microbiome balanced. You can get probiotics in fermented foods like yogurt and kefir, as well as in pill form and powder form.

Omega-3 Fatty Acids Omega-3 fatty acids are a form of PUFAs that play a crucial role in maintaining good health. Fatty fish like salmon, mackerel, and sardines are a good source of these acids, namely eicosapentaenoic acid (EPA) and docosahexaenoic acid (DHA). Anti-inflammatory characteristics and possible advantages to cardiovascular health, cognitive function, and joint health are among the many reasons omega-3 fatty acids have gained popularity. Fish oil capsules and plant-based equivalents made from algae or flaxseed are two common forms of omega-3 supplementation.

Antioxidants Antioxidants are chemicals that assist prevent cellular harm from free radicals by neutralizing those radicals. Unstable chemicals called free radicals have been linked to cellular damage and illness. Vitamins C and E, beta-carotene, selenium, and the many polyphenols found in fruits, vegetables, and tea are all examples of antioxidants. Antioxidants may aid in immune system function, lessen the likelihood of developing chronic illnesses, and encourage healthy aging. They may be gained from a diet high in fruits, vegetables, and whole foods, or as dietary supplements (Ahmad et al. 2011).

Fibre Supplements The digestive system relies on dietary fibre, which is why it's so important. It aids in weight management, improves cardiovascular health, and encourages regular bowel motions. Fruits, vegetables, and whole grains are the best sources of dietary fibre, but fibre supplements may help fill in the gaps if your diet is lacking. The fibre in these aids may be from psyllium husk, inulin, or glucomannan, or it may be insoluble fibre.

4 Sources of Nutraceuticals

Plants, animals, microbes, and even synthetic processes are all potential origins for nutraceuticals. These plants and animals offer the raw ingredients used to make nutraceutical goods, many of which include a wide variety of bioactive chemicals. This article delves into where various nutraceuticals come from and why they're so important to the creation of healthful goods (Verma et al. 2022).

(i) **Plant-Based Sources**

Nutraceuticals may be found in abundance in plant foods since they include many different bioactive components. Plants are the source of many of the active substances in nutraceuticals, including herbal extracts, botanical compounds, and phytochemicals. Some instances are:

- **Herbs and Spices:** Curcumin, gingerol, allicin, cinnamaldehyde, and flavonoids are just a few examples of the bioactive compounds found in herbs and spices including turmeric, ginger, garlic, cinnamon, and ginkgo biloba; these compounds demonstrate antioxidant, anti-inflammatory, antibacterial, and other health-promoting effects.

- **Fruits and Vegetables:** Vitamins, minerals, fibre, and phytochemicals are all easily obtained from a diet rich in fruits and vegetables. Antioxidants, polyphenols, carotenoids, and other bioactive substances are especially concentrated in berries, citrus fruits, cruciferous vegetables, leafy greens, and other colourful food.
- **Medicinal Plants:** For generations, people have relied on the healing qualities of several plants that are now considered traditional medicines. Herbal supplements like aloe vera, echinacea, chamomile, ashwagandha, and ginseng are all good examples.
- **Tea and Herbal Infusions:** Antioxidants, polyphenols, and other beneficial components may be found in beverages including green tea, black tea, white tea, and herbal infusions produced from plants like chamomile, peppermint, and hibiscus (Verma et al. 2022).

(ii) **Animal-Based Sources**

The unique bioactive chemicals found in animal sources of some nutraceuticals are one reason for their health benefits. Instances include:

- **Fish and Seafood:** Omega-3 fatty acids, especially EPA and DHA, found in fatty fish like salmon, mackerel, sardines, and trout have been linked to improved cardiovascular health, cognitive function, and inflammatory control. Omega-3 fatty acid supplements in the form of fish oil are quite popular.
- **Collagen:** Supporting joint health, skin elasticity, and tissue integrity, collagen is a structural protein present in the skin, bones, and connective tissues of animals.
- **Bee Products:** Honey, royal jelly, bee pollen, and propolis are all products of bees that have been studied for their possible health benefits due to the presence of bioactive chemicals such as antioxidants, antibacterial agents, and anti-inflammatory molecules.
- **Animal-Derived Enzymes:** Digestive enzymes including proteases, lipases, and amylases are only some of the enzymes that may be sourced from animals and employed in nutraceutical compositions (Verma et al. 2022).

(iii) **Microorganisms**

Nutraceuticals rely heavily on the work of microorganisms like bacteria, yeast, and algae. Instances include:

- **Probiotics:** Probiotics are good bacteria that occur mostly in the digestive tract and may be taken as dietary supplements. Probiotics are often made from bacteria like Lactobacillus and Bifidobacterium.
- **Yeast Extracts:** Bioactive substances, vitamins, and minerals are all provided by yeast extracts like *Saccharomyces cerevisiae*. They're a common ingredient in nutraceuticals since they're rich in useful nutrients..
- **Algae:** Due to their high protein, vitamin, mineral, and antioxidant content, algae like spirulina and chlorella are categorized as nutraceuticals (Holzapfel et al. 2001).

(iv) **Synthetic Processes**

To improve their efficacy or bioavailability, several nutraceuticals are manufactured by chemical procedures. Examples of such man-made dietary supplements include:

- **Synthetic Antioxidants:** Butylated hydroxytoluene (BHT) and butylated hydroxyanisole (BHA) are two man-made antioxidants often found in processed foods and dietary supplements.
- **Synthetic Vitamins:** Vitamins that are often utilized in nutraceutical formulations may be produced in the lab; they include vitamins C (ascorbic acid) and E (tocopherol).
- **Modified Bioactive Compounds:** Some bioactive substances may be made more stable, bioavailable, or effective by chemical modification or extraction methods (Holzapfel et al. 2001).

5 Potential of Nutraceuticals

The unique mix of nutritional and pharmacological qualities found in nutraceuticals has demonstrated promising results in improving health and well-being. The health advantages of these items extend well beyond those of simple diet since they are sourced from natural sources including plants, herbs, and bioactive substances. Nutraceuticals are gaining popularity among customers who want to improve their overall health and treat particular health issues because of the growing interest in preventative healthcare and the need for natural alternatives. Nutraceuticals have great promise since they may improve health by helping the body perform better in certain ways and by correcting nutritional imbalances. Nutraceutical supplements, for instance, may assist guarantee an adequate intake of nutrients including vitamins and minerals that are crucial to the body's correct functioning. The immune system, the cardiovascular system, and even mental performance may all get focused help from nutraceuticals. Nutraceuticals, with their own blend of bioactive components, provide a more all-encompassing strategy for improving health.

Nutraceuticals also have the potential to play a part in illness prevention and management. Preventive benefits of some nutraceuticals against chronic illnesses including cardiovascular disease, diabetes, cancer, and neurological disorders have been proposed by scientific investigations. Antioxidants in nutraceuticals, such as polyphenols and flavonoids, may help lower the risk of oxidative stress-related illnesses because of their anti-inflammatory and free radical-scavenging characteristics. The use of nutraceuticals may promote healthy aging and enhance quality of life in older adults. Conditions specific to old age and general health deterioration are commonly mentioned in conjunction with aging. Some dietary supplements, such as those containing resveratrol or coenzyme Q10, have been studied for their potential to slow aging, improve cell health, and increase energy levels. Omega-3 fatty acid-rich supplements, such as fish oil, have also been investigated for their possible positive effects on cognition and brain function.

Nutraceuticals have the potential to improve health and also have the benefit of being safer and well-tolerated than traditional medications. Nutraceuticals, being obtained from natural sources, are thought to have a decreased risk of unwanted effects and to be typically well-tolerated. This makes them appealing to those who prefer more natural remedies or who are wary of the possible negative effects of prescription medications.

The promise of nutraceuticals, however, has to be understood with some care. Despite the encouraging findings of several trials, further investigation into the drugs' mechanisms of action, appropriate doses, and possible drug interactions is required. Furthermore, nutraceutical items might vary in quality and uniformity, so it's important to choose trustworthy brands and verify accurate labelling and production procedures (Holzapfel et al. 2001).

The promise of nutraceuticals rests on their capability to go above and beyond the nutritional needs of the body. Bioactive chemicals have been the subject of substantial research for their potential therapeutic and preventative benefits against a wide range of ailments. The following Table 1.1 summarizes several commonly used nutraceuticals and the benefits they provide: (Mukherjee 2019).

Dolkar et al. (2017)

6 Regulatory Aspects of Nutraceuticals

Nutraceutical safety, quality, and effectiveness rely heavily on regulatory considerations. Since these items incorporate characteristics of both food and medicine, regulation systems vary by country and area. Classification, labelling, ingredient safety, production methods, and marketing claims are all common components of nutraceutical regulation. Categorization of products as a regulatory component of nutraceuticals. There may be variances in the language and concepts used to classify nutraceuticals between countries. They might be considered nutritional supplements, functional meals, or even traditional herbal remedies in certain nations. What kinds of labelling criteria apply, what kinds of claims are permissible, and so on for nutraceuticals are all determined by their categorization (Witkamp and van Norren 2018).

Compliance with labelling regulations is crucial for maintaining honesty in the marketplace and giving buyers reliable data. Ingredients, dosing directions, storage requirements, expiry dates, and allergy warnings are common information to find on a nutraceutical's label. Furthermore, rules frequently stipulate on what kinds of claims may be made on the label, with the goal of preventing false or misleading advertising and promoting instead the use of evidence-based practices. The regulation of the quality and security of ingredients is another important facet of nutraceuticals. Guidelines for the safety evaluation of substances used in nutraceutical products are often established by regulatory authorities. The toxicological profile,

Table 1.1 Potential of nutraceuticals properties

Nutraceuticals	Properties
Curcumin	Anti-inflammatory, antioxidant, anticancer
Omega-3 fatty acids	Cardiovascular health, brain function, anti-inflammatory
Probiotics	Gut health, immune support, digestive health
Green tea extract	Antioxidant, metabolism boost, weight management
Resveratrol	Antioxidant, anti-aging, cardiovascular health
Coenzyme Q10	Energy production, heart health, antioxidant
Vitamin D	Bone health, immune function, mood regulation
Cranberry extract	Urinary tract health, antioxidant
Garlic extract	Cardiovascular health, immune support
Ginkgo biloba	Cognitive function, circulation, antioxidant
Glucosamine	Joint health, cartilage support
Proanthocyanidins	Antioxidant, cardiovascular health, skin health
Lycopene	Antioxidant, prostate health, skin health
Astaxanthin	Antioxidant, eye health, skin health
Chondroitin sulfate	Joint health, cartilage support
Melatonin	Sleep regulation, antioxidant, immune support

drug-drug interactions, and maximum safe doses may all factor into these recommendations. Furthermore, several nations have regulatory bodies that either approve substances or set maximum contaminant levels.

Regulators keep an eye on production processes and quality assurance as well. Guidelines established by the Good Manufacturing Practices (GMP) program set the bar for the safe and reliable manufacture, distribution, and storage of nutraceuticals. Following Good Manufacturing Practices (GMP) guidelines aids in avoiding product contamination, guarantees truthful labelling, and boosts confidence in the product's safety and efficacy. Marketing claims for nutraceuticals are likewise governed by regulations. Consumers should not be misled, hence the claims must be backed up by science. Companies may be required by regulatory organizations to provide proof supporting the claims or to employ health claims that have already been authorized. This safeguards customers' ability to make educated purchases and avoids deceptive advertising (Witkamp and van Norren 2018).

The nutraceutical sector cannot survive without strict enforcement of laws and close monitoring of compliance. Inspections, audits, and random sampling are common methods used by regulatory organizations to verify whether a producer or retailer is following the rules. There may be repercussions, such as fines, product recalls, or even legal action, for failing to comply.

The worldwide nutraceutical business has a continuing challenge: harmonizing regulatory requirements. Disparities in national legislation may impede international commerce and restrict people's access to safe and effective goods. The worldwide commerce of nutraceuticals is being facilitated by efforts to harmonize regulatory frameworks, develop international norms, and encourage mutual acceptance of safety evaluations.

7 Nutraceuticals Versus Modern Medicines

There are benefits and drawbacks to both nutraceuticals and conventional medical treatment. Nutraceuticals are dietary supplements or functional foods obtained from natural sources, as opposed to conventional medications which are created after extensive scientific study and clinical trials. In this post, we'll look at how nutraceuticals and conventional pharmaceuticals vary in terms of research and development, effectiveness, safety, accessibility, and philosophy toward health care (Subbiah 2008).

(i) **Development Process**

Modern Medicines These days, the development of a new pharmaceutical follows a strict protocol that includes plenty of research, animal testing, and human trials. All of the pharmaceuticals go through rigorous testing for side effects, dosing, and safety, thanks to these procedures. Before a drug may be authorized and sold, regulatory bodies like the Food and Drug Administration (FDA) in the United States impose stringent restrictions and regulations on the pharmaceutical industry.

Nutraceuticals Nutraceuticals, on the other hand, come from organic materials including plants, animals, and microbes. Some of the substances in nutraceuticals have been tested in clinical experiments, while many others are based on anecdotal evidence and folklore. Nutraceutical research and development is often less standardized and controlled than those of conventional pharmaceuticals. Researching the effectiveness and safety of nutraceutical components, however, is becoming more popular.

(ii) **Efficacy**

Modern Medicines Today's drugs are developed with pinpoint accuracy to address individual illnesses and circumstances. In order to guarantee their effectiveness in clinical settings, they are typically subjected to extensive testing. Evidence from clinical trials and other forms of significant study confirms the effectiveness of today's medications. These drugs are designed to have predictable therapeutic results, making them useful in the control and treatment of a wide range of medical issues.

Nutraceuticals In contrast to traditional pharmaceuticals used to cure illness, nutraceuticals take the form of dietary supplements or functional foods. Nutraceutical effectiveness may range widely based on factors including active ingredient, formulation, and dose. Although certain nutraceuticals have been shown to have positive benefits in research, the proof for their effectiveness is typically scant and may not satisfy the same stringent requirements as conventional pharmaceuticals. Furthermore, genetics, nutrition, lifestyle, and general health may all play a role in how a person responds to nutraceuticals.

(iii) **Safety**

Modern Medicines In order to identify possible hazards and side effects, modern drugs go through thorough safety testing throughout the development process. The safety profile of medications is continuously monitored, and adverse effects are reported and examined, by regulatory bodies. However, even cutting-edge pharmaceuticals are not without potential hazards and unwanted consequences, particularly in situations of incorrect dosing, drug interactions, and individual differences in drug metabolism.

Nutraceuticals When produced from natural ingredients, nutraceuticals offer an even better safety profile. However, the purity, extraction technique, formulation, and dose of a nutraceutical may all affect its potential risk. Although many nutraceuticals have a good safety record, others may cause unwanted side effects due to their components or drug interactions. If you have a pre-existing ailment or are already on any kind of medicine, you should talk to your doctor before adding any kind of nutraceutical to your regular routine.

(iv) **Availability and Accessibility**

Modern Medicines Modern medications may be obtained either with a doctor's prescription or without one at most drugstores. Availability and quality of contemporary medications are ensured by stringent regulations on their production and delivery. However, variables including healthcare systems, pricing, and geography might affect how easily people can get their hands on contemporary drugs. In certain circumstances, especially in underdeveloped nations or outlying locations, access to modern medications may be restricted.

Nutraceuticals Nutraceuticals may be found in a wide variety of grocery shops, health food stores, and even online without the need for a doctor's prescription. Their availability is high, and there is a plethora of alternatives for buyers to consider. However, assuring the uniformity, purity, and regulated dose of these products may be difficult due to the absence of strong standards and quality control in the nutraceutical business.

(v) **Approach to Healthcare**

Modern Medicines Disease prevention and treatment are the main focuses of modern medicine. They treat symptoms, cure illnesses, and manage chronic disorders by focusing on certain pathways, chemicals, or receptors in the body. Science, evidence-based standards, and a reductionist mindset form the basis of today's medical practice, which seeks to find particular molecular targets for intervention.

Nutraceuticals Nutraceuticals are health supplements that adopt a more all-encompassing view of medicine by highlighting the connection between diet and illness prevention. They are often used as non-traditional or supplementary treatments. The purpose of nutraceuticals is to improve health by supplementing the diet with bioactive chemicals and nutrients that may be deficient. Personalized nutrition,

Table 1.2 A comparative table of Nutraceuticals and modern medicines

Aspect	Nutraceuticals	Modern medicines
Source	Derived from natural sources (plants, animals, minerals)	Synthesized chemically or derived from natural sources
Regulation	Less stringent regulations, often considered as dietary supplements	Stringent regulations, subjected to clinical trials and approval processes (U.S.F.D.A. 2022)
Mode of Action	Often work through multiple mechanisms, may have holistic effects	Typically target specific pathways or molecules (AlAli et al. 2021)
Side Effects	Generally fewer and milder side effects, although individual reactions vary	Can have more pronounced and varied side effects, often listed in product information (N.C.C.I.H 2020)
Accessibility	Widely available over the counter in health food stores or online	Typically available by prescription or over the counter at pharmacies
Cost	Can vary widely in cost, from inexpensive to relatively expensive	Can be costly due to research, development, and marketing expenses
Treatment Approach	Emphasizes preventive and holistic approaches to health	Often focuses on symptom management or targeted treatment of specific conditions
Research	Increasing scientific interest, but may lack comprehensive clinical trials	Extensively researched, with robust clinical trials supporting efficacy and safety (W.H.O. 2005)
Integration with Conventional Medicine	Often used as complementary or alternative treatments	Primary focus of conventional medical practice

in which consumers' dietary requirements and tastes are taken into account while developing a final product, is also linked to nutraceuticals (Table 1.2).

8 Current Scenario

The future of the nutraceutical business in India is bright. Products have expanded greatly in variety during the last decade, reflecting the market's explosive development. One positive effect of the thriving economy is that people's take-home pay has gone up. A rise in diet-related health problems has also been linked to the combination of bad eating habits and inactivity. However, there is a growing understanding of the connection between what we eat and our long-term health. These factors have helped create a conducive environment for the growth of India's nutraceutical business. Numerous factors, including India's highly educated workforce, cutting-edge R&D facilities, and abundant raw materials, put us at an international advantage. There is a potential USD 1 Billion in the Indian nutritional market. When compared to the worldwide market's CAGR of 7%, the Indian market's CAGR of

18% over the previous three years has been very impressive. However, India's untapped market is worth between $2 and $4 billion, or two to four times the size of the present industry, with around 148 million potential clients. In a USD 1 billion market, 54% is held by functional food, 32% by dietary supplements, and 14% by functional beverages. There are very few pure play nutraceutical enterprises in the Indian industry, which is dominated by the pharmaceutical and fast moving consumer goods industries. GlaxoSmithKline consumer healthcare, Dabur India, Cadila healthcare, EID Parry's, Zandu Pharmaceuticals, Himalaya herbal healthcare, Amway, Sami labs, Elder pharmaceuticals, and Ranbaxy are only few of the main firms marketing nutraceuticals in India.

The present state of the nutraceutical market is indicative of the trend away from reactive medicine and towards preventative medicine. Consumers are increasingly interested in using natural remedies to improve their health, manage chronic illnesses, and maintain an overall state of wellness, which has led to a rise in the demand for nutraceuticals. The potential, threats, and developments in the nutraceutical market are discussed in this chapter. The nutraceutical industry is growing as a result of rising interest in and demand for health and wellness-oriented goods. More and more people are realizing that what they eat directly affects how they feel and how long they live. They are on the lookout for goods that may help them achieve certain health goals and avoid illness. Due to customers' newfound willingness to spend money on wellness goods, the nutraceutical industry has seen explosive growth (Zelig and Rigassio Radler 2012).

Increases in the number of people suffering from conditions like diabetes, obesity, and cardiovascular disease have also contributed to the growth of the nutraceutical industry. A lack of exercise and unhealthy eating habits have been related to these diseases. Nutraceuticals are increasingly seen as a practical strategy for controlling and preventing chronic illnesses due to their ability to alleviate nutritional deficiencies and offer therapeutic effects. Because of this, doctors are starting to include nutraceutical supplements in patient care programs. The nutraceutical market has become more specialized to address various customer needs and preferences. As a simple and tasty approach to improve one's health, functional meals and drinks have become more popular. Bioactive substances including probiotics, antioxidants, fiber, and omega-3 fatty acids are often included to these products. Targeted nutritional supplementation in the form of capsules, pills, and powders is also seeing a surge in popularity.

The rise of individualized nutrition is another interesting development. Companies may now provide customized nutraceuticals according to customers' health profiles, genetic data, or biomarkers thanks to scientific developments and widespread access to genetic testing. By taking this route, individuals are better able to meet their unique dietary requirements and experience positive health benefits. Artificial intelligence and data analytics are predicted to become more integrated into the field of personalized nutrition, leading to more precise and individualized product recommendations. Research and development efforts are also advancing in the nutraceutical business. The effectiveness and mechanisms of action of different nutraceutical components are now being investigated in many research investigations. This study is crucial because it will help back up the health claims made for nutraceuticals. Innovative delivery technologies are also being developed to improve the bioavailability and stability of nutraceutical substances, leading to greater absorption and efficacy in the body.

Nutraceutical companies, however, have their own set of problems. Safety regulations and testing for potential hazards are a major issue. When it comes to nutraceuticals, the regulatory environment differs from country to country due to differences in registration procedures and health claim validity. Maintaining customer confidence and defending public health require that nutraceutical goods be tested for safety, effectiveness, and quality. To overcome these issues, it is crucial to implement stringent quality control methods, adhere to good manufacturing standards, and use clear labelling. There is a variety of nutraceutical goods on the market, making it difficult for customers to navigate and make educated decisions. Consumers' bewilderment and skepticism might be exacerbated by health claims that are either false or overblown. It is essential for consumers to be well-versed in nutraceuticals in order to make well-informed choices (Zelig and Rigassio Radler 2012).

9 Conclusion

In conclusion, nutraceuticals have become a promising new class of goods because of their ability to combine the best of both the food industry and the pharmaceutical industry. By fusing the dietary advantages of food with targeted bioactive molecules, they may promote health and well-being. Vitamins, minerals, herbs, botanicals, enzymes, and many more compounds are all considered nutraceuticals. These items are often used in functional meals or as dietary supplements. Nutraceuticals have become more popular due to rising interest in preventative healthcare and a need for all-natural remedies. Supplements and dietary changes are being sought out by many as strategies to improve general health and lower the danger of developing certain illnesses. Nutraceuticals are dietary supplements that promise extra health advantages on top of those already provided by healthy eating and regular exercise.

The chapters in this book provide an in-depth exploration of the role of nutrients and nutraceuticals in preventive medicine. It covers topics such as the prevention of cancer, metabolic syndrome, diabetes, chronic renal disease, chronic pancreatitis, chronic obstructive pulmonary disease, inflammatory autoimmune diseases, infertility, and suboptimal health. The chapters discuss how different nutrients can impact human health positively and negatively, emphasizing the importance of balanced nutrition for overall well-being. It also delves into the remedial effects of nutrients in various chronic diseases and the potential dangers of excessive nutrient intake on human health. Overall, the chapters highlight the significant role of nutrients in maintaining and improving health outcomes.

References

Ahmad MF, Ashraf SA, Ahmad FA, Ansari JA, Siddiquee MRA. Nutraceutical market and its regulation. Am J Food Technol. 2011;6(5):342–7.

AlAli M, Alqubaisy M, Aljaafari MN, AlAli AO, Baqais L, Molouki A, Abushelaibi A, Lai K-S, Lim S-HE. Nutraceuticals: transformation of conventional foods into health promoters/disease preventers and safety considerations. Molecules. 2021;26(9):2540. https://doi.org/10.3390/molecules26092540.

Andlauer W, Fürst P. Nutraceuticals: a piece of history, present status and outlook. Food Res Int. 2002;35(2–3):171–6.

Dolkar D, Bakshi P, Wali V, Sharma V, Shah R. Fruits as nutraceuticals. Ecol Environ Conserv. 2017;23:113–8.

Holzapfel WH, Haberer P, Geisen R, Björkroth J, Schillinger U. Taxonomy and important features of probiotic microorganisms in food and nutrition. Am J Clin Nutr. 2001;73(2):365s–73s.

Lunin N. Ueber die Bedeutung der anorganischen Salze für die Ernährung des Thieres. Z Physiol Chem. 1881;5(1):31–9.

Mukherjee PK. Quality control and evaluation of herbal drugs: evaluating natural products and traditional medicine. Elsevier; 2019.

National Center for Complementary and Integrative Health (N.C.C.I.H.). Dietary and herbal supplements: What You Need to Know. 2020. Available from: https://www.nccih.nih.gov/health/dietary-and-herbal-supplements

Padmavathi D. A general review on "Nutraceuticals": its golden health impact over human community. Int J Food Sci Nutr. 2018;3(2):214–7.

Pandey M, Verma RK, Saraf SA. Nutraceuticals: new era of medicine and health. Asian J Pharm Clin Res. 2010;3(1):11–5.

Sharma M, Majumdar PK. Occupational lifestyle diseases: an emerging issue. Indian J Occupat Environ Med. 2009;13(3):109.

Subbiah MR. Understanding the nutrigenomic definitions and concepts at the food–genome junction. Omics. 2008;12(4):229–35.

United States Food and Drug Administration (U.S.F.D.A.). Development & Approval Process. 2022. Available from: https://www.fda.gov/drugs/development-approval-process-drugs

Verma R, Pandey M, Saraf S. Nutraceuticals: new era of medicine and health. Asian J Clin Nutr. 2010;3

Verma NK, Singh AK, Chaurasiya AK, Singh A. IASR Journal of Medical and Pharmaceutical Science. 2022.

Witkamp RF, van Norren K. Let thy food be thy medicine…. when possible. Eur J Pharmacol. 2018;836:102–14.

World Health Organization. Quality control methods for medicinal plant materials. Geneva: World Health Organization; 2005.

Zelig R, Rigassio Radler D. Understanding the properties of common dietary supplements: clinical implications for healthcare practitioners. Nutr Clin Pract. 2012;27(6):767–76.

Chapter 2
Foundation of Nutraceuticals in Preventive Medicine

Pranav Pancham, Divya Jindal, and Manisha Singh

1 Introduction

The intricate relationship and interaction between nutrients and physiology form the foundation of our overall health and vitality. Their dynamic interaction is not static; it constantly adapts and responds to our body's changing needs. From the digestive system's role in absorbing, breaking down nutrients, and utilising the nutrients by cells to the metabolism's conversion of these nutrients into energy, every cellular process and organ system relies on the availability and balance of essential nutrients (Calder et al. 2020). By exploring this dynamic relationship, we can gain valuable insights into how to make informed dietary choices that promote our well-being and prevent nutritional deficiencies or imbalances (Chen et al. 2018). However, different nutrients have specific roles and functions – carbohydrates are a primary energy source, while proteins contribute to tissue growth and repair. Fats play a crucial role in insulation, hormone production, and the absorption of fat-soluble vitamins. Vitamins and minerals are co-factors in enzymatic reactions, supporting various physiological processes. Water is essential for maintaining hydration, facilitating

P. Pancham
School of Medicine, Western Sydney University, Campbelltown, NSW, Australia

D. Jindal
Indian Institute of Technology Bombay Monash Research Academy, Mumbai, India

M. Singh (✉)
Faculty of Health, Graduate School of Health, University of Technology Sydney, Sydney, NSW, Australia

ARCCIM, School of Public Health, Faculty of Health, University of Technology Sydney, Sydney, NSW, Australia

Department of Biotechnology, Jaypee Institute of Information Technology (JIIT), Noida, Uttar Pradesh, India

S. M. Jafari et al. (eds.), *Nutraceuticals: A New Millennium Approach for Preventive Medicine*, Food Bioactive Ingredients,
https://doi.org/10.1007/978-3-031-76481-3_2

nutrient transport, and regulating body temperature (Slavin et al. 2012; Gómez-Gallego and Salminen 2021). The intricate relationship between nutrients and physiology extends beyond primary metabolism. Also, the nutrients interact with specific organ systems, influencing their functions (calcium and vitamin D are essential for bone health, and iron is crucial for oxygen transport through the circulatory system), significantly impacting the immune system (Wallace and Murray 2020; Holick 2017). Further, it has also been seen that physical activity and exercise highlight the relationship between nutrients and cellular physiology. The body's energy demands increase during exercise, requiring additional nutrients for optimal performance and recovery. Proper pre- and post-exercise nutrition, including carbohydrates for energy and protein for muscle repair, can significantly impact exercise outcomes and our body's overall physiological functions (Bytomski 2018).

Therefore, understanding the dynamic relationship of essential nutrients required for balanced physiological processes forms the fundamental basis of nutraceuticals in preventive medicine. Nutraceuticals are bioactive compounds derived from food sources that offer potential health benefits beyond their essential nutritional value. The specific nutrients have bioactive properties that can positively impact physiological processes, including reducing inflammation, supporting antioxidant defences, modulating immune function, and regulating cellular signalling pathways (Russo 2019). In a regulated manner, these nutrients are formulated as nutraceuticals that help healthcare professionals in preventive medicine leverage the power of targeted nutritional interventions. This, in return, promotes overall wellness, reduces the risk of chronic diseases, is utilised to bridge dietary gaps and supports optimal physiological functioning (Biesalski and Tinz 2018). These interventions can range from personalised dietary recommendations to the use of specific nutrient supplements or functional foods fortified with bioactive compounds. The synergy between nutraceuticals and balanced physiology is exemplified in personalised nutrition. Advances in genetic testing and personalised medicine enable the identification of individual genetic variations that influence nutrient metabolism, absorption, and utilisation. This knowledge allows for developing tailored dietary plans and nutraceutical interventions that align with an individual's unique physiological needs, optimising health outcomes and disease prevention (Mayurasakorn et al. 2020). Moreover, nutraceuticals offer a promising avenue for addressing nutrient deficiencies or imbalances which may arise from poor dietary choices, restrictive diets, malabsorption disorders, or specific health conditions. The preventive potential of nutraceuticals extends to various chronic diseases, including cardiovascular diseases, diabetes, cancer, neurodegenerative disorders, and metabolic syndrome. Scientific evidence suggests that specific nutraceuticals, such as omega-3 fatty acids, polyphenols, curcumin, resveratrol, and probiotics, have shown promising effects in reducing the risk and progression of these diseases. Their mechanisms of action often involve modulating inflammatory pathways, improving oxidative stress management, enhancing immune function, and promoting cellular health (Calder 2017; Eichelmann et al. 2016). By leveraging the power of nutraceuticals, healthcare professionals can optimise health outcomes and empower individuals to take control of their well-being through informed dietary choices and targeted interventions. This book chapter examines the role of macronutrients and micronutrients in supporting optimal physiological functioning and their potential as nutraceutical agents. Also, it explores the interplay between nutraceuticals and physiological processes, highlighting the potential of nutraceuticals in the prevention and management of chronic diseases.

2 Macronutrients: Fuelling the Body's Functions

Macronutrients, comprising carbohydrates, proteins, and fats, serve as the primary sources of energy and play vital roles in fuelling the body's functions. Understanding their impact on physiology is crucial for developing effective nutraceutical interventions in preventive medicine. Carbohydrates, the body's primary energy source, are broken down into glucose and utilised by cells for energy production (Astrup et al. 2020). Additionally, carbohydrates contribute to synthesising glycogen, the storage form of glucose in the liver and muscles. Nutraceuticals derived from carbohydrates (dietary fibres, prebiotics, and resistant starches) have been linked to various health benefits, including improved glycemic control, weight management, and enhanced gut health (Crichton and Bryan 2015). Furthermore, proteins, composed of amino acids, are critical for the growth, repair, and maintenance of body tissues. They play vital roles in enzymatic reactions, immune function, hormone production, and transportation of molecules within the body (Bianchi and Gutierrez 2019). Dietary protein intake is essential for preserving muscle mass, healing wounds, and optimising immune responses (Schwingshackl et al. 2021). Nutraceuticals derived from proteins (whey protein, specific amino acids or protein isolates) have been studied for their potential to promote muscle protein synthesis, enhance athletic performance, and support healthy ageing (Myles 2014). Additionally, fats, often demonized in the past, are essential for various physiological functions. They serve as a concentrated energy source and aid in absorbing fat-soluble vitamins. They have a significant role in maintaining the structure and function of cell membranes, hormone synthesis, and inflammation regulation. The balance between fats (saturated, monounsaturated, and polyunsaturated fatty acids) is essential for maintaining cardiovascular health and preventing chronic diseases (Rezende et al. 2014; Calder 2015). Nutraceuticals derived from fats (omega-3 fatty acids) have shown promising effects in reducing inflammation, improving lipid profiles, and supporting cognitive function (Table 2.1).

3 Micronutrients: Catalysts for Cellular Processes

Micronutrients, including vitamins and minerals, act as catalysts for numerous cellular processes and have significance as potential nutraceutical agents in preventive medicine. Vitamins are organic compounds that act as coenzymes or cofactors,

Table 2.1 Listing the Essential Macronutrients needed in significant amounts by the body for proper functioning and health, along with their foods, dietary sources, and contributions to supporting various bodily functions and processes

S.No	Essential Macronutrients required by the human body	Sources of Macronutrients	Functions	References
1	Vitamin A	Obtain it from the diet either as preformed vitamin A or in the form of provitamin A carotenoids. Such as herbs, cereals, specific vegetable oils, carrots, spinach, and tomato.	Maintain epithelium integrity, Eye retina, Immune system, cell growth, etc.	Carazo et al. (2021)
2	Vitamin D	UV-B radiation, fatty fish, eggs and fish oil	Calcium homeostasis	Janoušek et al. (2022)
3	Vitamin E	Seeds, Green leafy vegetables, and fruits	Antioxidant, membrane stabilisation enzyme activity, inhibition of platelet coagulation	Niki and Abe (2019)
4	Vitamin K	spinach, chard, parsley, and various types of lettuce	γ-ccarboxylation of proteins	Mladěnka et al. (2021)
5	Vitamin C	star fruit, guava, black currant, kiwi, and strawberry cruciferous vegetables, especially broccoli, kale, and peppers	reduce ferric ions into ferrous ones, Enzymatic Cofactor, 2-Oxoglutarate-Dependent Dioxygenases	Doseděl et al. (2021)
6	Vitamin B1	whole grains, bread, meat (especially pork), and pulses, while milled wheat flour, polished rice, vegetables, and fruits	cofactor of several crucial enzymes, mitochondria function	Biosynthesis of vitamins in plants part a – vitamins A, B1, B2, b3, B5 (2011)
7	Vitamin B2	Eggs, legumes, nuts, mushrooms, and organ meat (liver)	Catalyze reduction-oxidation reactions	Riboflavin (1998)
8	Vitamin B6	highly fortified cereals; beef liver and other organ meats; and highly fortified, soy-based meat substitutes.	Aminotransferases, decarboxylases, racemases, and dehydratases.	Dietary reference intakes for thiamin, riboflavin, niacin, vitamin B6, folate, vitamin B12, pantothenic acid, biotin, and choline (1998)

facilitating enzymatic reactions in energy production, DNA synthesis, immune function, and antioxidant defence (De la Fuente et al. 2017). They are classified as either water-soluble (such as vitamin C and the B vitamins) or fat-soluble (such as vitamins A, D, E, and K). And nutraceuticals derived from both categories of vitamins have gained significant attention for their potential to promote health and prevent diseases (e.g., vitamin D has been studied for its role in bone health, immune function, and cancer prevention, while vitamin C is recognised for its antioxidant properties and support of collagen synthesis) (Fulgoni et al. 2011; Molfino et al. 2013). Likewise, minerals are inorganic elements essential for various physiological processes and serve as structural components, cofactors for enzymatic reactions, and regulators of cellular functions. Minerals (calcium, phosphorus, magnesium, and potassium) are vital for maintaining bone health, nerve function, muscle contraction, and fluid balance. Trace minerals (iron, zinc, copper, selenium, and iodine) are required in smaller amounts but play critical roles in immune function, energy metabolism, and antioxidant defence (Farag et al. 2023) (Table 2.2).

Nutraceuticals derived from minerals have shown potential in preventing deficiencies and optimising cellular processes. (Iron supplementation: treat iron-deficiency anaemia, zinc: enhanced immune function and wound healing) (Table 2.2). B vitamins and minerals are involved in energy metabolism, acting as cofactors for enzymes in critical metabolic pathways, including carbohydrate, protein, and lipid metabolism (Tardy et al. 2020). They also play a significant role in gene expression, supporting DNA synthesis, repair, and transcription. Furthermore, micronutrients contribute to antioxidant defence, helping to neutralise free radicals and protect cells from oxidative damage. Additionally, nutraceuticals derived from micronutrients, such as antioxidant supplements, have been investigated for their potential to reduce the risk of chronic diseases, including cardiovascular diseases, cancer, and age-related macular degeneration (Suzuki and Kunisawa 2015).

4 Phytonutrients: Antioxidant Powerhouses

Phytonutrients, or phytochemicals, are natural compounds found in plants and are antioxidant powerhouses. They encompass various compounds, including flavonoids, carotenoids, phenolic acids, and other plant-derived substances. These compounds are responsible for the vibrant colours, flavours, and aromas of fruits, vegetables, herbs, and spices (Zhang et al. 2015). Beyond their sensory attributes, phytonutrients exhibit remarkable antioxidant properties, protecting our cells from oxidative damage caused by free radicals (Rudrapal et al. 2022). Oxidative stress, a condition characterised by an imbalance between free radicals and antioxidants in the body, has been linked to various chronic diseases, including cardiovascular diseases, cancer, neurodegenerative disorders, and ageing-related conditions. Phytonutrients act as potent antioxidants, neutralising free radicals and reducing oxidative stress (Pizzino et al. 2017). They also support the body's endogenous antioxidant defence system, enhancing its ability to combat oxidative damage.

Table 2.2 Listing the essential micronutrients needed in significant amounts by the body for proper functioning and health, along with their foods, dietary sources, and contributions to supporting various bodily functions and processes

S.No	Essential Micronutrients required by the human body	Sources of macronutrients	Functions	References
1	Iodine	Fish (such as cod and tuna), seaweed, shrimp, Dairy products	Metabolism, differentiation, and fetus development	Opazo et al. (2020)
2	Zinc	Meat, liver, kidney, fish, chicken and cereals	Cell division and the synthesis of DNA and protein	Deshpande et al. (2013)
3	Iron	Red meat, miscellaneous cereals, nuts, and shellfish	Oxygen transport, synthesis of deoxyribonucleic acid (DNA) and electron transport	Rucker and Lanoue (2004)
4	Manganese	Water, spices, apricots, bananas, soybeans, nuts, green leafy vegetables, and whole grains	Bone growth and integrity	Shah (2017)
5	Calcium	Shellfish, fish, dark green leafy vegetables and milk	Adequate growth and bone development	Matkovic and Ilich (2009)
6	Magnesium	Nuts, cereals, and tea leaves	Enzymes activity	Yokel and Crossgrove (2004)
7	Selenium	Meat, cereals, grains, and dairy products like eggs	Antioxidant	Surai (2006)
8	Copper	Cereals, fruit, vegetables and, to a lesser extent, meat and animal products	Acts as a cofactor for redox reactions	Bost et al. (2016)

Some phytonutrients directly scavenge free radicals, preventing them from damaging cellular components such as DNA, proteins, and lipids. Others stimulate the body's antioxidant enzyme systems, enhancing the production of endogenous antioxidants like glutathione and superoxide dismutase (as listed in Table 2.3). Additionally, phytonutrients may have anti-inflammatory effects, contributing to their overall health benefits (Lobo et al. 2010; Sharifi-Rad et al. 2020). They have gained significant attention in preventive medicine due to their potential to reduce the risk of chronic diseases and ensure a broader range of health benefits. A diverse and colourful plant-based diet is essential for obtaining many phytonutrients. Different fruits, vegetables, herbs, and spices contain unique phytonutrient profiles and consume various plant-based foods. Nutraceuticals derived from phytonutrients, such as supplements or concentrated extracts, have become a convenient way to obtain specific phytonutrients with targeted health benefits (Nasri et al. 2014; Puri et al. 2022).

Table 2.3 This table shows the phytonutrients commonly recommended for treating pancreatitis. It's been followed with the different food sources, benefits, and daily intake recommendations (Böhm et al. 2020; U.S. Department of Health and Human Services n.d.; Hovenkamp et al. 2008; Jones 2007; Sacco et al. 2013; Mennen et al. 2005)

S.No	Phytonutrients	Food Sources	Antioxidant Benefits	Recommended Nutrients Intake (Daily)	Complications due to high consumption
1	Carotenoids	Carrots, Spinach, Kale, Pumpkin, Apricots, Red peppers, Mangoes, Papayas	Beneficial for eye health, immune function, and overall antioxidant support	6 μg of β-carotene, and 12 μg for other carotenoids with provitamin A activity.	Risk of lung cancer and death.
2	Phytosterols	Almonds, Walnuts, Sesame seeds, Whole wheat, Brown rice, Oats, Lentils, Avocado, Broccoli, Berries	Lowers cholesterol levels, reduces risk of cardiovascular diseases, anti-inflammatory, immune-modulating and anti-cancer properties	2 g/day	Dosages as high as 45 g/d were reported to be well tolerated without severe side effects
3	Polyphenols	Berries, Cherries, Grapes, Apples, Citrus fruits, Almonds, Flaxseeds, Sesame seeds, Soybeans, Barley	Anti-inflammatory effects improve heart health, reduce the risk of certain cancers and neurodegenerative diseases, regulate blood sugar levels, promote healthy skin by protecting against UV radiation damage, reduce signs of ageing, and promote collagen synthesis.	It is estimated that the average daily diet of 1.5 g	carcinogenic/ genotoxic effects
4	Anthocyanins	Blueberries, cherries, red cabbage	Protect cells from damage, support heart health	No established Recommended Dietary Reference Intake	Excessive intake may cause gastrointestinal discomfort
5	Flavonoids	Citrus fruits, tea, onions	Reduce inflammation and improve blood vessel function	209 to 1017 mg/day	Excessive intake may interact with certain medications
6	Resveratrol	Red grapes, red wine, peanuts	Protect against heart disease, promote longevity	30–150 mg	Excessive intake may lead to digestive issues

(continued)

Table 2.3 (continued)

S.No	Phytonutrients	Food Sources	Antioxidant Benefits	Recommended Nutrients Intake (Daily)	Complications due to high consumption
7	Isoflavones	Soybeans, tofu, chickpeas	Support hormonal balance, reduce cancer risk	30 and 50 mg	High consumption may affect thyroid function
8	Lycopene	Tomatoes, watermelon, pink grapefruit	Guard against prostate cancer, promote skin health	8–21 mg/day	Excessive intake may cause digestive upset
9	Curcumin	Turmeric, curry powder	Anti-inflammatory, supports brain health	0-3 mg/kg of body weight, or ~ 0 -1.4 mg/ pound	High doses may interact with blood-thinning medications
10	Quercetin	Apples, onions, berries	Allergy relief, cardiovascular support	500 to 1000 mg daily	Excessive intake may lead to kidney stones
11	Sulforaphane	Broccoli, Brussels sprouts, cauliflower	Detoxification, cancer prevention	400 mcg/day	High intake may interfere with thyroid function

5 Fiber: Promoting Gut Health

Fibre is crucial in maintaining a healthy diet and fostering optimal gut health. It actively governs the microbiome's composition, diversity, and richness, serving as a valuable source of substrates for fermentation reactions. These reactions are conducted by specific microorganisms that possess the enzymatic tools required to degrade the intricate carbohydrates found in fibre (Cronin et al. 2021). The recommended daily fibre intake for adult men is approximately 36 grams, while for women, it is around 28 grams (Anderson et al. 2009). Fibre-rich foods, including nuts, whole-grain flour, fruits, and vegetables, have been linked to several health benefits. These benefits encompass reduced levels of low-density lipoprotein (LDL) cholesterol, decreased insulin demand, increased stool bulk, softer faecal contents, improved laxative properties, and a positive impact on body weight regulation. Moreover, fibre has been associated with an increase in the production of short-chain fatty acid (SCFA) metabolites in the gut of mice. This increase in SCFA production is accompanied by an elevation in butyrate receptor levels and a significant reduction in polyposis (Bishehsari et al. 2018). A high dietary fibre intake has been consistently linked to significantly lower prevalence rates of coronary heart disease (CHD), stroke, and peripheral vascular disease (Liu et al. 1999). Higher consumption of whole grains is associated with a substantial 26% decrease in the prevalence of ischemic strokes. Specifically, data from four studies involving over 134,000 individuals demonstrate that individuals in the highest quintile of dietary fibre or whole grain intake have a relative risk of stroke of 0.74 compared to those in the lowest quintile. This indicates a significantly lower risk of stroke for individuals with higher intakes of dietary fibre or whole grains (Steffen et al. 2003).

5.1 Soluble Fiber

Soluble fibre is a type of fibre that can dissolve in water and form a gel-like substance in the digestive tract. It is commonly found in oats, barley, legumes, fruits, and vegetables. When consumed, soluble fibre absorbs water and undergoes fermentation in the colon. Soluble fibre can be categorised into viscous (gel-forming) and non-viscous. Viscosity, in the context of dietary fibre, refers to the ability of specific polysaccharides to thicken or form gels when mixed with fluids. This occurs due to physical entanglements among the polysaccharide constituents within the solution. Soluble non-viscous fibres, such as inulin and wheat dextrin, are rapidly and completely fermented in the gut. As a result, they have the potential to produce significant amounts of gas. On the other hand, soluble viscous fibres, like psyllium, and insoluble fibres, such as wheat bran, are relatively less fermentable, leading to lower production of gas (Chutkan et al. 2012).

5.2 Insoluble Fiber

Unlike soluble fibre, insoluble fibre does not dissolve in water but adds bulk to the stool. It is commonly found in whole grains, nuts, seeds, and the skin of fruits and vegetables. Insoluble fibre promotes regular bowel movements by increasing stool volume and facilitating passage through the intestines. Examples of insoluble fibre-rich foods include wheat and rice (Baky et al. 2022).

5.3 Prebiotic Fiber: Nourishing the Gut Microbiome

Prebiotic dietary fibres are specific types of compounds that have a significant impact on shaping the microbiota. They serve as a carbon source, promoting beneficial bacterial growth and inducing specific microbiome changes that contribute to the host's overall health and metabolism. Due to the numerous health benefits of these bacterial genera, they are often used as indicators of microbiota health and targeted for dietary interventions. Lactobacilli, for example, have been found to decrease mucosal inflammation in the gastrointestinal (GI) tract. They also assist in the digestion of lactose for lactose-intolerant individuals, alleviate constipation, improve symptoms of irritable bowel syndrome (IBS), and potentially help prevent traveller diarrhoea. On the other hand, *Bifidobacteria* naturally inhabit the GI tract

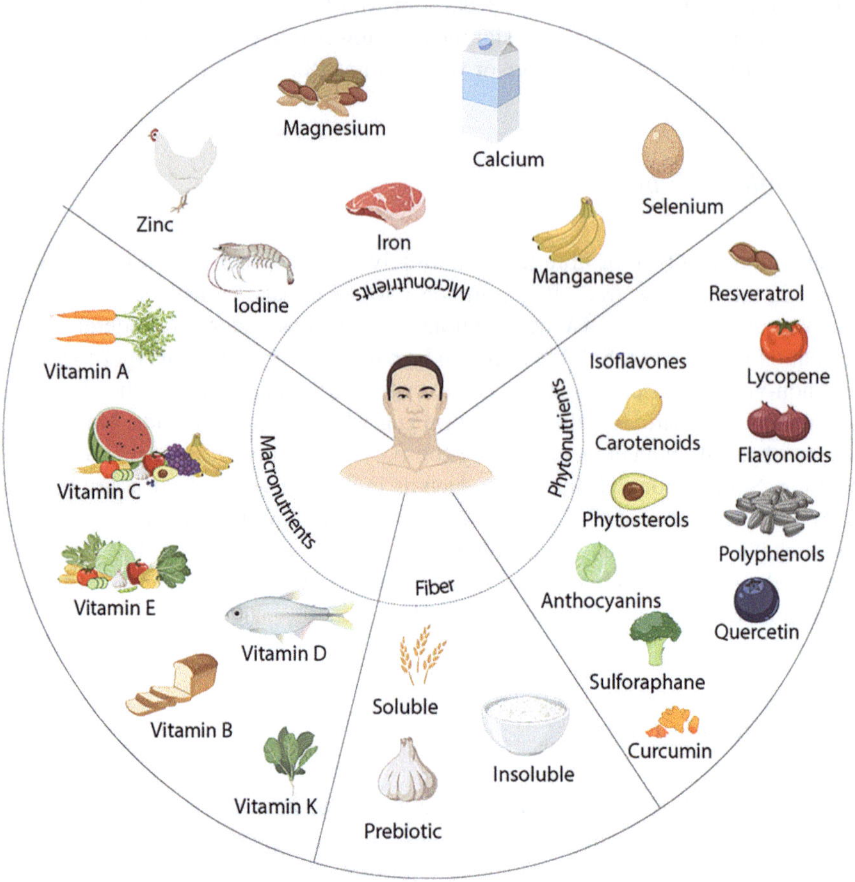

Fig. 2.1 Illustration showing the various nutrients and its sources

of healthy adults and possess a solid capacity to ferment specific oligosaccharides. This characteristic makes them a standard marker for evaluating prebiotic potential and effectiveness (Carlson et al. 2018; Slavin 2013) (Fig. 2.1).

6 Hydration and Cellular Function

To maintain a proper water and mineral balance, the body relies on a complex system involving detectors located throughout various sites in the body. These detectors are interconnected through neural pathways and communicate with integrative centres in the brain that process this information. These centres also respond to humoral factors, known as neurohormones, which regulate diuresis, natriuresis, and blood pressure. Such neurohormones include angiotensin, mineralocorticoids, vasopressin, and atrial natriuretic factor. In the case of water deficit, the extracellular compartment experiences an increase in ionic concentration, leading to water movement from the intracellular compartment and causing cells to shrink. Two types of brain sensors detect this shrinkage—one controlling drinking behaviour and the other regulating urine excretion. These sensors send signals to the kidneys, primarily by releasing the antidiuretic hormone vasopressin, producing a smaller volume of more concentrated urine. Conversely, when the body has an excess of water, the reverse processes take place. The lower ionic concentration in body fluids allows more water to enter the intracellular compartment. This leads to cell hydration, inhibiting the sensation of thirst, and prompts the kidneys to excrete more water (Popkin et al. 2010; Lorenzo et al. 2019).

Additionally, there is a link between exercise-related asthma and low fluid intake, which can contribute to bronchopulmonary disorders. Insufficient fluid intake during exercise can lead to dehydration, harming the airways and respiratory system. Dehydration can make the airways more sensitive and prone to irritation, increasing the risk of bronchospasm and exacerbating asthma symptoms during physical exertion. Individuals with exercise-related asthma need to maintain proper hydration by consuming adequate fluids before, during, and after exercise to help mitigate the potential impact on bronchopulmonary health (Kalhoff 2003).

7 Nutrient Absorption and Metabolism

Nutrient absorption is a vital process through which the body obtains essential macronutrients, micronutrients, and other necessary components from our food. This intricate process primarily occurs in the gastrointestinal (GI) tract, with the small intestine being the leading site for nutrient absorption. On the other hand, the colon mainly handles the transport of water and electrolytes (Physiology n.d.). In the small intestine, three main mechanisms contribute to the transport of sodium ions (Na+) across the apical membrane: (a) Nutrient-coupled Na + absorption: This mechanism involves the participation of various families of Na + -dependent nutrient transporters. For example, specific transporters for sugars and amino acids play a crucial role in absorbing these nutrients. (b) Electroneutral NaCl absorption: This process primarily occurs through the Na+/H+ exchange mechanism. It involves the exchange of Na + and H+ ions, resulting in the absorption of both sodium and chloride ions. (c) Electrogenic Na + absorption in the colon: This mechanism is predominantly found in the colon and involves the activity of epithelial Na + channels (ENaC), which facilitate the electrogenic absorption of sodium ions. These mechanisms collectively contribute to the efficient absorption of nutrients, maintaining electrolyte balance, and water transport throughout the GI tract (Feranchak 2003; Rajendran et al. 2018).

Efficient chloride absorption in the gut results in an average stool concentration of 10–15 millimoles per litre (mmol/L). However, in malabsorption conditions, such as congenital chloride diarrhoea, the chloride concentration in the stool can exceed 90 mmol/L. Chloride is absorbed from the intestinal lumen through three distinct mechanisms: (a) Paracellular (passive) pathway: Chloride can passively move between cells through the space between them, known as the paracellular pathway. (b) Electroneutral pathways: These pathways involve coupled transport of sodium (Na+) and hydrogen ions (H+) with chloride ions (Cl-) exchange. The movement of Na + and H+ ions drives the absorption of chloride. (c) Protein-dependent chloride absorption: Certain intestinal epithelium proteins facilitate chloride ions' active transport (Said et al. 2000). Water-soluble vitamins are essential for average growth and development as they participate in various metabolic processes. Unlike fat-soluble vitamins, humans cannot synthesise many water-soluble vitamins and must rely on their intake from external sources. An adequate supply of water-soluble vitamins through a balanced diet is essential to meet the body's requirements (Biochemistry n.d.).

Vitamin C, also known as ascorbic acid, is a water-soluble vitamin that humans cannot synthesise and must obtain from dietary sources. Its intestinal transport occurs through a carrier-mediated mechanism dependent on sodium ions (Na+). There are two known Na + -dependent carriers for vitamin C in humans: vitamin C transporter-1 (SVCT1) and SVCT2, encoded by the SLC23A1 and SLC23A2 genes. Biotin, another water-soluble vitamin, can be obtained from two sources in the human intestine: dietary sources and bacterial production. Biotin is absorbed through a carrier-mediated process dependent on sodium ions (Na+). The intestinal sodium-dependent multivitamin transporter (SMVT), encoded by the SLC5A6 gene, is responsible for biotin uptake throughout the intestinal tract (Padayatty and Levine 2016).

Cobalamin, or vitamin B12, is obtained from animal-derived foods and can be produced by the colonic microbiota. Dietary cobalamin binds to the protein haptocorrin (transcobalamin-1; TC1) in the stomach. Trypsin releases cobalamin from the TC1 complex in the small intestine, binding to gastric intrinsic factor (GIF). The B12/GIF complex binds to the multi-ligand heterodimeric receptor called cubam, composed of amnionless (AMN) and cubilin (CUBL), located at the apical membrane of ileal enterocytes. After internalisation, the complex undergoes recycling, and cobalamin is transported out of the enterocytes by the multi-drug resistant associated protein-1 (MRP1) at the basolateral membrane. Systemically, transcobalamin carries cobalamin for utilisation and storage (Kiela and Ghishan 2016). The intestinal transport of vitamin B2 (riboflavin) is believed to occur through riboflavin transporters known as RFVT1 and RFVT2, encoded by the SLC52A1 and SLC52A2 genes, respectively.

Thiamin, also known as vitamin B1, is widely available in dietary sources such as whole grains, nuts, and dried legumes. Intestinal absorption of thiamin is attributed to two thiamin transporters in humans: THTR1 and THTR2, encoded by the SLC19A2 and SLC19A3 genes, respectively. These transporters are expressed throughout the gastrointestinal tract (Mosegaard et al. 2020). Vitamin A, a fat-soluble vitamin, exists in the diet as retinol or ethyl esters (preformed) and as

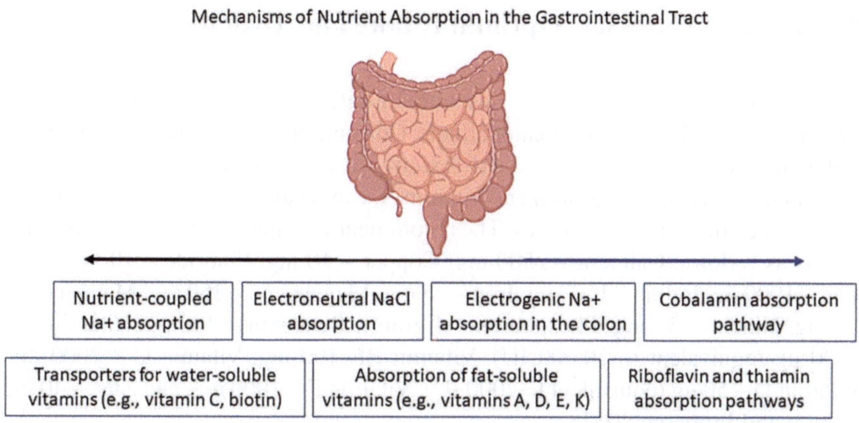

Fig. 2.2 Illustration showing the different mechanisms of nutrients absorption in GI tract

pro-vitamin A carotenoids. The absorption of vitamin A in the intestines depends on its dietary form. Pro-vitamin A carotenoids are converted to retinol within enterocytes. As a fat-soluble vitamin, retinol and retinol esters are re-esterified before incorporation into chylomicrons with other lipids. Chylomicrons enter the circulation via the lymphatics and are taken up by hepatocytes through apolipoprotein E-dependent internalisation. Within hepatocytes, retinol esters are stored primarily in the form of palmitate. Retinol-binding proteins transport vitamin A from the liver to peripheral tissues (Reboul 2013) (Fig. 2.2).

Vitamin D has two major forms: vitamin D2 (ergocalciferol) and vitamin D3 (cholecalciferol). Vitamin D3 is synthesised in the skin upon exposure to ultraviolet B (UVB) radiation. Pre-vitamin D3 is then converted to cholecalciferol, which is transported to the liver bound to a binding protein. In the liver, both vitamin D2 and D3 undergo hydroxylation to form 25-hydroxyvitamin D. Subsequently, in the kidneys, 1α-hydroxylation occurs to produce the biologically active form of vitamin D, known as 1,25-dihydroxyvitamin D3 or calcitriol. Vitamin D is involved in the regulation of intestinal absorption of calcium (Ca^{2+}) and phosphate (Pi) and their reabsorption in the kidneys (Ross 2011). Vitamin E encompasses a group of compounds known as tocopherols and tocotrienols. The main biological activity is attributed to α-tocopherol, predominant in most foods. Vitamin E is a lipid-soluble antioxidant, protecting cells from reactive oxygen species generated during lipid oxidation. Vitamin K1 is converted to vitamin K2 (menaquinones) by intestinal bacteria or can be synthesised de novo. Vitamin K2 is the primary storage form of vitamin K. Absorption of vitamin K occurs through the lipid pathway in the terminal ileum and is dependent on biliary and pancreatic enzymes. The level of gamma-carboxylated proteins in the blood indirectly measures vitamin K status (Szewczyk et al. 2021).

8 Nutrient Excesses: Optimal Intake and Toxicity

The requirement for each nutrient lies within a safe and desirable range, exceeding which poses a risk of adverse health effects. Consumption of any nutrient, including water, in excessive amounts can lead to toxicity, depending on the dosage. Overdoses of certain nutrients, such as iron and vitamin A, can result in poisoning, acute toxicity, and even death in severe cases. The recommended daily intake for various nutrients is as follows: Calcium – 2500 mg, Copper – 10 mg, Fluoride – 10 mg, Folic acid – 1000 µg, Iodine – 1100 µg, Iron – 45 mg, Magnesium – 350 mg, Manganese – 11 mg, Niacin – 35 mg, Phosphorus – 4 grams, Selenium – 400 µg, Vitamin A – 3000 µg (equivalent to 10,000 IU), Vitamin B6–100 mg, Vitamin C – 2000 mg, Vitamin D – 50 µg (equivalent to 2000 IU), Vitamin E – 1000 mg, and Zinc – 40 mg (Saylor and Probasco 2017).

9 Nutraceuticals and Nutrient Synergy

Food contains essential nutrients with specific active compounds with functional properties that can help prevent diseases. The immune system, which consists of various tissues, cells, organs, proteins, and chemicals, plays a crucial role in protecting the body. Therefore, the immune system's effectiveness can be seen as an indicator of proper nutrition (Natarajan et al. 2019). Garlic and honey exhibit a beneficial synergistic effect. When garlic is ingested and reaches the stomach, it stimulates the production of gastric juices, which play a vital role in iron absorption and enhance the circulation of nutrients in the bloodstream (Shoba et al. 1998). Almond skin, rich in flavonoids, provides significant health benefits and works synergistically with vitamin C and vitamin E to protect against the oxidation of LDL (low-density lipoprotein) (Chen et al. 2005). The salads containing tomatoes, carrots, and green leafy vegetables provide carotenoids such as lycopene, lutein, beta-carotene, alpha-carotene, and zeaxanthin. Additionally, egg yolks also contain zeaxanthin and lutein. Consuming boiled eggs together with the salad has been shown to significantly increase the absorption of carotenoids, ranging from 3 to 9 times higher than consuming the salad alone (Cook and Briggs 1986).

Lemon and green leafy vegetables are beneficial in addressing iron deficiency anaemia by increasing haemoglobin levels. They help improve the bioavailability of iron in the bloodstream. This is achieved through ascorbic acid or vitamin C, which enhances the absorption of non-heme iron from the diet. Vitamin C first converts ferric iron into a non-absorbable form and then facilitates its conversion back to ferrous iron, which is more easily absorbed by the mucosal cells, leading to better iron absorption (Lynch and Cook 1980). Yoghurt and bananas synergise, providing mutual benefits through probiotics and prebiotics. Probiotics introduce beneficial bacteria into the gut, while prebiotics serve as a source of nourishment for these good bacteria. The gut bacteria are supported and encouraged to thrive by consuming probiotics and prebiotics, improving gastrointestinal digestion and overall gut health (Krisnanda 2019).

Bananas contain inulin, a prebiotic fibre that is a fuel source for growing beneficial bacteria in yoghurt. This symbiotic relationship between inulin and probiotics supports the improvement of immune function and regulation of digestion. Additionally, when combined with probiotics, inulin helps to enhance the absorption and utilisation of calcium in the body, leading to increased levels of this essential mineral (Fernandez and Marette 2017).

10 Complementary Nutrients: Enhancing Bioavailability

Using complementary nutrient-rich whole food sources to address micronutrient and calorie malnutrition has gained significant attention. Recent studies have highlighted the potential of lentils and kale as complementary food sources to combat global malnutrition and calorie-related issues. These foods provide a diverse range of essential nutrients and offer the opportunity for further enrichment through prebiotic carbohydrates and genetic advancements. By exploring the carbohydrate profiles and genetic potential of lentils and kale, we can work towards developing sustainable and nutritious food systems that contribute to addressing the global challenge of malnutrition (Migliozzi et al. 2015).

A separate study focused on the potential of fortifying maise/soybean and peanut with *Moringa oleifera* leaf powder to create protein-rich complementary food products. This fortification approach enhanced nutrient quantity, quality, and availability in the fortified products. Specifically, feeding studies conducted with rats demonstrated improved protein quality indices, indicating the enhanced nutritional value of the fortified foods. This suggests that incorporating Moringa oleifera leaf powder into these food products can contribute to developing acceptable and nutritionally improved complementary foods (Shiriki et al. 2015).

11 Nutrient Interactions: Synergistic Effects

The relationship between infection and malnutrition has long been recognised as synergistic. While extensive research exists on the role of individual nutrients in immune function, the understanding of multiple nutrients and their interactions is less explored. The availability of one nutrient can influence the action of another nutrient in the immune system, as observed with vitamin E and selenium, vitamin E and vitamin A, zinc and copper, and dietary fatty acids and vitamin A. These nutrient-nutrient interactions can hurt immune function (Scrimshaw and SanGiovanni 1997).

The association between vitamin E and selenium was among the first reported interactions in mammalian metabolism. Schwarz and Foltz discovered that small amounts of selenium could prevent liver necrosis in rats deficient in vitamin E. This finding began a relationship between these two nutrients and sparked a further investigation into their interconnected roles (Kubena and McMurray 1996a). Another notable example of an interaction involving vitamin E is its relationship with vitamin C. An in vitro study demonstrated a synergistic effect of supplemental vitamins E and C on the inhibition of the release of arachidonic acid. This suggests that when combined, these two vitamins may significantly impact specific biochemical processes more than when consumed individually (ElAttar and Lin 1992).

Several studies have noted an antagonistic effect between vitamins E and A. While both vitamin E and A supplements were found to increase antibody production and phagocytosis, it was observed that when either one was increased, immune function was less pronounced. This effect may be attributed to the suppression of gastrointestinal absorption and tissue reserves of vitamin E caused by high vitamin A levels. The interaction between these two vitamins highlights the importance of maintaining a balance in their intake for optimal immune function (Tengerdy and Nockels 1975; Tengerdy and Brown 1977; Pudelkiewicz et al. 1964).

Vitamin A and vitamin D are both essential for optimal immune responses. Vitamin A or retinol deficiency can impair mitogen-induced proliferation in rats, reduce antibody responses to pneumococcal polysaccharides, and decrease natural killer (NK) cell and interferon-gamma (IFN-γ) activity in rats. Moreover, vitamin A deficiency has been associated with impaired resistance to infectious diseases in rodents and chickens. These findings highlight the importance of adequate vitamin A and vitamin D intake for maintaining a robust immune system and effective defence against infections (Kubena and McMurray 1996b).

12 Nutrient Requirements Across the Lifespan

During pregnancy, the energy and nutrient demands of the body increase significantly. The total energy cost of pregnancy is estimated to be around 80,000 kcal, which translates to an average daily requirement of 2500 kcal for women in the second and third trimester, or an additional 200–500 kcal per day, depending on their activity level. According to the Dietary Reference Intake (DRI) recommendations, pregnant women are advised to consume an extra 340 kcal during the second and 452 kcal during the third trimester. Additionally, they should aim for approximately 130 grams of carbohydrates per day and about 1.1 grams of protein per kilogram of body weight per day. The need for iron also significantly increases during pregnancy, nearly doubling to 18 to 27 mg per day. This is due to the 50% increase in the mother's blood volume. Adequate intake of long-chain polyunsaturated fatty acids, particularly docosahexaenoic acid (DHA), is crucial for fetal brain growth and visual development. It is recommended that pregnant women consume 300 mg of DHA per day. For lactating mothers, the energy requirement further increases. They are estimated to need about 640 additional kcal per day to support breast milk production. This ensures an adequate supply of nutrients for both the mother and the baby during breastfeeding (Ilich and Brownbill 2009; Nutrition during lactation 1996).

A woman's energy requirements increase during lactation to support breast milk production. Per the Dietary Reference Intake (DRI), the current recommendation suggests an additional 330 kcal/day intake for the first 6 months and 400 kcal/day for the second 6 months. This gradual increase allows for a healthy weight loss during lactation. For protein, the DRI recommends 13 grams per day for children between the ages of 1 and 3, and it increases to 19 grams per day for children aged 4 to 8. The DRI for carbohydrates remains constant at 130 grams daily for children of all ages. The recommended intake of fat should be 25–35% of the total energy intake for children between the ages of 4 and 18. Calcium is vital for children's growth, and the DRI suggests that children between the ages of 1 and 3 should consume 1–2 servings of dairy per day to meet the calcium requirement of 500 mg/day. For children between the ages of 4 and 8, 2–3 servings of dairy are recommended to meet the calcium requirement of 800 mg/day (Nutrition during Pregnancy and Lactation 2020).

Significant growth occurs during adolescence (8–17), with about 20% of adult height and 50% of adult weight gained. The DRI recommendations are divided into two age groups: 9–13 and 14–18. Both males and females in these age groups are advised to consume 56 grams and 46 grams of protein daily (Ilich and Brownbill 2009; Soliman et al. 2014). The DRI recommendations are divided into three age categories for adults: 19–30, 31–50, and over 50. The estimated energy requirements are 2403 kcal for females and 3067 kcal for males in all three age categories, with slight calorie reductions for each year above 19. The recommended protein intake for males and females aged 19 and over is 0.8 grams per kilogram of body weight per day. In older adults, the nutrient requirements change. For males over 50 years, the DRI for vitamin B6 increases to 1.7 mg/day, while for females over 50 years, it increases to 1.5 mg/day. The calcium requirement increases to 1200 mg/day for adults over 50 years. The iron requirement decreases to 8 mg/day for females over 50 years. Chromium and vitamin D requirements also change with age (Ilich and Brownbill 2009; National Research Council (US) 1989).

It's important to note that these recommendations provide a general guideline, and individual needs may vary based on factors such as activity level, health status, and specific nutritional requirements. Consulting with a healthcare professional or registered dietitian can provide personalised recommendations based on individual circumstances.

References

Anderson JW, Baird P, Davis RH Jr, Ferreri S, Knudtson M, Koraym A, Waters V, Williams CL. Health benefits of dietary fiber. Nutr Rev. 2009;67(4):188–205. https://doi.org/10.1111/j.1753-4887.2009.00189.x.

Astrup A, Geiker NRW, Magkos F. Effects of low-carbohydrate diets on weight loss and cardiometabolic health: a comprehensive review. Eur J Clin Nutr. 2020;74(3):324–33. https://doi.org/10.1038/s41430-019-0506-9.

Baky MH, Salah M, Ezzelarab N, Shao P, Elshahed MS, Farag MA. Insoluble dietary fibers: structure, metabolism, interactions with human microbiome, and role in gut homeostasis. Crit Rev Food Sci Nutr. 2022:1–15. https://doi.org/10.1080/10408398.2022.2119931.

Bianchi F, Gutierrez M. The role of dietary protein in human health. Nutrients. 2019;11(5):1012. https://doi.org/10.3390/nu11051012.

Biesalski HK, Tinz J. Multivitamin/mineral supplements: rationale and safety considerations. Int J Vitam Nutr Res. 2018;88(3–4):179–84. https://doi.org/10.1024/0300-9831/a000477.

Biochemistry, water soluble vitamins – statpearls – NCBI bookshelf. n.d.. Available at: https://www.ncbi.nlm.nih.gov/books/NBK538510/. Accessed 10 July 2023.

Biosynthesis of vitamins in plants part a – vitamins A, B1, B2, b3, B5. 2011. Advances in Botanical Research. https://doi.org/10.1016/c2010-0-66876-3.

Bishehsari F, Engen P, Preite N, Tuncil Y, Naqib A, Shaikh M, Rossi M, Wilber S, Green S, Hamaker B, Khazaie K, Voigt R, Forsyth C, Keshavarzian A. Dietary fiber treatment corrects the composition of gut microbiota, promotes SCFA production, and suppresses colon carcinogenesis. Genes. 2018;9(2):102. https://doi.org/10.3390/genes9020102.

Böhm V, Lietz G, Olmedilla-Alonso B, Phelan D, Reboul E, Bánati D, Borel P, Corte-Real J, de Lera AR, Desmarchelier C, Dulinska-Litewka J, Landrier J-F, Milisav I, Nolan J, Porrini M, Riso P, Roob JM, Valanou E, Wawrzyniak A, et al. From carotenoid intake to carotenoid blood and tissue concentrations—implications for dietary intake recommendations. Nutr Rev. 2020;79(5):544–73. https://doi.org/10.1093/nutrit/nuaa008.

Bost M, Houdart S, Oberli M, Kalonji E, Huneau J-F, Margaritis I. Dietary copper and human health: current evidence and unresolved issues. J Trace Elem Med Biol. 2016;35:107–15. https://doi.org/10.1016/j.jtemb.2016.02.006.

Bytomski JR. Fueling for performance. Sports Health. 2018;10(1):47–53. https://doi.org/10.1177/1941738117743913. Epub 2017 Nov 27. PMID: 29173121; PMCID: PMC5753973.

Calder PC. Functional roles of fatty acids and their effects on human health. J Parenter Enter Nutr. 2015;39(1_suppl):18S–32S. https://doi.org/10.1177/0148607115595980.

Calder PC. Omega-3 fatty acids and inflammatory processes: from molecules to man. Biochem Soc Trans. 2017;45(5):1105–15. https://doi.org/10.1042/BST20160474.

Calder PC, Carr AC, Gombart AF, Eggersdorfer M. Optimal nutritional status for a well-functioning immune system is critical to protect against viral infections. Nutrients. 2020;12(4):1181. https://doi.org/10.3390/nu12041181.

Carazo A, Macáková K, Matoušová K, Krčmová LK, Protti M, Mladěnka P. Vitamin A update: forms, sources, kinetics, detection, function, deficiency, therapeutic use and toxicity. Nutrients. 2021;13(5):1703. https://doi.org/10.3390/nu13051703.

Carlson JL, Erickson JM, Lloyd BB, Slavin JL. Health effects and sources of prebiotic dietary fiber. Curr Dev Nutr. 2018;2(3) https://doi.org/10.1093/cdn/nzy005.

Chen C-Y, Milbury PE, Lapsley K, Blumberg JB. Flavonoids from Almond skins are bioavailable and act synergistically with vitamins C and E to enhance hamster and human LDL resistance to oxidation. J Nutr. 2005;135(6):1366–73. https://doi.org/10.1093/jn/135.6.1366.

Chen Y, Michalak M, Agellon LB. Importance of nutrients and nutrient metabolism on human health. Yale J Biol Med. 2018;91(2):95–103. PMID: 29955217; PMCID: PMC6020734.

Chutkan R, Fahey G, Wright WL, McRorie J. Viscous versus nonviscous soluble fiber supplements: mechanisms and evidence for fiber-specific health benefits. J Am Acad Nurse Pract. 2012;24(8):476–87. https://doi.org/10.1111/j.1745-7599.2012.00758.x.

Cook F, Briggs GM. The nutritive value of eggs. Egg Sci Technol. 1986:141–63. https://doi.org/10.1007/978-1-349-09142-3_7.

Crichton GE, Bryan J. Dietary carbohydrates, glycemic index, and cognition. Crit Rev Food Sci Nutr. 2015;55(11):1547–60. https://doi.org/10.1080/10408398.2012.661324.

Cronin P, Joyce SA, O'Toole PW, O'Connor EM. Dietary fibre modulates the gut microbiota. Nutrients. 2021;13(5):1655. https://doi.org/10.3390/nu13051655.

De la Fuente M, Hernanz A, Guayerbas N. Vitamin C intake, leukocyte vitamin C concentration, and circulating inflammatory markers in humans. Adv Nutr. 2017;8(6):947–56. https://doi.org/10.3945/an.117.015206.

Deshpande J, Joshi M, Giri P. Zinc: the trace element of major importance in human nutrition and health. Int J Med Sci Public Health. 2013;2(1):1. https://doi.org/10.5455/ijmsph.2013.2.1-6.

Dietary reference intakes for thiamin, riboflavin, niacin, vitamin B6, folate, vitamin B12, pantothenic acid, biotin, and choline. 1998. https://doi.org/10.17226/6015.

Doseděl M, Jirkovský E, Macáková K, Krčmová L, Javorská L, Pourová J, Mercolini L, Remião F, Nováková L, Mladěnka P. Vitamin C—sources, physiological role, kinetics, deficiency, use, toxicity, and determination. Nutrients. 2021;13(2):615. https://doi.org/10.3390/nu13020615.

Eichelmann F, Schwingshackl L, Fedirko V, Aleksandrova K. Effect of plant-based diets on obesity-related inflammatory profiles: a systematic review and meta-analysis of intervention trials. Obes Rev. 2016;17(11):1067–79. https://doi.org/10.1111/obr.12439.

ElAttar TMA, Lin HS. Effect of vitamin C and vitamin E on prostaglandin synthesis by fibroblasts and squamous carcinoma cells. Prostaglandins Leukot Essent Fat Acids. 1992;47(4):253–7. https://doi.org/10.1016/0952-3278(92)90194-n.

Farag MA, Abib B, Qin Z, Ze X, Ali SE. Dietary macrominerals: updated review of their role and orchestration in human nutrition throughout the life cycle with sex differences. Curr Res Food Sci. 2023;6:100450. https://doi.org/10.1016/j.crfs.2023.100450. PMID: 36816001; PMCID: PMC9932710.

Feranchak AP. Ion channels in digestive health and disease. J Pediatr Gastroenterol Nutr. 2003;37(3):230–41. https://doi.org/10.1097/00005176-200309000-00006.

Fernandez MA, Marette A. Potential health benefits of combining yogurt and fruits based on their probiotic and prebiotic properties. Adv Nutr. 2017;8(1) https://doi.org/10.3945/an.115.011114.

Fulgoni VL, Keast DR, Bailey RL, Dwyer J. Foods, fortificants, and supplements: where do Americans get their nutrients? J Nutr. 2011;141(10):1847–54. https://doi.org/10.3945/jn.111.142257.

Gómez-Gallego C, Salminen S. Nutritional interventions and gut microbiota in older people: a literature review. J Funct Foods. 2021;80:104467. https://doi.org/10.1016/j.jff.2020.104467.

Holick MF. The vitamin D deficiency pandemic and consequences for nonskeletal health: mechanisms of action. Mol Asp Med. 2017;57:28–40. https://doi.org/10.1016/j.mam.2016.10.001.

Hovenkamp E, Demonty I, Plat J, Lütjohann D, Mensink RP, Trautwein EA. Biological effects of oxidized phytosterols: a review of the current knowledge. Prog Lipid Res. 2008;47(1):37–49. https://doi.org/10.1016/j.plipres.2007.10.001.

Ilich JZ, Brownbill RA. Nutrition through the life span: needs and health concerns in critical periods. In: Handbook of stressful transitions across the lifespan, pp. 625–641; 2009. https://doi.org/10.1007/978-1-4419-0748-6_31.

Janoušek J, Pilařová V, Macáková K, Nomura A, Veiga-Matos J, Silva DD, Remião F, Saso L, Malá-Ládová K, Malý J, Nováková L, Mladěnka P. Vitamin D: sources, physiological role, biokinetics, deficiency, therapeutic use, toxicity, and overview of analytical methods for detection of vitamin D and its metabolites. Crit Rev Clin Lab Sci. 2022;59(8):517–54. https://doi.org/10.1080/10408363.2022.2070595.

Jones PJH. Ingestion of phytosterols is not potentially hazardous. OUP Academic; 2007, November 1. https://academic.oup.com/jn/article/137/11/2485/4664490.

Kalhoff H. Mild dehydration: a risk factor of broncho-pulmonary disorders? Eur J Clin Nutr. 2003;57(S2) https://doi.org/10.1038/sj.ejcn.1601906.

Kiela PR, Ghishan FK. Physiology of intestinal absorption and secretion. Best Pract Res Clin Gastroenterol. 2016;30(2):145–59. https://doi.org/10.1016/j.bpg.2016.02.007.

Krisnanda R. Vitamin C helps in the absorption of iron in iron deficiency anemia. Jurnal Penelitian Perawat Profesional. 2019;2(3) https://doi.org/10.37287/jppp.v2i3.137.

Kubena KS, McMurray DN. Nutrition and the immune system. J Am Diet Assoc. 1996a;96(11):1156–64. https://doi.org/10.1016/s0002-8223(96)00297-0.

Kubena KS, McMurray DN. Nutrition and the immune system. J Am Diet Assoc. 1996b;96(11):1156–64. https://doi.org/10.1016/s0002-8223(96)00297-0.
Liu S, Stampfer MJ, Hu FB, Giovannucci E, Rimm E, Manson JE, Hennekens CH, Willett WC. Whole-grain consumption and risk of coronary heart disease: results from the Nurses' Health Study. Am J Clin Nutr. 1999;70(3):412–9. https://doi.org/10.1093/ajcn/70.3.412.
Lobo V, Patil A, Phatak A, Chandra N. Free radicals, antioxidants and functional foods: impact on human health. Pharmacogn Rev. 2010;4(8):118–26. https://doi.org/10.4103/0973-7847.70902. PMID: 22228951; PMCID: PMC3249911.
Lorenzo I, Serra-Prat M, Yébenes JC. The role of water homeostasis in muscle function and frailty: a review. Nutrients. 2019;11(8):1857. https://doi.org/10.3390/nu11081857.
Lynch SR, Cook JD. Interaction of vitamin C and iron. Ann N Y Acad Sci. 1980;355(1 Micronutrient):32–44. https://doi.org/10.1111/j.1749-6632.1980.tb21325.x.
Matkovic V, Ilich JZ. Calcium requirements for growth: are current recommendations adequate? Nutr Rev. 2009;51(6):171–80. https://doi.org/10.1111/j.1753-4887.1993.tb03097.x.
Mayurasakorn K, Williams FE, Ten Eyck RP, Vaag AA, Færgeman NJ, Lyssiotis CA. Targeting physiologic nutrient signalling for the metabolic treatment of insulin-requiring diabetes. Mol Metabol. 2020;42:101067. https://doi.org/10.1016/j.molmet.2020.101067.
Mennen LI, Walker R, Bennetau-Pelissero C, Scalbert A. Risks and safety of polyphenol consumption. Am J Clin Nutr. 2005;81(1) https://doi.org/10.1093/ajcn/81.1.326s.
Migliozzi M, Thavarajah D, Thavarajah P, Smith P. Lentil and kale: complementary nutrient-rich whole food sources to combat micronutrient and calorie malnutrition. Nutrients. 2015;7(11):9285–98. https://doi.org/10.3390/nu7115471.
Mladěnka P, Macáková K, Kujovská Krčmová L, Javorská L, Mrštná K, Carazo A, Protti M, Remião F, Nováková L. Vitamin K—sources, physiological role, kinetics, deficiency, detection, therapeutic use, and toxicity. Nutr Rev. 2021;80(4):677–98. https://doi.org/10.1093/nutrit/nuab061.
Molfino A, Gioia G, Rossi Fanelli F, Muscaritoli M. The role for dietary omega-3 fatty acids supplementation in older adults. Nutrients. 2013;5(8):3152–63. https://doi.org/10.3390/nu5083152.
Mosegaard S, et al. Riboflavin deficiency—implications for general human health and Inborn Errors of Metabolism. Int J Mol Sci. 2020;21(11):3847. https://doi.org/10.3390/ijms21113847.
Myles IA. Fast food fever: reviewing the impacts of the Western diet on immunity. Nutr J. 2014;13:61. https://doi.org/10.1186/1475-2891-13-61.
Nasri H, Baradaran A, Shirzad H, Rafieian-Kopaei M. New concepts in nutraceuticals as alternative for pharmaceuticals. Int J Prev Med. 2014;5(12):1487–99. PMID: 25709784; PMCID: PMC4336979.
Natarajan TD, Ramasamy JR, Palanisamy K. Nutraceutical potentials of synergic foods: a systematic review. J Ethnic Foods. 2019;6(1) https://doi.org/10.1186/s42779-019-0033-3.
National Research Council (US). Subcommittee on the tenth edition of the recommended dietary allowances. Recommended dietary allowances: 10th Edition. Washington, DC: National Academies Press (US); 1989. 3, Energy. Available from: https://www.ncbi.nlm.nih.gov/books/NBK234938/
Niki E, Abe K. Chapter 1. vitamin E: Structure, properties and functions. Vitamin E; 2019, pp. 1–11. https://doi.org/10.1039/9781788016216-00001.
Nutrition during lactation. National Academy Press. 1996.
Nutrition during Pregnancy and Lactation. 2020. https://doi.org/10.3390/books978-3-03928-055-1.
Opazo MC, Coronado-Arrázola I, Vallejos OP, Moreno-Reyes R, Fardella C, Mosso L, Kalergis AM, Bueno SM, Riedel CA. The impact of the Micronutrient Iodine in health and diseases. Crit Rev Food Sci Nutr. 2020;62(6):1466–79. https://doi.org/10.1080/10408398.2020.1843398
Padayatty S, Levine M. Vitamin C: the known and the unknown and goldilocks. Oral Dis. 2016;22(6):463–93. https://doi.org/10.1111/odi.12446.
Physiology, digestion – statpearls – NCBI bookshelf. (n.d.). https://www.ncbi.nlm.nih.gov/books/NBK544242/
Pizzino G, Irrera N, Cucinotta M, Pallio G, Mannino F, Arcoraci V, Squadrito F, Altavilla D, Bitto A. Oxidative Stress: Harms and Benefits for Human Health. Oxid Med Cell Longev.

2017;2017:8416763. https://doi.org/10.1155/2017/8416763. Epub 2017 Jul 27. PMID: 28819546; PMCID: PMC5551541.

Popkin BM, D'Anci KE, Rosenberg IH. Water, hydration, and health. Nutr Rev. 2010;68(8):439–58. https://doi.org/10.1111/j.1753-4887.2010.00304.x.

Pudelkiewicz WJ, Webster L, Matterson LD. Effects of high levels of dietary vitamin A acetate on tissue tocopherol and some related analytical observations. J Nutr. 1964;84(2):113–7. https://doi.org/10.1093/jn/84.2.113.

Puri V, Nagpal M, Singh I, Singh M, Dhingra GA, Huanbutta K, Dheer D, Sharma A, Sangnim T. A comprehensive review on nutraceuticals: therapy support and formulation challenges. Nutrients. 2022;14:4637. https://doi.org/10.3390/nu14214637.

Rajendran VM, Schulzke J-D, Seidler UE. Ion channels of the gastrointestinal epithelial cells. In: Physiology of the Gastrointestinal Tract; 2018. p. 1363–404. https://doi.org/10.1016/b978-0-12-809954-4.00058-x.

Reboul E. Absorption of vitamin A and carotenoids by the enterocyte: focus on transport proteins. Nutrients. 2013;5(9):3563–81. https://doi.org/10.3390/nu5093563.

Rezende LF, Saconato H, Soares BGO, Atallah ÁN. Efficacy and safety of probiotics in the treatment of constipation: a systematic review and meta-analysis of randomised controlled trials. Neurogastroenterol Motil. 2014;26(8):1252–9. https://doi.org/10.1111/nmo.12386.

Riboflavin. Vitamin analysis for the health and food sciences; 1998. https://doi.org/10.1201/9781420050165.ch8.

Ross AC. Dietary reference intakes for calcium and vitamin D. National Academies Press; 2011.

Rucker RB, Lanoue L. Trace minerals: metabolism and deficiency (copper, zinc, selenium, manganese). In: Encyclopedia of gastroenterology; 2004. p. 502–7. https://doi.org/10.1016/b0-12-386860-2/00689-4.

Rudrapal M, Khairnar SJ, Khan J, Dukhyil AB, Ansari MA, Alomary MN, Alshabrmi FM, Palai S, Deb PK, Devi R. Dietary polyphenols and their role in oxidative stress-induced human diseases: insights into protective effects, antioxidant potentials and mechanism(s) of action. Front Pharmacol. 2022;13:806470. https://doi.org/10.3389/fphar.2022.806470.

Russo GL. Nutraceuticals and human health: the food-to-supplement paradigm. Food Frontiers. 2019;1(1):58–68. https://doi.org/10.1002/fft2.5.

Sacco SM, Horcajada M-N, Offord E. Phytonutrients for bone health during ageing. Br J Clin Pharmacol. 2013;75(3):697–707. https://doi.org/10.1111/bcp.12033.

Said HM, Rose R, Seetharam B. Chapter 2 intestinal absorption of water-soluble vitamins: cellular and molecular aspects. Curr Top Membr. 2000:35–75. https://doi.org/10.1016/s1063-5823(00)50004-0.

Saylor D, Probasco JC. Nutritional deficiencies. Oxford Medicine Online; 2017. https://doi.org/10.1093/med/9780199937837.003.0184.

Schwingshackl L, Bogensberger B, Hoffmann G. Diet quality as assessed by the healthy eating index, alternate healthy eating index, dietary approaches to stop hypertension score, and healthful plant-based diet index about cardiovascular disease and mortality risk: a systematic review and meta-analysis of prospective studies. J Acad Nutr Diet. 2021;121(8):1296–1315.e7. https://doi.org/10.1016/j.jand.2021.01.017.

Scrimshaw N, SanGiovanni J. Synergism of nutrition, infection, and immunity: an overview. Am J Clin Nutr. 1997;66(2) https://doi.org/10.1093/ajcn/66.2.464s.

Shah A. Hypomagnesemia/Hypermagnesemia. Oxford Medicine Online [Preprint]; 2017. https://doi.org/10.1093/med/9780190226459.003.0039.

Sharifi-Rad M, Anil Kumar NV, Zucca P, Varoni EM, Dini L, Panzarini E, Rajkovic J, Tsouh Fokou PV, Azzini E, Peluso I, Prakash Mishra A, Nigam M, El Rayess Y, Beyrouthy ME, Polito L, Iriti M, Martins N, Martorell M, Docea AO, Setzer WN, Calina D, Cho WC, Sharifi-Rad J. Lifestyle, oxidative stress, and antioxidants: Back and forth in the pathophysiology of chronic diseases. Front Physiol. 2020;11:694. https://doi.org/10.3389/fphys.2020.00694.

Shiriki, D., Igyor, M. A., Gernah, D. I. (2015). Nutritional evaluation of complementary food formulations from maize, soybean and peanut fortified with Moringa oleifera leaf powder Food Nutr Sci, 06(05), 494–500. https://doi.org/10.4236/fns.2015.65051.

Shoba G, Joy D, Joseph T, Majeed M, Rajendran R, Srinivas P. Influence of piperine on the pharmacokinetics of curcumin in animals and human volunteers. Planta Med. 1998;64(04):353–6. https://doi.org/10.1055/s-2006-957450.

Slavin J. Fiber and prebiotics: mechanisms and health benefits. Nutrients. 2013;5(4):1417–35. https://doi.org/10.3390/nu5041417.

Slavin JL, Lloyd B, Statement HCSSA. Health benefits of fruits and vegetables. Adv Nutr. 2012;3(4):506–16. https://doi.org/10.3945/an.112.002154.

Soliman A, De Sanctis V, Elalaily R, Bedair S. Advances in pubertal growth and factors influencing it: can we increase pubertal growth? Indian J Endocrinol Metabol. 2014;18(7):53. https://doi.org/10.4103/2230-8210.145075.

Steffen LM, Jacobs DR, Stevens J, Shahar E, Carithers T, Folsom AR. Associations of whole-grain, refined-grain, and fruit and vegetable consumption with risks of all-cause mortality and incident coronary artery disease and ischemic stroke: the atherosclerosis risk in communities (ARIC) study. Am J Clin Nutr. 2003;78(3):383–90. https://doi.org/10.1093/ajcn/78.3.383.

Surai PF. Selenium in nutrition and health. Nottingham University Press; 2006.

Suzuki H, Kunisawa J. Vitamin-mediated immune regulation in the development of inflammatory diseases. Endocr Metab Immune Disord Drug Targets. 2015;15:212–5. https://doi.org/10.2174/1871530315666150316122128.

Szewczyk K, Chojnacka A, Górnicka M. Tocopherols and tocotrienols—bioactive dietary compounds; what is certain, what is doubt? Int J Mol Sci. 2021;22(12):6222. https://doi.org/10.3390/ijms22126222.

Tardy AL, Pouteau E, Marquez D, Yilmaz C, Scholey A. Vitamins and minerals for energy, fatigue and cognition: a narrative review of the biochemical and clinical evidence. Nutrients. 2020;12(1):228. https://doi.org/10.3390/nu12010228. PMID: 31963141; PMCID: PMC7019700.

Tengerdy RP, Brown JC. Effect of vitamin E and A on humoral immunity and phagocytosis in E. coli infected chicken. Poult Sci. 1977;56(3):957–63. https://doi.org/10.3382/ps.0560957.

Tengerdy RP, Nockels CF. Vitamin E or vitamin A protects chickens against E. coli infection. Poult Sci. 1975;54(4):1292–6. https://doi.org/10.3382/ps.0541292.

U.S. Department of Health and Human Services. Dietary supplement fact sheets. NIH Office of Dietary Supplements. n.d.. https://ods.od.nih.gov/factsheets/list-all/.

Wallace TC, Murray R. Zoning in on the role of bioactives in dietary patterns for preventing chronic diseases. Nutrients. 2020;12(3):758. https://doi.org/10.3390/nu12030758.

Yokel RA, Crossgrove JS. Manganese toxicokinetics at the blood-brain barrier. Boston: Health Effects Institute; 2004.

Zhang YJ, Gan RY, Li S, Zhou Y, Li AN, Xu DP, Li HB. Antioxidant phytochemicals for the prevention and treatment of chronic diseases. Molecules. 2015;20(12):21138–56. https://doi.org/10.3390/molecules201219753. PMID: 26633317; PMCID: PMC6331972.

Chapter 3
Nutraceuticals in Metabolism and Xenobiotics

Hussaina Banu Buhari Malkhan Ali

1 Introduction

1.1 Xenobiotics

Xenobiotics are toxic chemical substances from exogenous sources, such as pesticides, food additives, pollutants, pharmaceutical drugs, cosmeceuticals and carcinogens, that enters the body through the food eaten, air breathed in or exposure to toxic chemicals through the skin to reach the blood stream resulting in toxicity and harmful effects. Certain xenobiotics are toxic in their parental form or as their metabolite when acted upon by the metabolizing enzymes of the body (Zhao et al. 2021).

1.2 Fate of Xenobiotics & Endogenous Substrates

Most of the xenobiotics are lipophilic in their chemical nature with better solubility in lipids or fats than in aqueous solutions, which limits their absorption and may accumulate in adipose tissues and the phospholipids of the cell, wherein the accumulated toxic chemical can have a harmful effect at the site of exposure or at the site of its distribution. However, there are a number of drug or xenobiotic metabolizing enzymes involved in the prevention of bioaccumulation of toxic chemicals in the tissues by mediating the biotransformation of lipophilic xenobiotics to more polar or water-soluble substrates that can be further metabolized into a suitable form that can be readily eliminated by the body (Hrubý and Anzenbacher 1997; Amacher 2010).

H. B. B. M. Ali (✉)
College of Health Sciences, Abu Dhabi University, Abu Dhabi, UAE

S. M. Jafari et al. (eds.), *Nutraceuticals: A New Millennium Approach for Preventive Medicine*, Food Bioactive Ingredients,
https://doi.org/10.1007/978-3-031-76481-3_3

1.3 Absorption, Distribution, Metabolism and Excretion (ADME)

Although, the metabolism of most of the toxic chemicals is mediated by the combined action of both phase I and phase II metabolizing enzymes leading to the detoxication of the exogenous substrate, the biotransformation can be mainly mediated by the Phase I enzymes. They play a potential role in the biotransformation of the lipophilic xenobiotic into a more water soluble form by introducing certain polar groups to form a reactive site that helps in the accessibility by the phase II enzymes, which facilitates the elimination or excretion of the toxic chemicals by subjecting the hydrophilic toxic chemicals to various chemical reactions such as conjugation (Rendic and Guengerich 2014). The drug uptake transporters remove the toxic chemicals from the blood by mediating their uptake in liver and kidney; and the drug efflux transporters mediate the excretion of the metabolite into the bile and/or the urine (Yaqoob et al. 2022).

1.3.1 Phase I Metabolism

Phase I metabolizing enzymes are primarily associated with the functionalization of toxic compounds, mainly through the oxidation, reduction and hydrolysis reactions, which introduces the polar functional groups to the toxic chemicals, rendering them hydrophilic in nature. Phase I enzymes are the first line of defence by the body, where a broad spectrum of toxic lipophilic substrates including the xenobiotics, hormones, steroids or pharmaceutical drugs are introduced with polar moieties such as hydroxyl (-OH), carboxyl (-COOH) or amino (-NH_2) to form a reactive hydrophilic site for the phase II metabolising enzymes.

Phase I enzymes mainly encompass the cytochrome P450s superfamily of monooxygenases, which are predominant players of xenobiotic metabolism. The CYP-P450 isoforms are heme containing enzymatic proteins catalysing the oxidative reactions such as S-oxidation, N-oxidation, C-H bond hydroxylation, epoxidation, N-dealkylation, N-hydroxylation, and O-dealkylation (Anzenbacher and Anzenbacherová 2001; Hodges and Minich 2015; Wiśniewska and Mazerska 2009). Other phase I metabolizing enzymes include xanthine oxidases, flavin monooxygenases, alcohol and aldehyde dehydrogenases, and monoamine oxidases.

In humans, there are about 57 cytochrome isoforms identified, mainly distributed in the liver and intestinal tissues. Among the large family of cytochrome P450 enzymes, only a few of the enzymes belonging to the CYP1, 2 & 3 families are found to be the most abundant isozyme involved in the metabolism of a wide range of exogenous and endogenous substrates in the liver and intestine (Monostory and Vereczkey 1995; Pang et al. 2021).

The most common cytochromes are the membrane bound enzymes in the liver cell endoplasmic reticulum, which includes CYP3A4 (most common), CYP2C9, CYP2C8, CYP2E1 and CYP1A2; whereas the CYP3A5, CYP2A6, CYP2B6,

CYP2D6, and CYP2C19 are the less common forms; and the enzymes CYP2J2, CYP1A1 and CYP1B1 are the most common extra hepatic cytochrome isoforms in tissues such as lung, kidney, intestine, skin and adipose tissue (Zanger and Schwab 2013).

1.3.2 Phase II Metabolism

Phase II metabolism involves the conjugation reactions followed by the phase I chemical reactions that generate the reactive intermediates with hydrophilic reactive site readily available for the phase II metabolizing enzymes to further conjugate an endogenous hydrophilic substrate making it suitable for enhanced excretion in urine and/or bile. Several metabolic products of the Phase I enzymes, which are water soluble intermediates but remain active and toxic, are acted upon easily by the phase II enzymes that mainly intends to further increase the water solubility of the substrate by catalysing several conjugation reactions rendering the conversion of endogenous and xenobiotic toxic chemicals into water soluble inactive substances followed by the elimination from the body through the urine or bile (Sevior et al. 2012).

The most common endogenous substrates for the conjugation are the glutathione, glucuronic acid, methyl groups and sulphates that makes the metabolite readily excretable. The key enzymes of the phase II metabolism are glutathione S-transferase (GST), UDP-glucuronosyltransferases (UGT), sulfotransferases, N-acetyl transferases (NAT), acyl transferase and methyl transferase that uses glutathione, glucuronic acid, sulphate, acetyl co-enzyme A, acyl co-enzyme A, S-adenosyl methionine as conjugating agents (Almazroo et al. 2017; Meyer 1996).

The endogenous or exogenous substrates that already have the polar functional groups such as -OH, $-NH_2$ or COOH groups, have the innate potential to bypass the phase I metabolism and directly become hydrophilic substrates for the phase II metabolism and can be subjected to the conjugation reactions for elimination through urine or bile (Phang-Lyn and Llerena 2020).

Inadequacy of expression of phase II metabolism results in intracellular oxidative damage resulting in inflammation and carcinogenesis. The expression of the phase II metabolizing enzymes GST and UGT, in particular GSTM1, GSTA1, GSTP1, UGT1A8 and UGT1A10, are found to be associated with the detoxification of various carcinogens.

1.3.3 Drug Uptake and Efflux Transporters

Besides the metabolizing enzymes, the uptake and efflux transporters of drugs play a significant role in the elimination of the toxic chemicals from the body. Most of the toxic metabolites or the reactive intermediates have differing water solubility or charges cannot freely cross the phospholipid membrane of cells; and hence the interplay between the drug transporters across the organs greatly influences the mobility of the metabolites across the blood, liver, intestine and kidneys, mediating detoxification reactions.

The drug uptake transporters, that are mainly in the liver and the kidney, remove the toxic chemicals from the blood by increasing the absorption of the toxic chemicals in the intestinal lumen; and the drug efflux transporters mediate the pumping of the processed toxic metabolite into the bile or the urine for excretion. The transporter proteins mainly belong to the super families of ABC (ATP binding cassette) and SLC (solute carrier), which are involved in mediating the efflux and influx of toxic chemicals in the cells (Takano et al. 2006). The ABC family includes seven families of about 20 carrier proteins, which functions as ATP powered solute pumps; and the SLC comprises 52 families of drug uptake transporters. Most of the transporter proteins are distributed in the liver tissue and extra-hepatic tissues such as lung and kidneys (Döring and Petzinger 2014).

Drug uptake transporters include mainly the human organic anion uptake transporter proteins (OATP), including OATP1B1, OATP1B3, OATP2B1 and Na^+ (sodium ion) taurocholate co-transporting polypeptide (NTCP) in liver cells; and OAT1, OAT3 and OAT4 in the kidney cells. These transport proteins are mainly associated with the uptake of glucuronide and sulfate metabolites produced by the phase II metabolizing enzymes. The drug efflux transporters include major transporter of organic ions (particularly MRP2 & MRP3), BCRP (Breast Cancer Resistance Protein), BSEP (Bile Salt Export Pump), MATE (Multidrug Export Transport Pump) and p-glycoprotein (Nakanishi and Tamai 2015; Patel et al. 2016).

1.3.4 Transcription Factors

The expression of phase I and phase II metabolism as well as the drug uptake and efflux pumps are regulated by certain transcription factors such as Nrf2 (Nuclear factor erythroid-2-related factor 2), hepatic nuclear factor 1, CCAAT-enhancer binding protein-β, PPAR (Peroxisome Proliferator-Activated Receptors), and PXR (Pregnane X nuclear Receptors) and FXR (Farnesoid X Receptor). These transcription factors are activated in response to the xenobiotic intracellular oxidative stresses and are primarily involved in the regulation of expression of drug-metabolizing enzymes and drug transporters.

The activation of every single transcription factor is found to be associated with a specific signalling pathway that regulates the modulation of transcription factors through ligand-protein or protein-protein interactions in response to the oxidative stress signals. For instance, the Nrf2 and Antioxidant responsive elements (ARE) are widely responsible for the detoxification and excretion of several common xenobiotics that are found to be involved in the transactivation of phase II metabolizing enzymes, particularly GST, heme oxygenases-1, UGT, NQO-1 and γ-glutamylcysteine synthetase and also various organic anion transporters.

AREs are specific DNA consensus sequence that are electrophile responsive elements in nature. They are the enhancer regions located in the promoter regions of genes encoding the detoxification enzymes such as GST, NAD(P)H quinone oxidoreductase 1 (NQO1), UDP-glucuronosyltransferase, gamma-glutamate cysteine ligase and hemeoxygenase-1, as well as some of the oxidative stress enzyme genes rendering them protection against xenobiotics and other endogenous substrates that causes genotoxic damage to cells. The reactive oxygen species, which are the oxidative stress inducers generated during metabolic reactions while processing toxic chemicals trigger the ARE regions resulting in the ARE-mediated chemoprotective gene expression (Pool-Zobel et al. 2005; Lee et al. 2009).

In response to the xenobiotic and oxidative stress, Nrf2 transcription factor dissociates from the bound cytoplasmic Kelch-like ECH-associated protein 1 (Keap1), allowing the translocation of Nrf2 to into the nucleus after phosporylation, wherein Nrf2 binds with Maf protein to form a heterodimer triggering the ARE-mediated chemoprotective gene expression of several detoxification and cytoprotective enzymes (Nguyen et al. 2003; Rushmore et al. 1991).

2 Nutraceuticals in Xenobiotic Metabolism

Dietary nutrients play a critical role in the regulation of xenobiotic metabolizing enzymes by altering their expression and function. The dietary nutrients mainly include the polyphenolic compounds, alkaloids, organosulfur compounds, nitrogen containing compounds, terpenoids, essential fatty acids and oils, vitamins, minerals, and other protecting dietary agents such as bioactive peptides, butyrates and probiotics that plays a pivotal role in the metabolism and elimination of endogenous and exogenous toxic substances.

2.1 Role in Detoxification

Previous studies on xenobiotic metabolism and its association with the intake of dietary nutrients have made it evident that some of the nutritional deficiencies results in deficiency of clearance of xenobiotics and endogenous substrates, leading to the induction of tumor and cancer growth. In order to achieve chemoprevention, the main role of the metabolizing enzyme is to target the detoxification and elimination of the pre-carcinogen rather than the active conversion of pre-carcinogen to carcinogen to initiate the cancer growth. Studies have shown that there are several nutraceutical agents that play a critical role in the elimination of toxins by inducing the phase I enzymes to metabolize the toxic substrate facilitating the xenobiotic clearance from the body.

Studies have shown that there are several nutraceutical agents that play a critical role in the elimination of toxins by inducing the phase II enzymes that catalyzes the various conjugation reactions forcing the xenobiotic to eliminate from the body. Excessive activity of phase I enzymes like CYP1A superfamily without adequate expression of the phase II enzymes might result in the progression of cancers due to the toxic procarcinogens. There are a number of studies available on the increased expression of the phase II metabolizing enzymes by nutraceutical agents or antioxidants in animal studies rendering protection against cancer causing chemicals (Pool-Zobel et al. 2005).

Researches show that several of the phenolic antioxidants and aromatic phytochemicals present in the food (e.g. beta-naphthoflavone and 3-methylcholanthrene) have immense potential in the activation of transcription factors triggering the synthesis of phase II metabolizing enzymes that renders the elimination of toxic intermediates processed by phase I metabolizing enzymes (Rushmore et al. 1991).

2.2 *Chemoprevention & Biotransformation of Xenobiotics*

During the phase I metabolism, several xenobiotics are converted into its soluble metabolite, which are in most cases a reactive electrophilic species that has the potential to induce oxidative stress in cells resulting in carcinogenic bio-activation. The phase I enzymes have an immense interindividual variability in genetic expression; and many of the pro-carcinogenic chemicals formed in phase I enzymatic reactions covalently or non-covalently attach to the DNA or RNA to initiate cancer growth in cells. The excessive genetic expression and enzymatic activity of phase I enzymes are associated with cancers, as they result in rapid formation of the highly reactive intermediates in the cells without much stimulation of the phase II enzymes (Paine 1981).

For instance, the genetic polymorphisms of the cytochrome P450 family 1, which constitutes mainly CYP1A1, CYP1A2 and CYP1B1, have been frequently associated with drug resistance, and the process of initiation and progression of cancer. The CYP1A family is mainly involved in the detoxification of common and complex chemical substrates such as poly aromatic hydrocarbons (PAH), heterocyclic amines and polychlorinated biphenyls. Increased expression of the CYP1 enzymes have been associated with cancers due to their potential to bio-transform toxic procarcinogens to carcinogens having reactive electrophilic groups that forms carcinogenic DNA adducts leading to cancer progression (Božina et al. 2009).

2.3 *Nutraceuticals as Bifunctional Modulators*

Nutraceuticals play a vital role in the chemoprevention or protection against the induction of cancers by various carcinogenic agents mediated by the inhibition of phase I enzymes involved in the oxidation of xenobiotics into several carcinogenic substrates that forms DNA adducts triggering the carcinogenesis process (Reed et al. 2018). Genetic variability in the enzymes of the phase I metabolism is associated with both the induction and inhibition of cancers. CYP3A4 has been found to be downregulated in colorectal tissue of patients in various stages of colorectal cancers, whereas the expression of CYP2C9 was higher in colon and rectal cancer patients in comparison with the healthy mucosal tissues (Beyerle et al. 2020; Mansoor et al. 2023; Pool-zobel et al. 2005).

Excessive expression of the enzymes of phase I metabolism with lesser support from the phase II detoxification system leads to carcinogenesis due to the rapid metabolism of xenobiotics to carcinogenic DNA adducts paving way to genotoxicity. Hence, the rate of phase I metabolism is limited and the elimination by the phase II metabolism is favored by the body. Thus, the nutraceuticals can selectively modulate the expression of the detoxification systems rendering the protection from intracellular oxidative stress induced by the xenobiotic.

Several nutraceuticals play an important role in increasing the bioavailability of the drugs in the blood stream by inhibiting some of the cytochrome P450 enzyme mediated drug uptake mechanism, for instance CYP3A4 and CYP2C9 isozymes, mainly in the liver (Kimura et al. 2010a; Bailey et al. 1998). However, there are some of the natural metabolites that inhibits certain cytochrome P450 expression in the intestine and not in the liver, enhancing the bioavailability of the drug. Research by Lown et al. (1997) have shown that the grape fruit juice inhibits the CYP3A protein of intestine promoting the oral bioavailability of the anti-hypertensive drug felodipine in healthy individuals.

3 Phytochemicals in Diet

3.1 Bioactive Natural Products

There are several classifications of phytonutrients present in the diet (Bellik et al. 2012; Martinez et al. 2017). The phytochemicals of the diet mainly include the polyphenolic compounds, which fall under the category of flavonoid phenolic compounds (includes flavones, isoflavones, flavanones, anthocyanins, flavonols, and flavan-3-ols); and the non-flavonoid phenolic compounds (includes coumarins, phenolic acids, lignans, stilbens, curcumin, garcinol and tannins). The phenolic compounds are aromatic compounds that possesses multiple hydroxyl (-OH) groups and are the bioactive moieties that exhibits multiple health benefits, particularly strong antioxidants pertaining to their iron-chelating and free radical scavenging ability (Surh YJ. 2003). The dietary nutrients also encompass vitamins, minerals, sterols, bioactive peptides, essential fats and oils, alkaloids, organosulfur compounds, terpenoids and other protecting bioactive dietary agents such as butyrate from fermented plant product.

3.2 *Flavonoid Polyphenols*

3.2.1 Flavonols

The flavonols have the 3-hydroxyflavone backbone, which are flavonoids with a ketone group; and the class includes quercetin, quercetin-3O-glucuronide, naringenin, kaempferol, myricetin, rutin, and isorhamnetin. There have been several researches on the pharmaceutical effectiveness of flavonols as antioxidants and are found to have a strong influence on the biological activity of diverse enzymes in detoxification.

Study by Warwick et al. (2012) on dietary inclusion of phytochemicals in infants to understand their potential to induce phase II metabolism, rendering protection from exposure to xenobiotics in later life. The phytochemicals sulforaphane, quercetin and catechin were screened for their efficacy and a significant induction of mRNA of glutathione S-transferase A1 (GSTA1) and NAD(P)H:quinone oxidoreductase 1 (NQO1) was noticed. Among the tested phytochemicals, sulforaphane exhibited a strong induction of the phase II metabolizing enzymes followed by catechin and then quercetin.

The polyphenolic compound in diet, particularly kaempferol from ginseng, quercetin, hypericin, allicin from garlic, silibinin from milk thistle, and hyperforin from St. John's wort, were tested in vitro for their efficacy to modulate the CYP3A4 and p-glycoprotein. The Caco-2 cells grown with 1-alpha,25-dihydroxy vitamin D3 and multidrug-resistant 1 (MDR1) transfected MDCK cells was tested for the CYP3A4-mediated metabolism with cortisol; and through P-gp-mediated efflux of the human immunodeficiency virus (HIV) protease inhibitor ritonavir. The investigation showed that the phytonutrients kaempferol, quercetin and hypericin, significantly inhibited P-gp-mediated efflux of ritonavir in Caco-2 cells and also the CYP3A4 activity; allicin exhibited inhibition of the drug ritonavir efflux on MDR1-MDCK cells, while hyperforin and silibinin did not influence the cellular uptake of the metabolite. The phytonutrient hypericin inhibited reduced nicotinamide adenine dinucleotide phosphate (NADPH)-dependent metabolism of cortisol; and kaempferol and quercetin significantly inhibited cortisol metabolism (Patel et al. 2004).

The effect of the flavonol quercetin on the expression of N-acetyl transferase 2 (NAT2), CYP1A2, CYP2A6, and xanthine oxidase (XO) was tested in vivo in 12 humans and the results showed that quercetin inhibited CYP1A2, whereas it enhances the effects of CYP2A6, NAT2 and XO (Chen et al. 2009). The effect of quercetin on hepatic cytochrome P450 isozymes of rats had shown that the flavonoid quercetin was one of the most potent inducer of CYP2B isozyme and a less potent inducer of CYP1A isozyme (Rahden-Staron et al. 2001). Valerio et al. (2001) have studied the effect of quercetin on phase II detoxification by screening NQO1 activity in the human MCF-7 breast cancer cell line; and have reported a 2 fold and 3–4 fold increase in the protein and mRNA expression, respectively.

3.2.2 Flavan-3-ols

The flavanols (flavan-3-ols), commonly referred to as catechins, include catechin, epicatechin, epicatechin gallate, gallo-catechin, procyanidin, and benzotropolones. The flavanols comprise the basic skeleton of a 2-phenyl-3,4-dihydro-2H-chromen-3-ol that has no double bonds between the second and third carbon, and has two chiral carbons. There have been several studies done on catechins and its derivatives that signifies the effectiveness of catechins in the phase I as well as phase II xenobiotic metabolism (Satoh et al. 2016; Srinivasan et al. 2008; Tan et al. 2010; Warwick et al. 2012).

3.2.3 Flavanones

The flavanones have the flavanoid skeleton devoid of the double bond between the carbons 2 and 3, unlike the flavones and flavonols (Zuiter 2014). Flavanones, in particular, the methoxy flavanones have been extensively studied to understand their modulation of the xenobiotic metabolizing enzymes; and it includes hesperidin, hesperetin, morin, naringin, eriodictyol, naritutin, didimin and naringenin.

Famurewa et al. (2022) have studied the efficacy of the flavanones hesperidin and hesperetin in blocking the toxicity associated with heavy metals and have shown that the flavanones have the potential to modulate the signalling networks associated with the heavy metal induced dysregulations such as Nrf2/HO-1/ARE. The ability of hesperidin in the hepatic transformation of xenobiotics in the breast cancer induced by 7,12-dimethylbenz(a)anthracene was investigated and the tests showed that hesperidin restored all of the altered biochemical pathways, especially by the induction or inhibition of phase I and II metabolism in breast and liver, and modulation of the lysosomal and TCA cycle enzymes (Nandakumar et al. 2011).

Narirutin is a dihydroxy flavonone commonly found in citrus fruits that exhibits several health preventive effects. Narirutin is one of the effective protective agent against drug induced liver injury caused by acetaminophen overdosage. Acetaminophen or para-acetyl-amino-phenol is the active constituent of the commonly administered analgesic paracetamol for fever and mild to moderate pain control; and increased usage of acetaminophen is one of the common causes of drug induced hepatotoxicity.

Studies by Fang et al. have shown that the narirutin administration in mice alleviated the oxidative stress and hepatotoxicity by increasing the expression of antioxidant enzymes and by activating the macrophagy/autophagy lysosomal pathway (Fang et al. 2023). Flavonoids such as narirutin were also found to inhibit the intestinal CYP3A4 resulting in increased bioavailability of drugs without being cleared by the drug metabolizing enzymes (Kondža et al. 2021).

Recent study by Singh et al. have shed light on the ability of narirutin in micro-RNA targeted chemoprevention and cancer therapy, and predicted that narirutin could be an antagonist of oncogenic microRNAs in the body (Singh et al. 2022). Conversely, there are also in vitro studies that have shown that the flavonoids like narirutin and naringin has the potential to lower the concentration of medications in blood by upregulating the CYP3A4 and p-glycoprotein expression (Okada et al. 2017).

Naringenin is also a flavonoid of the citrus family that has been investigated extensively for its therapeutic efficacy. Previous studies have shown that naringenin is found to possess chemo-preventive role as it reduced the metabolic activation of benzopyrene by inhibiting the aryl hydrocarbon (benzo(a)pyrene) hydroxylase in vitro and protected against the pro-toxicants activated by the cytochrome P450 1A2 monooxygenase (Ueng et al. 1999).

3.2.4 Flavones and Isoflavones

The flavones and isoflavones are chromone ring containing polyphenolic compounds, which are one of the predominant dietary flavanoids and include apigenin, chrysin, baicalein, luteolin, diosmin, chrysin, rhoifolin, daidzein, genistein, daidzin, genistin, formononetin, tangeritin, glycitein and chrysoeriol.

Tangeretin, which is a pentamethoxy flavone present in the tangerine juice, has the potential to stimulate cytochrome P450 3A4 (CYP3A4) and found to be an effective regioselective stimulator of midazolam 1′-hydroxylation in the in vitro studies using human liver microsomes. However, the effect of tangeretin on CYP3A4 expression in the randomized cross over study in 8 healthy humans did not show any consistent results (Backman et al. 2000).

Takemura et al. have shown that the methoxyflavonoids such as chrysoeriol (flavone) and isorhamnetin (flavonol) selectively inhibited CYP1B1 that is associated with the conversion of 17-beta-estradiol (E(2)) to carcinogenic 4-hydroxy-E(2), and has a potential role in halting the processes of carcinogenesis and cancer progression (Takemura et al. 2010).

Apigenin is a 4′, 5, 7,-trihydroxyflavone and widely known for its potent antioxidant and chemopreventive properties. Svehlikova (2004) examined the influence of the two phytochemicals sulforaphane and apigenin on phase II metabolising enzymes in the CaCo-2 cells and the experiments showed that apigenin and sulforaphane induced the expression of UGTA1 by 4 folds and 3.7 folds; and whereas only sulforophane induced GSTA1 expression by 2.8 folds and apigenin did not have any significant effect on the GSTA1expression. However, synergistically sulforaphane and apigenin exhibited an induction of UGT1A1 of up to 12-folds.

The anti-oxidant and chemo-preventive ability of the flavones apigenin and luteolin to induce NRF2-mediated gene expression associated with xenobiotic induced oxidative stress and inflammation was studied by Paredes-Gonzalez et al. (2015) in HepG2 cells; and the tests showed that both the flavones at a low doses (1.56–6.25 μm) did not show any significant influence on the NRF2-mediated gene expression. However, the higher doses of the phytochemicals significantly activated the PI3K/Nrf2/ARE system marked by the suppression of lipopolysaccharide-induced biochemical markers.

3.2.5 Anthocyanidins

The most common anthocyanidins are malvinidin, petunidin, delphinidin, pelargonidin, peonidin, and cyanidin. Anthocyanidins have been found to be associated the control of inflammation and in regulating the scavenging of intracellular reactive oxygen species (Bonetti et al. 2017).

3.3 *Non-flavonoid Polyphenols*

3.3.1 Phenolic Acids

The phenolic acids contain a phenol ring linked to a carboxyl group. They are classified into two groups, namely the benzoic acid derivatives and the cinnamic acid derivatives. The hydroxycinnamic acid of fruits mainly includes caffeic, ferulic, coumaric, sinapic, chlorogenic, cinnamic and rosamarinic acid; and the hydroxybenzoic acid mainly includes gallic, ellagic, vanillic, syringic and para-hydroxy benzoic acid.

The chemopreventive effect of sinapic acid was tested against dimethylhydrazine induced carcinogenesis in rat colon; and the expression of drug-metabolizing enzymes of phase I (cytochrome P450 and P4502E1) were enhanced and the levels of phase II enzymes (GST, DT-diaphorase and UGT) were reduced in response to the xenobiotic in the liver and mucosa of colon cells; and the effects were reversed upon sinapic acid supplementation (Balaji et al. 2014). In a similar study induced by the same xenobiotic in rats, p-methoxy cinnamic acid was effective in modulating the enzymes associated with inflammation and cancer (Sivagami et al. 2012).

3.3.2 Coumarins

Coumarins are benzopyrone derivatives that comprises fused benzene and α-pyrone rings. They comprise simple coumarin, furanocoumarin, pyranocoumarin, and phenyl-coumarin. A great deal of researches have been done recently on the influence of various coumarins in regulating the expression of detoxification enzymes and the chemoprevention.

In vitro studies by Kimura et al. (2010a, b) on the efficacy of about sixty polyphenols for their modulation of CYP3A4 and CYP2C9 activity and found that among the tested phytonutrients, three of the coumarins and twelve other flavonoids exhibited a marked inhibition of both CYP3A4 and CYP2C9 activities. Furanocoumarins such as isopimpinellin, imperatorin and trioxsalen have been found to be a potent chemopreventive agents as they mediated the inhibition of CYP1A2 mediated procarcinogen activation (Kleiner et al. 2001, 2003). Also, the furanocoumarins found in several fruits and vegetables such as bergamottin and 6′,7′-dihydroxybergamottin are found to be inhibitors of some of the phase I metabolising enzymes (Masuda et al. 2017).

3.3.3 Lignans and Flavolignans

Dietary lignans are phenyl propanoid dimers and one of the large group of the polyphenolic compounds. Flavanolignans are a unique class of lignans that comprises a phenylpropanoid moiety and a flavanoid moiety. The dietary lignans and flavanolignans comprise pinoresinol, matairesinol, lariciresinol, sesamol, sesamin and silibin.

Lee et al. (2009) have studied the therapeutic efficacy of nine lignans isolated from *Schisandra chinensis*, commonly known as five flavoured berries that has a sweet, sour, pungent, salty and bitter taste, in HepG2 human hepatocarcinoma cells. The investigation shows that the lignans possess an intense chemopreventive activity mediated by the induction of the phase II detoxification enzyme NQO1 through nuclear accumulation of Nrf2 through the Nrf2-ARE pathway, especially the lignans tigloylgomisin H and angeloylgomisin H were found to be a potent chemopreventive agent as they significantly induced the quinone reductase activity.

Silymarin is a mixture of polyphenolic flavonolignans, particularly silibinins, extracted from the plant Silybum marianum, commonly known as milk thistle. Silymarin is traditionally used to cure liver and digestive tract problems (Wadhwa et al. 2022). The effect of silymarin on mucositis, which is a cancer chemotherapy mediated gastrointestinal tract disorder, was tested on mucosal barrier injury in mice induced by epirubicin. The findings showed that the silymarin effectively reduced the condition of gastrointestinal mucositis and may prevent the gastrointestinal damage caused by epirubicin-induced toxicity. During this study, various gene expressions were studied including Bax, Bcl-2, p53 and particularly cytP450 expression. The results presented stronger expression for cyt P450 in the gastrointestinal tissues, demonstrating the intense drug detoxification. (Sasu et al. 2015).

Kiruthiga et al. (2013) investigated the effect of silymarin on the polyaromatic hydrocarbon representative benzo(a)pyrine toxicity in Wistar rats and found that the levels of hepatic phase I (CYP1A1) enzymes, which converts the xenobiotic into a reactive toxic intermediate, are inhibited; and the phase II enzymes, namely GST, epoxide hydroxylases and UGT, are modulated. Similar studies have shown that Nrf2 and pregnane X receptor (PXR) were the key players associated with the activation of detoxifying genes against the formation of B[a]P-7,8-dihydrodiol-9,10-epoxide DNA adducts (Jee et al. 2020).

Studies on chemoprotective activity of silibinin against colon cancer in rats against the carcinogenesis mediated by the xenobiotic 1,2 di-methylhydrazine (DMH) have shown that the test group administered with silibinin modulated the xenobiotic metabolizing enzymes leading to the DMH clearance, whereas the rats exposed to only DMH exhibited increased activities of cytochrome b5 reductase,

Fig. 3.1 Milk thistle rich in polyphenolic flavolignan silibin

cytochromeP450 reductase, CYP2E1 and decreased the activities of UGT, GST and DT-diaphorase in both the liver and colon cells (Fig. 3.1).

3.3.4 Garcinol

Garcinol is an active constituent of the Genus *Garcinia*, including *Garcinia indica* and *Garcinia cambogia*, which are indigenous plants of South East Asia, America, Australia and Africa. Garcinol is a polyisoprenylated benzophenone that has been found to play a key role in the chemoprevention as it exhibited the potential to strongly inhibit human leukaemia HL-60 cells with an IC50 value of 9.42 μM by inducing apoptosis via releasing cytochrome c into the cytosol, processing procaspase-9, activating caspases 2 & 3, inducing the degradation of poly(ADP-ribose) polymerase (PARP), and mediating DNA fragmentation (Pan et al. 2001). Garcinol also prevented colonic aberrant crypt foci in male F344 rats when fed with experimental diet containing garcinol indicating its effectiveness in inhibiting colon tumorigenesis (Tanaka et al. 2000a). Kopytko et al. (2021) emphasizes the role of garcinol in epigenetic regulation by inhibition of histone acetyl transferases and deregulation of miRNA associated with the development and progression of cancer cells.

Study by Bolla L in rats was intended to evaluate the inhibitory action of garcinol on cytochrome P450 isozymes when co-administered with the other drugs and the investigation showed strong inhibition of CYP1A2, CYP2C9, CYP2B6, CYP2D6 and CYP3A4 with IC50 values of 7.6, 8.0, 2.1, 9.5 and 5.1 μM, respectively; and also moderate inhibition of CYP2C19 and CYP2E1 with IC50 values of 16.4 and 19.0 μM, respectively (Bolla et al. 2021) (Fig. 3.2).

Fig. 3.2 Garcinol extracted from *Garcinia indica* (left) and *Garcinia cambogia* (right)

3.3.5 Stilbens

The stilbens include resveratrol, matteucinol, pinosylvin, and piceatonnol. Extensive researches have been performed on resveratrol that is a trihydroxy-trans-stilbene of grapes and peanuts. Numerous studies have shown that the resveratrol has immense potential to inhibit phase I metabolism playing a remarkable role in chemoprevention and to induce phase II enzymes (Guthrie et al. 2017; Canistro et al. 2009; Piver et al. 2001; Ji et al. 2021). Also, resveratrol has been found to play a crucial role in the inflammation prevention through the downregulation of expression of adhesion molecules (Calabriso et al. 2015).

3.3.6 Tannins

The tannins are polyphenolic compounds widely distributed in tea leaves (e.g. epigallocatechin gallate) and include mainly the hydrolysed and condensed tannins. The protective effect of tannic acid on murine phase I and phase II enzymes were explored and the results showed that there was a considerable reduction in the hepatic CYP2E1 levels, whereas the GST from kidney is elevated and reduced in livers, while the hepatic and renal NQO1 is inhibited (Krajka-Kuźniak 2003).

3.3.7 Curcuminoids

There are several pharmaceutically active constituents of the roots of *Curcuma longa*, commonly known as turmeric, which is a very common ingredient of several Asian cuisines that gives the yellow colour to the food and is also used in cosmetics traditionally and Ayurvedic medicine. Curcumin is a polyphenol (diferuloyl methane) that belongs to a class of curcuminoids comprising curcumin, demethoxy curcumin and bis-demethoxy curcumin, among which curcumin is widely known for its various health benefits including anti-oxidant, anti-inflammatory and anti-carcinogenic activity (Jurenka 2009; White et al. 2019). It is a potent inhibitor of NF-κB and its downstream genes, several growth factors and cell adhesion molecules associated with cancers and also an initiator of several inflammatory and immunogenic factor activation involved in cancers (Wilken et al. 2011).

Curcumin is also well known for its protective effects against toxic chemicals in liver, brain and cardiovascular tissues; and is also found to be associated with the prevention and treatment of various cancers. For instance, the hepatoprotective activity of curcumin in mercuric chloride toxicity was studied in mice and the results have shown that curcumin significantly increased the nuclear translocation of Nrf2 that is associated with the increase of cells to oxidants and electrophiles resulting in the induction of detoxifying gene expressions, particularly cyp450 signalling to eliminate mercuric chloride toxicity (Li et al. 2021; Pouremamali et al. 2022).

In order to understand the turmeric mediated chemoprevention, in vitro studies on rat liver microsomes using curcumin, dimethoxy curcumin and bis-demethoxy curcumin against benzo(a)pyrene and 4-methylnitrosamino-1-(3-pyridyl)-1-butanone chemicals and the results showed that the curcumin is the most effective in inhibiting the activation of carcinogens, metabolized by the CYP1A1, CYP1A2 and CYP2B1 cytochrome P450 enzymes (Thapliyal and Maru 2001). Iqbal M et al. studied the cancer chemopreventive action on curcumin in mice by subjecting them to dietary curcumin at 2%, w/v for 30 days and found that curcumin significantly increased the activities of the xenobiotic metabolizing enzymes glutathione peroxidase, glutathione reductase, GST, glucose-6-phosphate dehydrogenase, quinone reductase and catalase in liver and kidney cells of mice (Iqbal et al. 2003) (Fig. 3.3).

3.4 *Other Bioactive Dietary Agents*

Besides polyphenol, vitamins, minerals, sterols, bioactive peptides, essential fats and oils, alkaloids, organosulfur compounds, terpenoids and other protecting bioactive dietary agents such as butyrate from fermented plant or animal product, immense potential in the elimination of toxic chemicals and in the prevention and control of cancers.

Fig. 3.3 *Curcuma longa* (turmeric) rich in curcuminoids

3.4.1 Organosulfur Compounds

The organosulfur compounds include indole, allyl sulphur compounds and isothiocyanates. Numerous studies have reported the anti-inflammatory, antioxidant and chemopreventive potential of organosulfur compounds. They are found to modulate the expression of detoxification systems of cytochromes of phase I metabolism such as CYP2B, CYP3A2 and CYP1A2; and the expression of several phase II metabolizing enzymes such as quinone reductase, and GST (Navarro et al. 2009; Wark 2004; Yoxall et al. 2005).

Tan et al. (2010) studied the chemoprotective property of the phytochemicals sulforaphane, epigallocatechin gallate, green tea and broccoli sprout extracts, phenethyl isothiocyanate and benzyl isothiocyanate by inducing the expression of glutathione S-transferase P1 (GSTP1) and NQO1; and the tests showed that, among the tested phytochemicals, the extracts of green tea and broccoli sprout, and sulforaphane were effective in the induction of the phase II metabolizing enzymes.

3.4.2 Terpenoids

The terpenoids will include carotenoid terpenoids ($\alpha/\beta/\gamma$-carotene, lutein, lycopene, astaxanthin, canthaxanthin and β-cryptoxanthin) and the non-carotenoid terpenoids (phytosterols, phytostanols, saponins, limonene, ursolic acid and perillyl alcohol).

The effect of 4-oxocarotenoids, namely canthaxanthin and astaxanthin, and also lutein and lycopene on liver metabolizing enzymes were studied in rats by screening the levels of phase I cytochrome P4501A1 & 1A2 enzymes, and phase II enzymes, particularly 4-nitrophenol (4-NP) UDP-GT and quinone reductase (QR). The oxocartotenoid canthaxanthin induced the P4501A1 and 1A2 enzymes, which was evident by the significant increase of some of the P450-dependent markers, especially ethoxy-resorufin O-deethylase (EROD), methoxyresorufin O-demethylase (MROD), pentoxy-resorufin O-deethylase (PROD) and benzoxyresorufin O-dealkylases (BROD), and also co-induced 4NP-UGT and QR enzymes. Conversely, the canthaxanthin decreased nitrosodimethylamine N-demethylase (NDMAD) activity and did not influence the erythromycin N-demethylase (ERDM) activities. Inclusion of canthaxanthin in the diet of rats at a dose of about 10 ppm clearly showed the increase of EROD, MROD and 4NP-UGT, whereas the induction of QR was detectable in doses of greater than or equal to 100 ppm.

Astaxanthin also exhibited the same pattern of enzyme induction similar to canthaxanthin, but to a lesser extent. The effect of phase I enzymes and 4NP-UGT were observed in diets greater than or equal to 100 ppm of astaxanthin, and did not exhibit any influence on QR activity. Lutein did not have any influence on the effect of both phase I and phase II enzymes, and lycopene was found to only decrease the NDMAD activity, like canthaxanthin (Gradelet et al. 1996).

Limonoid is a non-carotenoid terpernoid commonly found in the citrus fruits and some of the limonoid compounds such as nomilin, isoobacunoic acid and deacetyl nomilin are found to induce the GSTs of stomach and intestine triggering the detoxification systems and favouring the elimination of the toxic substances (Perez et al. 2010).

Sinapic acid, p-methoxy cinnamic acid, luteolin, and phytosterol β-sitosterol exhibited a protective action against 1,2-dimethyl hydrazine-induced depletion of enzymes associated with oxidative stress such as inflammation and cancers such as catalase, superoxide dismutase, glutathione peroxidase, glutathione reductase, GST in rat colon and liver (Balaji et al. 2014; Sivagami et al. 2012; Baskar et al. 2012; Manju et al. 2005).

3.4.3 Alkaloids

There are several alkaloids present in food and it include pyrollidines, pyridines, pyrrolidizines and quinolines. There are several studies continuously made on their chemopreventive nature of the alkaloids.

4 Role of Dietary Nutrients in ADME

Several researches in the past had investigated the role of food-based components in the prevention and treatment of chronic diseases, delaying ageing and promotion of overall health and well-being. In comparison with the chemical drugs having a specific pharmaceutical activity, the dietary nutraceuticals and other phytochemicals provide a safety health profile with very few undesirable side effects and also can be as effective as a chemical drug with a diverse array of wellness. Hence, nutraceuticals have gained significant attention for protecting health and prevention or treatment of a disease or a disorder.

Most notably, functional foods have been widely investigated for their potential in elimination of toxic endogenous and exogenous substrates by the modulation of metabolic pathways associated with the biotransformation of toxic chemicals into water soluble and non-toxic forms readily eliminated in the urine or bile. The exposure and accumulation of the toxic chemical chemicals in the body eventually leads to chronic diseases like cancer, obesity, type 2 diabetes and cardiovascular disorders. The bioactive dietary agents such as polyphenolic compounds, particularly flavonoids and non-flavonoids, dietary fats and oils, bioactive peptides, vitamins and minerals play a pivotal role in modulating the detoxification pathways.

The pharmaceutically active phytochemicals, that largely influences the ADME and mediates the xenobiotic clearance from the body, includes functional foods such as vegetables (cruciferous, apiaceous, allium and nightshade vegetables), fruits (tropical, citrus, other subtropical fruits and temperate), spices, green tea polyphenols, green and herbs, whole grains, legumes, bioactive peptides, dietary fiber, probiotics and gut fermentation products.

4.1 Vegetables in ADME

4.1.1 Cruciferous Vegetables

The cruciferous vegetables belong to the Brassicaceae family, which includes several leafy vegetables and sprouts including the very common dietary vegetables like broccoli, cauli flower, Brussels sprouts, cabbage, kale, turnip and radish. They are rich in several potential antioxidants and cancer protecting agents such as polyphenols, carotenoids, sulphur containing glucosinolates and S-methyl sulfoxide (Soundararajan and Kim 2018; Zeng et al. 2023).

The effect of dietary consumption of broccoli on CYP1A2 and CYP2A6, enzymatic activities was studied and the results showed that the broccoli consumption significantly increased the expression of CYP1A2 and CYP2A6 in all the 10 volunteers (Hakooz and Hamdan 2007). Previous researches on the phytonutrients of cruciferous vegetables have shown the induction of CYP1A1 and CYP1A2 in humans and induction of CYP1B1 in animals (Peterson et al. 2009).

Several studies have been done on the indoles, which are the active constituents of cruciferous vegetables, in the estrogen detoxification through cytochrome P450 enzymes to check their chemopreventive efficacy against the estrogen responsive tumours. Studies by Michnovicz and Bradlow (1990) have shown that the hepatic estradiol-3-hydroxylation mediated by the cytochrome P450 enzymes was induced in rats. A similar study by Horn et al. 2002 on the metabolism of estrogen have shown an increase in the expression of mRNA levels of cytochrome P450 in liver and mammary glands.

Leibelt et al. in 2003 has demonstrated that the indole-3-carbinol and 3,3′-diindolylmethane phytochemicals found in the cruciferous vegetables have the potential to induce CYP1A1 (liver and colon) and CYP1A2 (liver) in rats; and induce CYP3A2 only in female rats. Thus, indole-3-carbinol is an effective inducer of expression of several cytochrome P450 enzymes when compared to the indole-3-carbinol (Leibelt et al. 2003).

A similar research on consumption of cruciferous vegetables by providing daily diet rich in 250 grams each of Brussels sprouts and broccoli on the metabolism of heterocyclic amine (2-amino-1-methyl-6-phenylimidazo[4,5-b]pyridine) present in cooked meat; and the results showed that the consumption of cruciferous vegetable induces both the phase I and II metabolism of heterocyclic amines with increased urinary clearance of the xenobiotic during the phase of vegetable consumption in comparison to the control phases with no inclusion in diet (Walters 2004).

Isothiocyanates are also one of the active constituents of many of the cruciferous vegetables. Studies by Navarro et al. (2009) in 33 men and 34 women on cruciferous vegetable supplementation have shown that the consumption has increased the phase II metabolizing enzyme GST noted by the induction of the GST variants GSTA1/2. Similar studies by Kall et al. (1996) in an in vivo experiment in 16 humans have shown that the intake of dietary broccoli influences human CYP1A2 and other drug metabolizing enzymes resulting in the prevention of carcinogenesis.

Similarly, sulforaphane present in the cruciferous vegetables were found to decrease the CYP2B and CYP3A2 apoprotein levels and upregulate the CYP1A2 levels in rats. With respect to the phase II metabolizing enzymes, sulforophane was found to stimulate quinone reductase in a dose-dependent manner, and did not have any marked influence the expression of GST, epoxide hydrolase and glucuronosyl transferase enzymes (Yoxall et al. 2005). Wark (2004) had reported that the glucosinolates and dithiolthiones of the Brassica family induced the human rectal GST detoxification enzymes'.

Chinese cabbage widely consumed in East Asia, rich in polyphenols like quercetin and kaempferol, was tested for the efficacy against stress related pathological conditions in endothelial cells of human umbilical vein; and it was found that the phytonutrients could suppress tumor necrosis factor (TNF)-α-induced inflammatory response resulting in anti-adhesive activity mediated by the downregulation of vascular cell adhesion molecule-1 (VCAM-1) via an Nrf-2 pathway resulting in the expression of ARE triggered expression of the cell protecting enzymes (Joo et al. 2017). Kale or leaf cabbage also did exhibit a similar reduction of TNF-α induced cell adhesion molecules E-selectin, ICAM-1 and VCAM-1 mRNA expression (Kuntz and Kunz 2014) (Fig. 3.4).

Fig. 3.4 Cruciferous vegetables – Cabbage, Rutabaga, Broccoli, Cauli flower and Brussels sprout

4.1.2 Apiaceous Vegetables

Vegetables like carrots, celery, parsley, dill, fennel, cilantro and parsnips that belong to the *Apiaceae* family are rich sources of flavonoid polyphenols and found to have a profound effect on the detoxification pathways. Three of the furanocoumarins from apiaceous vegetables, namely isopimpinellin, imperatorin and trioxsalen, were tested for their chemopreventive activity and the researches have shown that the furanocoumarins inhibited CYP1A2 mediated procarcinogen activation, particularly isopimpinellin exhibited a longer inhibition of the CYP1A2 enzyme (Kang et al. 2011). Similar studies on the inhibition of tumor initiation in mice and human breast MCF-7 adenocarcinoma cell lines by imperatorin and isopimpinellin have been reported by Kleiner et al. (2001, 2003) to be inhibiting the DNA adduct formation by blocking the cytochrome P450 mediated bioactivation of the polycyclic aromatic hydrocarbon carcinogens, namely benzo[a]pyrene and 7,12-dimethylbenz[a]anthracene.

Peterson et al. (2006) studied eight of the phytochemicals that are active constituents of the apiaceous vegetables, namely psoralen, 5-methoxy psoralen, 8-methoxy psoralen, apigenin, quercetin, narigenin, caffeic acid and chlorogenic acid, to inhibit CYP2C1 activity, particularly chemoprevention by inhibition of CYP1A2-mediated mutagenicity of aflatoxin B1 in yeast cells. The results showed that psoralen, 5-methoxy psoralen and 8-methoxy psoralen significantly reduced the aflatoxin B1 mediated mutagenicity by inhibiting the CYP1A2 mediated methoxyresorufin O-demethylase activity and exhibited improved cell survival in comparison with other phytochemicals. Also, quercetin showed inhibition against CYP1A2 and apigenin significantly reduced the mutagenicity caused by aflatoxin B1 in the yeast cells (Peterson et al. 2006) (Fig. 3.5).

Fig. 3.5 Apiaceous vegetables – Celery, Dill, Carrots and Parsnips

4.1.3 Allium Vegetables

Garlic, onion, shallots, green onions, chives and leeks are the vegetables that belong to the *Allium* family and contain several organosulfur compounds, polyphenols and saponins such as kaempferol, ferulic acid, quercetin and its derivatives, isorhamnetin, delphinidin, gallic acid, protocatechuic acid, anthocyanins, naringenin, hesperidin, luteolin, apigenin, chrysoeriol, balcalein, fisetin, morin, catechin, epicatechin and other organosulfur compounds such as allicin, alliin, diallyl sulphide/disulphide/trisulfide, S-allyl-cysteine, ajoene, and propin (Fossen and Andersen 2003; Fredotović et al. 2017; Kothari et al. 2020; Shang et al. 2019).

Many of the phytochemicals of the Allium family, particularly the lipophilic sulphur containing compounds have found to have an effective role in the detoxification reactions. Research by Ip and Lisk (1997) have shown that selenium enriched garlic diet in rats inhibited the initiation of cancer by the cancer-causing chemical dimethyl benz[a]anthracene (DMBA). Rats fed with selenium and garlic containing diet for two weeks resulted in significant reduction of the DMBA adducts in liver and mammary gland. No significant changes were observed in the expression of phase I metabolising enzymes, particularly CYP1A1, CYP1A2, CYP2B1, CYP2E1 and CYP3A4. But then, the phase II metabolising enzymes, in particular, GST and UGT activities were elevated to about 2 to 2.5 folds in the cells of liver and kidney.

Zeng et al. (2009) have shown that the garlic oil is effective in inhibiting the hepatic CYP2E1 and CYP1A2; and did not produce any marked changes in the expression of CYP3A indicating that it did not exhibit any interaction with the metabolism of concomitant pharmaceutical drugs. Similar study by Ho (2010) on the effect of garlic extract on human hepatocytes have shown that the exposure of hepatocytes to garlic extract for about 4 days resulted in the reduction of CYP2C9 activity by about 90%, but had no effects on CYP3A4 enzymatic activity.

Onion is specifically rich in alkyl polysulphides, glycosides and flavonols and play a crucial role in inhibiting the chemical induced cancers and inducing the detoxification of toxic chemicals. Teyssier et al. (2001) investigated the role of onion powder in the detoxification pathways in rats. Rats fed with a diet containing 20% onion powder for 9 days have shown remarkable induction of CYP1A and CYP2B activities, whereas the CYP2E1 activity is reduced. In addition to these changes, the detoxification enzymes UGT and GST were induced. Together, the onion powder inhibited the cytochrome p450 enzymes associated with the carcinogen activation and also enhanced the phase II metabolising enzymes involved in the elimination of the toxic compounds (Fig. 3.6).

4.1.4 Nightshade Vegetables

Tomato, capsicum (bell pepper), chilly, potato and eggplant (brinjal) belong to the family of Solanaceae, which are referred to as the nightshade vegetables. They are nutritious vegetables that possesses several vitamins, minerals, dietary fibers and

Fig. 3.6 Vegetables of *Allium* family – Chives, Shallots, Scallions, Garlic and Onions

flavonoids like anthocyanins; and have been the staple food for centuries worldwide. They also possess various alkaloids in minute amounts and are often associated with allergies, inflammation and some intestinal disorders (Kuang et al. 2023).

Bhuvaneswari et al. (2005) have studied the chemoprotective effects of garlic and tomato on hamster buccal pouch carcinogenesis induced by 7,12-dimethylbenz[a]anthracene (DMBA). The results showed that the combination tomato and garlic significantly decreased the phase I enzymes involved in the cancer initiation; and increased the phase II enzymes in the pouch and liver eventually leading to the clearance of the toxin from the body (Fig. 3.7).

Fig. 3.7 Night shade vegetable – Chillies, Egg plants, Capsicums & Potatoes

4.2 Spices in ADME

Spices are the essential ingredients in most of the global cuisines, as it enhances aroma, colour, flavour and palatability of foods. Kimura et al. studied 55 spices widely used in the culinary preparations for their inhibitory potential against the most common cyp isoforms, CYP3A4 and CYP2C9. Among the tested spices, about 36 spices exhibited more than 50% inhibition, particularly 11 of the spices showed marked inhibition of the cyp isoforms, predicted through testosterone 6β-hydroxylation and diclofenac 4′-hydroxylation. The spices white pepper, mace, black pepper, nutmeg, ginger, turmeric, cinnamon (Chinese) and cinnamon (Srilankan) exhibited an IC50 value of 1.0, 3.9, 4.1, 4.2, 5.1, 17.0, 24.0 and 30.0 μg/ml against CYP3A4 (Japanese pepper, Sichuan pepper & sage not tested); and the spices mace, nutmeg, white pepper, ginger, black pepper, cinnamon (Chinese), turmeric, cinnamon (Srilankan), sage, Sichuan pepper and Japanese pepper exhibited

Fig. 3.8 Dietary spices rich in several polyphenols and alkaloids

an IC50 value of 1.1, 2.6, 3.2, 10.0, 12.1, 12.4, 14.8, 15.2, 29.0, 32.0 and 36.0 μg/ml against CYP3A4 isoform (Kimura et al. 2010b) (Fig. 3.8).

4.3 Green Tea Polyphenols in ADME

Green tea comprises several polyphenolic compounds, particularly the flavanols epigallocatechin, epigallocatechin-3-gallate, epicatechin, and epicatechin-3-gallate. The green tea catechins have shown a strong modulatory effects on the enzymes and transporters associated with the metabolism of drugs and the exogenous and endogenous toxic chemicals.

Srinivasan et al. studied the influence of green tea polyphenols through the modulation of various enzymes of detoxification metabolism against 4-nitroquinoline

1-oxide induced oral cancer in mice. The phase I and phase II enzymes of study, namely cytochrome b5 and P450 reductase, aryl hydrocarbon hydroxylase, DT-diaphorase, GST and UGT, was screened for their influence in chemoprevention; and the results showed a significant inhibition of the phase I enzymes due to the immense protective nature of the polyphenol to deactivate the carcinogen and also induction of the phase II enzymes mediating the clearance of the toxic chemical (Srinivasan et al. 2008).

Satoh et al. (2016) studied the inhibitory effect of 8 green tea catechin on the cytochrome enzymes CYP1A2, CYP2C9, CYP2D6 and CYP3A4; and found that all gallated catechin exhibited inhibition except CYP2D6, particularly (−)-catechin-3-O-gallate inhibited CYP2C9 with an IC50 value of 7.60 μM; and epigallocatechin-gallate inhibited CYP1A2 with an IC50 value of 8.93 μM.

Orientin is a glycosyl flavonoid found abundantly in rooibos tea that is found to be potent in protecting oxidative stress relating conditions when tested on rats by modulating the activities of the phase I and phase II drug metabolizing enzymes and favours the detoxification of carcinogen (Thangaraj et al. 2018) (Fig. 3.9).

Fig. 3.9 Green tea rich in several tannins, specifically catechins, and polyphenols

4.4 Dietary Oils & Fats in ADME

4.4.1 Dietary Oils & Fats

The effect of dietary oils, namely fish oil, olive oil, hazel nut oil and corn oil, and the influence of dietary supplements, particularly 1% cholesterol or DL-tocopherol acetate, and beta carotene on liver microsome membrane fluidity and the concentration of the metabolizing enzymes, cyt P450, aniline hydroxylase (AH), aminopyrine-N-dimethylase (AND) and UDP-glucuronyltransferase (UDP-GT), was tested on rats. The results showed that the consumption of fish oil had the overall effect of increasing the membrane fluidity of liver microsomes and increased content of the metabolizing enzymes, particularly Cyt P450 and UDP-GT (Lutz et al. 1998). Studies on dietary cholesterol in humans and rats have shown that the intake of cholesterol plays a significant role in altering liver microsomal fluidity and influencing the xenobiotic metabolic pathways (Hietanen et al. 1982; Lutz et al. 1999; Muriana et al. 1992a, b; Siess et al. 1983).

4.4.2 Poly Unsaturated Fatty Acids

Several studies had been conducted on the roles of dietary fats on xenobiotic metabolism. For instance, studies by Yoo et al. on consumption of dietary corn oil in rats revealed alteration of constitutive and induced levels of GST isozyme and some of the P450 enzymes; and dietary intake of poly unsaturated fatty acid studies by Kravchenko et al. have shown that the activity of liver metabolizing enzymes was similar in rats fed with diets of lard or corn oil, while the activities of epoxide hydrolase and UDP-GT were considerably higher fish oil diet group (Yoo et al. 1990; Kravchenko et al. 1992).

The increase in membrane fluidity was also observed in hazel nut oil in comparison with olive oil or corn oil, whereas the consumption of cholesterol decreased the fluidity in all dietary groups. The beta-carotene was found to exhibit higher Cyt P450 activity in combination with fish oil, corn oil or olive oil. In particular, the microsomal fluidity and AH activity was induced in beta-carotene and olive oil group, the UDP-GT activity was high in beta-carotene and corn-oil group, the enzyme AND activity increased with beta-carotene and hazel nut oil group, and the AH activity increased with beta-carotene and fish oil group (Lutz et al. 1998) (Fig. 3.10).

4.5 Fruits in ADME

Based on the geographical location, fruits can be classified as tropical, sub-tropical and temperate fruits. Tropical fruits grow in the warm tropical climates and will not be able to withstand any cold temperatures. Sub-tropical fruits grow in regions

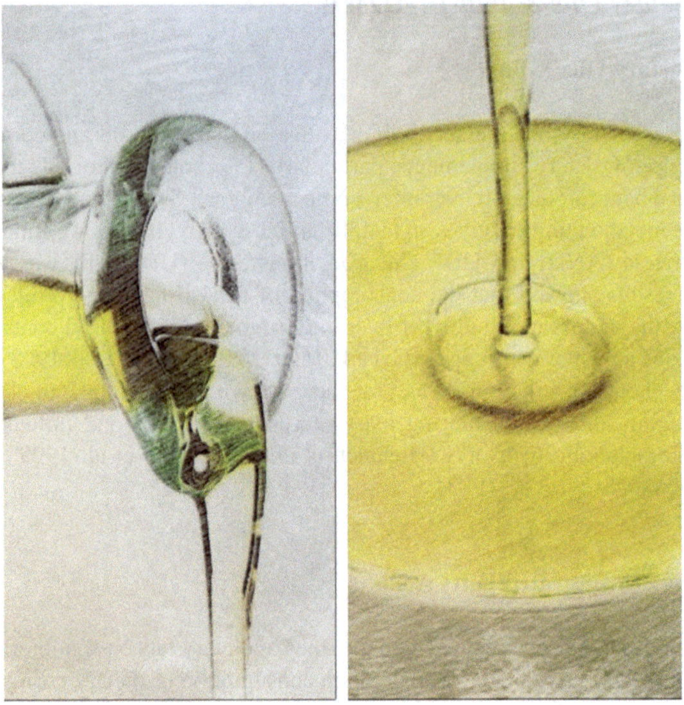

Fig. 3.10 Dietary oils, particularly polyunsaturated fatty acids

adjacent to the tropics that also requires warm temperature for fruiting and also can tolerate mild frost. These fruits are adapted to the temperature in the mid-latitude and requires chill temperatures for fruiting. The following section covers the different types of fruit juices and their stimulatory or inhibitory actions on the metabolizing enzymes that gives us an insight of the effect on nutraceuticals on drug metabolism.

Several researches are done on the concomitant intake of fruit juices with various drugs resulting in the modulating effects of the drug metabolizing enzymes and other transporter proteins (Mallhi et al. 2015). The fruits encompass several carotenoids, sterols and stanols, phenolic compounds including the flavonoids and non-flavonoids and other nutraceuticals that has several beneficial health effects, and has a strong influence on xenobiotic metabolism and multi-drug transporters (Lan et al. 2013; Sayago-Ayerdi et al. 2021).

4.5.1 Citrus Fruits

The citrus fruits are rich in essential nutrients, minerals and antioxidants such as vitamin C; and include orange, grape fruit, lemon, lime, mandarin and tangerine. They are one of the predominant and oldest domesticated crop, liked all around the world for their fresh fruits.

There have been several researches on the citrus triterpenoids in inducing the phase II metabolizing enzymes supporting the xenobiotic clearance from the body (Tanaka et al. 2000b; Takanaga et al. 2000). Studies by Perez et al. (2010), in an excised tissue from a mouse model, have shown that the citrus bioactive limonoids (sour orange), namely nomilin, isoobacunoic acid and deacetyl nomilin, induced stomach GST by 58%, 25% and 19%, respectively, against the xenobiotic 1-chloro-2,4-dinitrobenzene, whereas the GST in the liver is reduced by limonoid and isoobacunoic acid mixture. Nomilin as well as the limonoid mixture induced GST activity in both intestine and stomach, while deacetyl nomilin induced only the intestinal GST. Besides GST, the other potential phase II detoxification enzymes, namely NAD(P)H: quinone reductase (QR) activity was also induced by the triterpenoids.

The effect of major citrus fruit polyphenols, namely nobiletin, sinensetin, tangeretin, rutinosides, didymin, hesperidin and narirutin, on human organic anion transporting polypeptides 1B1, 1B3 and 2B1, have shown that all of the tested flavonoids strongly inhibit OATP2B1 with an IC_{50} of 1.6–14.2 μM, but only the polymethoxyflavones nobiletin, sinensetin, and tangeretin inhibit OATP1B1 and 1B3 at 2.1–21 μM (Bajraktari-Sylejmani and Weiss 2020).

Citrus fruits, especially grape fruit juice have been reported to be the major cytochrome P450 inhibitors. Fujita et al. (2008) have reported that the citrus species, especially the citrus extracts rich in naringin, bergamottin and 6′,7′-dihydroxybergamottin are reported to be potential inhibitors of CYP3A4, one of the significant phase I enzyme involved in the generation of reactive intermediates triggering carcinogenesis. Besides grape fruit, also other citrus fruits such as lime fresh and sweetie peel that are rich in furanocoumarins, especially bergamottin and 6′,7′-dihydroxybergamottin are strong inhibitors of cytochrome P450 enzymes (Masuda et al. 2017).

On the contrary, some of the citrus fruit juices do not inhibit the cytochrome enzymes, rather modulate the drug transporters enhancing the detoxification process. For instance, the methoxy flavones from the orange juice was tested in Caco-2 cells for their modulation of xenobiotic metabolism. The citrus compounds heptamethoxyflavone (HMF), tangeretin and nobiletin enhanced the drug uptake by the inhibition of the p-glycoprotein that mediates the drug efflux, and does not inhibit the CYP3A4 (Takanaga et al. 2000).

The citrus fruit juices from grapefruit and orange juice inhibited the organic anion-transporting polypeptide B (OATP2B1) expressed in the intestinal epithelial cells thereby blocking the intestinal absorption of anionic drugs via the inhibition of OATP-B (Satoh et al. 2005). Another study by Shirasaka et al. (2013) on grape fruit, orange and apple juice, having the flavonoids, particularly naringin, naringenin, hesperidin, hesperetin, phloridzin, phloretin, quercetin, and kaempferol, inhibited OATP2B1-mediated estrone-3-sulfate uptake with IC_{50} values of 0.222%, 0.807% and 2.27% for grape fruit, orange and apple juice respectively (Fig. 3.11).

Fig. 3.11 Citrus fruits, namely oranges, lemon, lime and grape fruit

4.5.2 Subtropical Fruits

Resveratrol (RSV) is a 3,5,4-trihydroxy-trans-stilbene that is commonly found in grapes, peanuts and wine. There are several studies that emphasises the potential of RSV in chemoprevention by the inhibition of bioactivation of phase I metabolizing enzymes and also induction of detoxifying phase II enzymes (Guthrie et al. 2017). In CD1 male mice, the effect of resveratrol on cytochrome P450 and phase II enzyme activity in liver and lung was investigated. The cytochrome P450 enzymes were significantly suppressed by up to 61% and 97% loss for the CYP3A1/2-linked 6 beta-hydroxylation of testosterone in liver and 2 alpha-hydroxylase in lung, respectively. With respect to the phase II enzymes, GST was significantly inhibited in lung (about 76% loss), whereas the UGT was significantly induced by about 83% in liver and reduced in the lung (of about 83% loss) (Canistro et al. 2009).

The role of RSV (about 10 μM in red wine) and red wine solids (RWS) in the cancer cell initiation and progression was tested in microsomes from rat liver, human liver or cells containing cDNA-expressed cytochromes by their ability to inhibit the metabolism of testosterone, chlorzoxazone, and ethoxyresorufin. Both RSV and RWS were irreversible inhibitors for CYP3A4 and reversible inhibitors for CYP2E1 (Piver et al. 2001). The enantiomers of resveratrol significantly inhibited CYP2C9, CYP3A, CYP1A2, CYP2B6, CYP2C19, CYP2E1 and uridine 5′-diphosphoglucuronosyl transferases (UGT1A1) in human liver microsomes (Ji et al. 2021).

Also, studies show that the red wine polyphenolic extract has been associated with the inhibition of vascular inflammation by downregulating the expression of adhesion molecules ICAM-1, VCAM-1, E-Selectin, MCP-1 and M-CSF (Calabriso et al. 2015).

Mulberry fruit extract, rich in cyanidin-3-glucoside, plays a significant role in chemoprevention and favours the detoxification of xenobiotics. Mulberry extract in response to the PAH benzo[a]pyrene, mediates the suppression of B[a]P derived DNA adduct formation in human keratinocytes mediated by the repression of aryl hydrocarbon receptor signalling (Woo et al. 2017). Chen et al. (2017) have studied the cytoprotective property of the mulberry extract against ethyl carbamate induced cytotoxicity, wherein the ethyl carbamate treatment resulted in glutathione (GSH) depletion and collapse of the mitochondrial membrane potential, whereas the mulberry extract could significantly inhibit the GSH depletion and restore mitochondrial membrane function.

Pomegranate is found to be a potent inhibitor of CYP3A and CYP2C9 mediated metabolism of xenobiotics into procarcinogens; and has potent phytochemicals such as tannins (ellagitannin, punicalagin and punicalin), flavonoids such as anthocyanins, flavan-3-ols, and flavonols. The administration of pomegranate juice was found to increase the absorption and bioavailability of the drugs buspirone, nitrendipine, metronidazole, saquinavir, and sildenafil by reducing the intestinal CYP3A4 and CYP2C9 (Mansoor et al. 2023). Pomegranate juice exhibited the ability to inhibit CYP3A metabolism of carbamazepine, particularly enteric and not hepatic CYP3A (Hidaka et al. 2005); and also inhibit CYP2C9 metabolism of tolbutamide, again enteric and not hepatic CYP2C9 (Nagata et al. 2006) (Fig. 3.12).

4.5.3 Other Tropical, Subtropical & Temperate Fruits

Among the tested dietary phytochemicals against IQ genotoxicity, the black, green and rooibos tea exhibited the highest inhibitory effect with an IC50 value of 0.8–0.9% followed by the sweet cherry juice with 0.17%, whereas mild inhibition was observed with an IC50 value of 0.48–0.71% for kiwi, plum and blueberry juices; and 0.42–0.54% for spinach and onion juices. The extracts from the fruits red apple, blackberry, strawberry, black currant and watermelon, and the vegetables beetroot, red-cabbage, broccoli, cauliflower, sweet pepper, tomato, and chard

Fig. 3.12 Sub-tropical fruits like pomegranate, grapes, fig, jack fruit and mulberry

moderately suppressed the IQ genotoxicity, whereas the grapefruit, red and white wine, sour cherry, red currant and pineapple juices were only weakly active; and Granny Smith green apple, orange and cucumber were ineffective. With respect to the 2-amino-1-methyl-6-phenylimidazo[4,5-b]pyridine (PhIP) carcinogen, inhibition of PhIP genotoxicity was much lesser than IQ genotoxicity.

In a similar study by Edenharder et al. (2002), a group of dietary phytochemicals were tested for their efficacy to protect against the genotoxicity induced by the heterocyclic aromatic amine PhIP derived from cooked red meat in comparison with the carcinogen 2-acetyl aminofluorene on genetically engineered V79 Chinese hamster fibroblasts that expresses human CYP1A2 and rat sulfotransferase. In this study, kiwi, blueberries, blackberries, watermelon, green tea, red wine, red grapes, parsley and spinach strongly reduced the genotoxic activity of PhIP, while the beer, coffee, black as well as rooibos tea, morellos, black-currants, plums, chives, red beets, and broccoli were less active. Similarly, the genotoxic activity of 2-acetyl aminofluorene was strongly reduced by the fruits and vegetables, morellos, black-currants, kiwi, watermelon, and spinach, and also green, black, and rooibos tea, and red wine, whereas the plums, red beets, and broccoli were less potent; and white grapes, white wine, bananas, and strawberries exhibited no activity.

In a study by Platt et al. (2010), the effect of 15 fruit juices (sweet cherry, kiwi, plum, blueberry, watermelon, blackberry, strawberry, black currant, red apple, sour cherry, grapefruit, red currant, pineapple, Granny Smith green apple and orange), 11 vegetable extracts (beetroot, sweet pepper, spinach, onion, broccoli, cauliflower, tomato, chard, red-cabbage and cucumber), 2 wines (red and white wines) and three teas (black, green and roiboos tea) to protect against the genotoxicity mediated by heterocyclic aromatic amines (namely 2-amino-3-methylimidazo[4,5-f]quinoline (IQ) and PhIP, which are potential carcinogens generated during the cooking of red meat and fish) was tested on genetically engineered V79 Chinese hamster fibroblasts that expresses human xenobiotic metabolizing enzymes CYP1A2, N(O)-acetyltransferase and sulfotransferase.

The effect of consumption of tropical fruits banana, guava, mangosteen, pineapple, and ripe mango and papaya on cytochrome P450 activities (CYP1A1, CYP1A2, CYP2E1 and CYP3A11) were studied in mice and the study had shown that the pine apple exhibited the strongest inhibition followed by mangosteen, guava, ripe mango, ripe papaya and banana (Chatuphonprasert and Jarukamjorn 2012) (Figs. 3.13 and 3.14).

4.6 Greens & Herbs

Centella asiatica has been one of the widely known green included in the diet of some of the countries like India, China, Srilanka and Southeast Asian countries; and the pharmaceutically active constituents include asiaticoside, asiatic acid and madecassic acid. Pan et al. (2010) tested the modulatory potential of plant extracts and commercially available asiaticoside, asiatic acid and madecassic acid on the three major cytochrome P450 isoforms. The results showed that the asiatic acid and madecassic acid selectively inhibited CYP2C9 and CYP2D6 than CYP3A4, while the extract exhibited an inhibition of CYP2C9 and not CYP2D6 and CYP3A4. Also, studies by Wright et al. (2020) showed that *Centella asiatica* extract had minimal induction or inhibition of several cytochrome P450 enzymes, including CYP3A4.

Carnosol is a natural product, which is a diterperne polyphenol, extracted from the Rosemary and Sage. The cytoprotective effects on carnosol in HepG2 cells have shown that the rosemary essential oil and carosol increased the nuclear accumulation of Nrf2 to stimulate the ARE to increase the intracellular GSH levels promoting the clearing of toxic substances from the body (Chen et al. 2017).

Fig. 3.13 Tropical fruits like banana, papaya, pineapple, kiwi and mango

Fig. 3.14 Temperate fruits like apples, cherries, plums, pears and peaches

Dasgupta et al. (2004) studied the potential of basil leaves (*Ocimum basilicum*) on xenobiotic metabolism and their antioxidant nature in Swiss albino mice liver cells. The results showed that the basil leaves were able to induce hepatic GST, DT-diaphorase, glutathione reductase, superoxide dismutase and catalase.

Neem flowers (*Azadirachta indica var. siamensis*), Thai fruits, Chinese bitter gourd (*Momordica charantia Linn.*) and sweet basil leaves (*Ocimum basilicum Linn*) were screened for their potential to modulate phase I and II enzymes in rat liver; and the neem flowers and Thai bitter gourd contain phase II enzyme (GST) inducers; and repressors of monooxygenases, whereas the sweet basil leaves are bifunctional inducers (inducing phase I and phase II enzymes) and Chinese bitter gourd fruits represses only the monooxygenases (Kusamran et al. 1998).

Sasaki et al. (2005) investigated the effects of thyme (*Thymus vulgaris* L.), rich in phenolic compounds carvacrol and thymol, on the efficacy of xenobiotic-metabolizing enzymes such as 7-ethoxycoumarin O-deethylase (ECOD), GST and quinone reductase (QR). The administration of thyme (2%), thymol (200 mg/kg) and carvacrol (200 mg/kg) resulted in significantly higher ECOD, GST and QR activities of 1.1–1.4-fold, 1.3–1.9-fold and 1.3–1.7-fold, respectively (Fig. 3.15).

4.7 Whole Grain, Legumes & Bioactive Peptides

Rice protein hydrolysates are the sources of various bioactive peptides that are found to increase the glutathione levels and gamma-glutamyl cysteine synthetase, regulated by the Nrf2 associated with the expression of heme oxygenase-1 antioxidant enzyme (Morita et al. 2017). Kawakami et al. (2022) studied the protective effects of the rice-derived peptides (RP) on hydrogen peroxide induced oxidative stress in HepG2 cells and reported that three of the peptides reduced the oxidative stress through the Nrf2 downstream antioxidant enzyme heme oxygenase – 1.

Helsby et al. (2000) investigated the role of wheat bran diet in protecting cancers in rats. The administration of wheat bran diet resulted in significant increase in CYP3A2 expression and reduced the CYP1A1/A2 expression in hepatic cells. Also, the GST and glucuronosyl transferase phase II metabolising enzymes were reduced in hepatic cells. However, the expression of CYP3A, CYP1A and the glutathione transferase isozymes (GSTA1 & A2) were increased in the colon cells, suggesting the antimutagenic role of dietary wheat bran.

A cancer study in rats induced by the xenobiotic dimethylhydrazine to test the efficacy of the phenolic acid p-methoxy cinnamic acid from rice bran was done and the tests show that the exposure to the xenobiotic decreased the levels of superoxide dismutase, catalase, glutathione peroxidase, glutathione reductase; and the effects were reversed upon the administration of the phenolic acid (Sivagami et al. 2012). The study showed that p-methoxy cinnamic acid has a major influence on prevention of inflammation and cancers (Fig. 3.16).

Fig. 3.15 Herbs – Basil, thyme, rosemary and sage

4.8 Probiotics, Dietary Fibers & Fermentation Products

Butyrates are short chain fatty acids produced by the gut microbiome when digesting food such as dairy products, legumes, vegetables & fruits, whole grains and resistant starches from rice and potatoes, which are rich in dietary fibers. Butyrate is a by-product that is generated in the large intestine of animals and humans

Fig. 3.16 Whole cereals and legumes rich in dietary nutrients

during the fermentation of soluble fibers. The enteric commensal bacterium, in particular, *Clostrdium butyricum*, *Anaerobutyricum hallii* and *Clostridium beijerincki* have been associated with butyrate production during fermentation of dietary fibers.

Researches have shown the butyrate is a significant luminal product that plays a crucial role in promoting detoxication by inducing the expression of the GSTP1 protein resulting in clearance of the toxic chemical (Ebert 2003). Zapletal et al. (2019) studied the chemopreventive action of butyrate in the colon epithelium against benzo[a]pyrene, which is a common dietary carcinogen; and have observed increased induction of N-acetyltransferase (in particular NAT2), NQO1 and UGTs by BaP in HCT-116 and HT29 cells, while the enzymatic activity of only NQO1 and UGTs was increased in differentiated Caco-2 cells (Fig. 3.17).

Fig. 3.17 Butyrates derived from gut fermentation

5 Conclusion

Over the past few decades, a great attention has been paid on functional foods rich in bioactive phytochemicals found to play a significant role in protecting the cells from oxidative stress induced by the exogenous and endogenous toxic chemical. The emphasis is on the influence of dietary nutrients on the modulation of the metabolizing enzymes, drug uptake and efflux pumps, and transcriptional regulators associated with processing and eliminating toxic chemicals from the body. The xenobiotic metabolizing enzymes of phase I are cytochrome P450 enzymes involved in the introduction of polar moieties via oxidation, reduction and hydrolysis; and the phase II enzymes, such as glutathione S-transferase and UDP-glucuronosyltransferases, carry out the conjugation reactions.

Several food nutraceuticals play a pivotal role in modulating the activity of the metabolizing enzymes or their receptors, either by inducing them resulting in intense detoxification to eliminate the toxic substances, or by inhibiting the biotransformation of the xenobiotics into harmful cancer-causing agents that covalently attaches to the DNA resulting in DNA damage leading to the development of cancer. Concurrently, the nutraceuticals play a dual role of inhibiting the formation of carcinogenic DNA adduct formation by inhibiting the drug metabolizing phase I enzymes; and conversely stimulating the phase II enzymes intending to eliminate the toxic substances from the body preventing the initiation of cancers. Additionally, the nutraceuticals play a critical role in improving the bioavailability of the drugs by inhibiting some of the enzymes in liver and/or intestine.

In summary, this chapter reviews the current knowledge on the potential of several dietary phytochemicals in the prevention of health conditions related to the xenobiotic induced oxidative stress and injury resulting in adverse health complications such as inflammation and cancers.

References

Almazroo OA, Miah MK, Venkataramanan R. Drug metabolism in the liver. Clin Liver Dis [Internet]. 2017;21(1):1–20. Available from: https://pubmed.ncbi.nlm.nih.gov/27842765/

Amacher DE. The effects of cytochrome P450 induction by xenobiotics on endobiotic metabolism in pre-clinical safety studies. Toxicol Mech Methods [Internet]. 2010 [cited 2019 Oct 31];20(4):159–166. Available from: https://www.ncbi.nlm.nih.gov/pubmed/20218941

Anzenbacher P, Anzenbacherová E. Cytochromes P450 and metabolism of xenobiotics. Cell Mol Life Sci. 2001;58(5):737–47.

Backman JT, Mäenpää J, Belle DJ, Wrighton SA, Kivistö KT, Neuvonen PJ. Lack of correlation between in vitro and in vivo studies on the effects of tangeretin and tangerine juice on midazolam hydroxylation. Clin Pharmacol Ther [Internet]. 2000 [cited 2023 Jul 10];67(4):382–390. Available from: https://pubmed.ncbi.nlm.nih.gov/10801247/.

Bailey DG, Malcolm J, Arnold O, David Spence J. Grapefruit juice–drug interactions. Br J Clin Pharmacol [Internet]. 1998;46(2):101–10. Available from: https://www.ncbi.nlm.nih.gov/pmc/articles/PMC1873672/

Bajraktari-Sylejmani G, Weiss J. Potential risk of food-drug interactions: citrus polymethoxyflavones and flavanones as inhibitors of the organic anion transporting polypeptides (OATP) 1B1, 1B3, and 2B. Eur J Drug Metab Pharmacokinet. 2020;45(6):809–15.

Balaji C, Muthukumaran J, Nalini N. Chemopreventive effect of sinapic acid on 1,2-dimethylhydrazine-induced experimental rat colon carcinogenesis. Hum Exp Toxicol. 2014;33(12):1253–68.

Baskar AA, Al Numair KS, Gabriel Paulraj M, Alsaif MA, Muamar MA, Ignacimuthu S. β-Sitosterol prevents lipid peroxidation and improves antioxidant status and Histoarchitecture in rats with 1,2-Dimethylhydrazine-induced colon cancer. J Med Food. 2012;15(4):335–43.

Bellik Y, Boukraâ L, Alzahrani H, Bakhotmah B, Abdellah F, Hammoudi S, et al. Molecular mechanism underlying anti-inflammatory and anti-allergic activities of phytochemicals: an update. Molecules. 2012;18(1):322–53.

Beyerle J, Holowatyj AN, Haffa M, Frei E, Gigic B, Schrotz-King P, et al. Expression patterns of xenobiotic-metabolizing enzymes in tumor and adjacent normal mucosa tissues among patients with colorectal cancer: the ColoCare study. 2020 [cited 2023 Jul 11];29(2):460–9. Available from: https://aacrjournals.org/cebp/article/29/2/460/71995/Expression-Patterns-of-Xenobiotic-Metabolizing.

Bhuvaneswari V, Abraham SK, Nagini S. Combinatorial antigenotoxic and anticarcinogenic effects of tomato and garlic through modulation of xenobiotic-metabolizing enzymes during hamster buccal pouch carcinogenesis. Nutrition. 2005;21(6):726–3.

Bolla L, Srivastava P, Velayutham Ravichandiran, N. Satheeshkumar. Cytochrome P450 and P-gp mediated herb-drug interactions and molecular docking studies of garcinol. 2021 [cited 2023 Jun 25];11(12):992–992. Available from: https://www.ncbi.nlm.nih.gov/pmc/articles/PMC8705362/.

Bonetti F, Brombo G, Zuliani G. Chapter 19 - Nootropics, functional foods, and dietary patterns for prevention of cognitive decline [Internet]. In: Watson RR, editor. ScienceDirect. Academic; 2017 [cited 2023 Jul 17]. p. 211–232. Available from: https://www.sciencedirect.com/science/article/abs/pii/B9780128053768000198

Božina N, Bradamante V, Lovrić M. Genetic polymorphism of metabolic enzymes P450 (CYP) as a susceptibility factor for drug response, toxicity, and cancer risk. Arch Ind Hyg Toxicol. 2009;60(2):217–42.

Calabriso N, Scoditti E, Massaro M, Pellegrino M, Storelli C, Ingrosso I, et al. Multiple anti-inflammatory and anti-atherosclerotic properties of red wine polyphenolic extracts: differential role of hydroxycinnamic acids, flavonols and stilbenes on endothelial inflammatory gene expression. Eur J Nutr. 2015;55(2):477–89.

Canistro D, Bonamassa B, Pozzetti L, Sapone A, Abdel-Rahman SZ, Biagi GL, et al. Alteration of xenobiotic metabolizing enzymes by resveratrol in liver and lung of CD1 mice. Food Chem Toxicol [Internet]. 2009 [cited 2023 Jul 12];47(2):454–456. Available from: https://pubmed.ncbi.nlm.nih.gov/19101601/.

Chatuphonprasert W, Jarukamjorn K. Impact of six fruits-banana, guava, mangosteen, pineapple, ripe mango and ripe papaya-on murine hepatic cytochrome P450 activities. J Appl Toxicol. 2012;32(12):994–100.

Chen YS, Xiao P, Ouyang DS, Fan L, Guo D, Wang YN, et al. Simultaneous action of the flavonoid Quercetin on cytochrome P450 (CYP) 1A2, CP2A6,N-acetyltransferase and xanthine oxidase activity in healthy volunteers. Clin Exp Pharmacol Physiol. 2009;36(8):828–33.

Chen W, Li Y, Bao T, Gowd V. Mulberry fruit extract affords protection against ethyl carbamate-induced cytotoxicity and oxidative stress. Oxidative Med Cell Longev. 2017;2017:1–12.

Dasgupta T, Rao AR, Yadava PK. Chemomodulatory efficacy of basil leaf (Ocimum basilicum) on drug metabolizing and antioxidant enzymes, and on carcinogen-induced skin and forestomach papillomagenesis. Phytomedicine. 2004;11(2-3):139–151. https://doi.org/10.1078/0944-7113-00289.

Döring B, Petzinger E. Phase 0 and phase III transport in various organs: combined concept of phases in xenobiotic transport and metabolism. Drug Metab Rev. 2014;46(3):261–82.

Ebert MN. Expression of glutathione S-transferases (GSTs) in human colon cells and inducibility of GSTM2 by butyrate. Carcinogenesis. 2003;24(10):1637–44.

Edenharder R, Sager JW, Glatt H, Muckel E, Platt KL. Protection by beverages, fruits, vegetables, herbs, and flavonoids against genotoxicity of 2-acetylaminofluorene and 2-amino-1-methyl-6-phenylimidazo[4,5-b]pyridine (PhIP) in metabolically competent V79 cells, Mutation Research/Genetic Toxicology and Environmental Mutagenesis. 2002;521(1–2):57–72. https://doi.org/10.1016/S1383-5718(02)00212-7.

Famurewa AC, Renu K, Eladl MA, Chakraborty R, Myakala H, El-Sherbiny M, et al. Hesperidin and hesperetin against heavy metal toxicity: insight on the molecular mechanism of mitigation. Biomed Pharmacother. 2022;149:112914.

Fang Z, Xu Y, Liu G, Shao Q, Niu X, Tai W, et al. Narirutin activates TFEB (transcription factor EB) to protect against Acetaminophen-induced liver injury by targeting PPP3/calcineurin. Autophagy [Internet]. 2023 [cited 2023 Jun 25];1–17. Available from: https://pubmed.ncbi.nlm.nih.gov/36779633/.

Fossen T, Andersen ØM. Anthocyanins from red onion, Allium cepa, with novel aglycone. Phytochemistry. 2003;62(8):1217–20.

Fredotović Ž, Šprung M, Soldo B, Ljubenkov I, Budić-Leto I, Bilušić T, et al. Chemical composition and biological activity of allium cepa L. and allium × cornutum (Clementi ex Visiani 1842) methanolic extracts. Molecules. 2017;22(3):448.

Fujita T, Kawase A, Niwa T, Tomohiro N, Masuda M, Matsuda H, et al. Comparative evaluation of 12 immature citrus fruit extracts for the inhibition of cytochrome P450 isoform activities. Biol Pharm Bull. 2008;31(5):925–30.

Gradelet S, Astorg P, Leclerc J, Chevalier J, Vernevaut MF, Siess MH. Effects of canthaxanthin, astaxanthin, lycopene and lutein on liver xenobiotic-metabolizing enzymes in the rat. Xenobiotica. 1996;26(1):49–63.

Guthrie AR, Chow H-HS, Martinez JA. Effects of resveratrol on drug- and carcinogen-metabolizing enzymes, implications for cancer prevention. Pharmacol Res Perspect. 2017;5(1):e00294.

Hakooz N, Hamdan I. Effects of dietary broccoli on human in vivo caffeine metabolism: a pilot study on a Group of Jordanian Volunteers. Curr Drug Metab. 2007;8(1):9–15.

Helsby NA, Zhu S, Pearson AE, Tingle MD, Ferguson LR. Antimutagenic effects of wheat bran diet through modification of xenobiotic metabolising enzymes. Mutat Res [Internet]. 2000 [cited 2023 Jul 11];454(1–2):77–88. Available from: https://pubmed.ncbi.nlm.nih.gov/11035162/

Hidaka M, Okumura M, Fujita K, Ogikubo T, Yamasaki K, Iwakiri T, et al. Effects of pomegranate juice on human cytochrome P450 3A (CYP3A) and carbamazepine pharmacokinetics in rats. Drug Metab Dispos. 2005;33(5):644–8.

Hietanen E, Ahotupa M, Heikelä A, Laitinen M, Nienstedt W. Dietary cholesterol-induced changes of xenobiotic metabolism in liver. I. Influence of xenobiotic administration on hepatic membrane structure. Drug-Nutr Interact [Internet]. 1982 [cited 2023 Jul 2];1(4):279–292. Available from: https://pubmed.ncbi.nlm.nih.gov/6926835/

Ho BE. Effects of garlic on cytochromes P450 2C9- and 3A4-mediated drug metabolism in human hepatocytes. Sci Pharm. 2010;78(3):473–8.

Hodges RE, Minich DM. Modulation of metabolic detoxification pathways using foods and food-derived components: a scientific review with clinical application. J Nutr Metabol [Internet]. 2015;2015(760689):1–23. Available from: https://www.ncbi.nlm.nih.gov/pmc/articles/PMC4488002/

Horn TL, Reichert MA, Bliss RL, Malejka-Giganti D. Modulations of P450 mRNA in liver and mammary gland and P450 activities and metabolism of estrogen in liver by treatment of rats with indole-3-carbinol. Biochem Pharmacol. 2002;64(3):393–404.

Hrubý K, Anzenbacher P. [Cytochrome P450--its significance in the biotransformation of xenobiotics and interspecies comparisons]. Ceskoslovenska Fysiologie [Internet]. 1997 [cited 2023 Jun 25];46(1):34–9. Available from: https://pubmed.ncbi.nlm.nih.gov/9221213/.

Ip C, Lisk DJ. Modulation of phase I and phase II xenobiotic-metabolizing enzymes by selenium-enriched garlic in rats. Nutr Cancer [Internet]. 1997 [cited 2023 Jul 10];28(2):184–188. Available from: https://pubmed.ncbi.nlm.nih.gov/9290126/.

Iqbal M, Sharma SD, Okazaki Y, Fujisawa M, Okada S. Dietary supplementation of curcumin enhances antioxidant and phase II metabolizing enzymes in ddY male mice: possible role in protection against chemical carcinogenesis and toxicity. Pharmacol Toxicol. 2003;92(1):33–8.

Jee SC, Kim M, Sung JS. Modulatory effects of silymarin on Benzo[a]pyrene-induced hepatotoxicity. Int J Mol Sci [Internet]. 2020 [cited 2023 Jan 14];21(7):2369. Available from: https://www.ncbi.nlm.nih.gov/pmc/articles/PMC7177818/

Ji SB, Park SY, Bae S, Seo HJ, Kim SE, Lee GM, et al. Comprehensive investigation of stereoselective food drug interaction potential of resveratrol on nine P450 and six UGT isoforms in human liver microsomes. Pharmaceutics [Internet]. 2021 [cited 2023 Jul 11];13(9):1419. Available from: https://pubmed.ncbi.nlm.nih.gov/34575495/.

Joo HK, Choi S, Lee YR, Lee EO, Park MS, Lim YP, et al. Ethanol extract of Brassica rapa ssp. pekinensis suppresses tumor necrosis factor-α-induced inflammatory response in human umbilical vein endothelial cells. J Med Food [Internet]. 2017 [cited 2023 Jul 13];20(5):511–518. Available from: https://pubmed.ncbi.nlm.nih.gov/28504909/

Jurenka JS. Anti-inflammatory properties of curcumin, a major constituent of curcuma longa: a review of preclinical and clinical research [Internet]. Alternat Med Rev. 2009. Available from: https://pubmed.ncbi.nlm.nih.gov/19594223/.

Kall MA, Vang O, Clausen J. Effects of dietary broccoli on human in vivo drug metabolizing enzymes: evaluation of caffeine, oestrone and chlorzoxazone metabolism. Carcinogenesis. 1996;17(4):793–9.

Kang AY, Young LR, Dingfelder C, Peterson S. Effects of furanocoumarins from apiaceous vegetables on the catalytic activity of recombinant human cytochrome P-450 1A2. Protein J. 2011;30(7):447–56.

Kawakami K, Moritani C, Hatanaka T, Tsuboi S. Isolation of the hemeoxygenase-1 inducer from rice-derived peptide. J Clin Biochem Nutr. 2022;71:41–7.

Kimura Y, Ito H, Ohnishi R, Hatano T. Inhibitory effects of polyphenols on human cytochrome P450 3A4 and 2C9 activity. Food Chem Toxicol [Internet]. 2010a;48(1):429–35. Available from: https://pubmed.ncbi.nlm.nih.gov/19883715/

Kimura Y, Ito H, Hatano T. Effects of Mace and Nutmeg on human cytochrome P450 3A4 and 2C9 activity. Biol Pharm Bull. 2010b;33(12):1977–82.

Kiruthiga P, Karthikeyan K, Archunan G, Pandian SK, Devi KP. Silymarin prevents benzo(a) pyrene-induced toxicity in Wistar rats by modulating xenobiotic-metabolizing enzymes. Toxicol Ind Health. 2013;31(6):523–4.

Kleiner HE, Vulimiri SV, Miller L, Johnson WH, Whitman CP, DiGiovanni J. Oral administration of naturally occurring coumarins leads to altered phase I and II enzyme activities and reduced DNA adduct formation by polycyclic aromatic hydrocarbons in various tissues of SENCAR mice. Carcinogenesis [Internet]. 2001 [cited 2023 Jul 10];22(1):73–82. Available from: https://pubmed.ncbi.nlm.nih.gov/11159744/

Kleiner HE, Reed MJ, DiGiovanni J. Naturally occurring coumarins inhibit human cytochromes P450 and block benzo[a]pyrene and 7,12-dimethylbenz[a]anthracene DNA adduct formation in MCF-7 cells. Chem Res Toxicol [Internet]. 2003 [cited 2023 Jul 10];16(3):415–422. Available from: https://pubmed.ncbi.nlm.nih.gov/12641443/

Kondža M, Bojić M, Tomić I, Maleš Ž, Rezić V, Ćavar I. Characterization of the CYP3A4 enzyme inhibition potential of selected flavonoids. Molecules. 2021;26(10):3018.

Kopytko P, Piotrowska K, Janisiak J, Tarnowski M. Garcinol-A natural histone acetyltransferase inhibitor and Nnew anti-cancer epigenetic drug. Int J Mol Sci [Internet]. 2021 [cited 2022 Mar 7];22(6):2828. Available from: https://pubmed.ncbi.nlm.nih.gov/33799504/

Kothari D, Lee WD, Kim SK. Allium Flavonols: health benefits, molecular targets, and bioavailability. Antioxidants. 2020;9(9):888.

Krajka-Kuźniak V. The effects of tannic acid on cytochrome P450 and phase II enzymes in mouse liver and kidney. Toxicol Lett. 2003;143(2):209–16.

Kravchenko LV, Kuz'mina EE, Avren'eva LI, Kulakova SN, Levachev MM, Tutel'ian VA. Xenobiotic-metabolizing enzymes in rat liver and small intestinal mucosa with varying combinations of polyunsaturated omega-6 and omega-3 fatty acids in the diet. Voprosy Meditsinskoi Khimii [Internet]. 1992 [cited 2023 Jul 2];38(6):53–56. Available from: https://pubmed.ncbi.nlm.nih.gov/1298135/.

Kuang R, Levinthal DJ, Ghaffari AA, Ramos C, Tansel A, Binion DG. Nightshade vegetables: a dietary trigger for worsening inflammatory bowel disease and irritable bowel syndrome? 2023 May 19.

Kuntz S, Kunz C. Extracts from Brassica oleracea L. convar. acephala var. sabellica inhibit TNF-α stimulated neutrophil adhesion in vitro under flow conditions. Food Funct. 2014;5(6):1082.

Kusamran WR, Ratanavila A, Tepsuwan A. Effects of neem flowers, Thai and Chinese bitter gourd fruits and sweet basil leaves on hepatic monooxygenases and glutathione S-transferase activities, and in vitro metabolic activation of chemical carcinogens in rats. Food Chem Toxicol. 1998;36(6):475–84.

Lan K, Xie G, Jia W. Towards polypharmacokinetics: pharmacokinetics of multicomponent drugs and herbal medicines using a metabolomics approach. Evid-Based Complement Alternat Med

[Internet]. 2013 [cited 2019 Sep 8];2013:1–12. Available from: https://www.hindawi.com/journals/ecam/2013/819147/

Lee SB, Kim CY, Lee HJ, Yun JH, Nho CW. Induction of the phase II detoxification enzyme NQO1 in hepatocarcinoma cells by lignans from the fruit of Schisandra chinensis through nuclear accumulation of Nrf2. Planta Med [Internet]. 2009 [cited 2023 Jul 16];75(12):1314–1318. Available from: https://pubmed.ncbi.nlm.nih.gov/19452436/

Leibelt DA, Hedstrom OR, Fischer KA, Pereira CB, Williams DE. Evaluation of chronic dietary exposure to indole-3-carbinol and absorption-enhanced 3,3'-diindolylmethane in sprague-dawley rats. Toxicol Sci. 2003;74(1):10–21. https://doi.org/10.1093/toxsci/kfg103

Li S, Wang X, Xiao Y, Wang Y, Wan Y, Li X, et al. Curcumin ameliorates mercuric chloride-induced liver injury via modulating cytochrome P450 signaling and Nrf2/HO-1 pathway. Ecotoxicol Environ Safe [Internet]. 2021 [cited 2023 Jul 8];208:111426. Available from: https://pubmed.ncbi.nlm.nih.gov/33096358/

Lown KS, Bailey DG, Fontana RJ, Janardan SK, Adair CH, Fortlage LA, et al. Grapefruit juice increases felodipine oral availability in humans by decreasing intestinal CYP3A protein expression. [Internet] wwwjciorg; 1997. Available from: https://www.jci.org/articles/view/119439/pdf.

Lutz M, Alvarado J, Barraza P, Bonilla S, Luna L. Effects of dietary oils and cholesterol supplement on fluidity and enzyme activities of liver microsomes in the rat. Lipids [Internet]. 1998 [cited 2023 Jul 2];34 Suppl:S119. Available from: https://pubmed.ncbi.nlm.nih.gov/10419116/

Lutz, M., Alvarado, J., Barraza, P. et al. Effects of dietary oils and cholesterol supplement on fluidity and enzyme activities of liver microsomes in the rat. Lipids 34 (Suppl 1), S119 (1999). https://doi.org/10.1007/BF02562256

Mallhi TH, Sarriff A, Adnan AS, Khan YH, Qadir MI, Hamzah AA, et al. Effect of fruit/vegetable-drug interactions on CYP450, OATP and p-glycoprotein: a systematic review. Trop J Pharm Res. 2015;14(10):1927.

Manju V, Balasubramaniyan V, Nalini N. Rat colonic lipid peroxidation and antioxidant status: the effects of dietary luteolin on 1,2-dimethylhydrazine challenge. Cell Mol Biol Lett [Internet]. 2005 [cited 2023 Jul 16];10(3):535–551. Available from: https://pubmed.ncbi.nlm.nih.gov/16217561/

Mansoor K, Bardees R, Alkhawaja B, Mallah E, AbuQatouseh L, Schmidt M, et al. Impact of pomegranate juice on the pharmacokinetics of CYP3A4-and CYP2C9-mediated drugs metabolism: A preclinical and clinical review. Molecules [Internet]. 2023 [cited 2023 Jul 11];28(5):2117. Available from: https://pubmed.ncbi.nlm.nih.gov/36903363/

Martinez KB, Mackert JD, McIntosh MK. Chapter 18 - Polyphenols and intestinal health [Internet]. Watson RR, editor. ScienceDirect. Academic; 2017 [cited 2023 Jul 15]. p. 191–210. Available from: https://www.sciencedirect.com/science/article/abs/pii/B9780128053768000186

Masuda M, Watanabe S, Tanaka M, Tanaka A, Araki H. Screening of furanocoumarin derivatives as cytochrome P450 3A4 inhibitors in citrus. J Clin Pharm Ther. 2017;43(1):15–20.

Meyer UA. Overview of enzymes of drug metabolism. J Pharmacokinet Biopharm. 1996 Oct;24(5):449–59.

Michnovicz JJ, Bradlow HL. Induction of estradiol metabolism by dietary Indole-3-carbinol in humans. JNCI [Internet]. 1990;82(11):947–9. Available from: https://academic.oup.com/jnci/article/82/11/947/905664

Monostory K, Vereczkey L. Role of human cytochrome P-450 enzymes in the metabolism of xenobiotics. Acta Pharma Hungarica [Internet]. 1995 [cited 2023 Jun 25];65(5):147–156. Available from: https://pubmed.ncbi.nlm.nih.gov/7484161/

Morita K, Nishibori N, Kishibuchi R, Itoh M, Horie Y, Nemoto H. Fermented Brown Rice extract stimulates BDNF gene transcription in C6 glioma cells: possible connection with HO-1 expression. J Diet FANSuppl [Internet]. 2017 [cited 2023 Jul 14];14(2):214–228. Available from: https://pubmed.ncbi.nlm.nih.gov/27560964/

Muriana FJG, Ruiz-Gutierrez V, Vazquez CM. Influence of dietary cholesterol on 0polyunsaturated fatty acid composition, fluidity and membrane-bound enzymes in liver —microsomes of rats fed olive and fish oil. Biochimie. 1992a;74(6):551–6.

Muriana FJ, Vazquez CM, Ruiz-Gutierrez V. Fatty acid composition and properties of the liver microsomal membrane of rats fed diets enriched with cholesterol. J Biochem [Internet]. 1992b [cited 2023 Jul 2];112(4):562–567. Available from: https://pubmed.ncbi.nlm.nih.gov/1491012/

Nagata M, Hidaka M, Sekiya H, Kawano Y, Yamasaki K, Okumura M, et al. Effects of pomegranate juice on human cytochrome P450 2C9 and Tolbutamide Pharmacokinetics in rats. Drug Metab Dispos. 2006;35(2):302–5.

Nakanishi T, Tamai I. Interaction of drug or food with drug transporters in intestine and liver. Curr Drug Metab. 2015;16(9):753–64.

Nandakumar N, Haribabu L, Perumal S, Balasubramanian MP. Therapeutic effect of hesperidin with reference to biotransformation, lysosomal and mitochondrial TCA cycle enzymes against 7,12-dimethylbenz(a)anthracene-induced experimental mammary cellular carcinoma. Biomed Aging Pathol. 2011;1(3):158–68.

Navarro SL, Chang JL, Peterson S, Chen C, King IB, Schwarz Y, et al. Modulation of human serum glutathione S-transferase A1/2 concentration by cruciferous vegetables in a controlled feeding study is influenced by GSTM1 and GSTT1 genotypes. Cancer Epidemiol Biomarkers Prev. 2009;18(11):2974–8.

Nguyen T, Sherratt PJ, Pickett CB. Regulatory mechanisms controlling gene expression mediated by the antioxidant response element. Annu Rev Pharmacol Toxicol. 2003;43(1):233–60.

Okada N, Murakami A, Urushizaki S, Matsuda M, Kawazoe K, Ishizawa K. Extracts of immature Orange (Aurantii fructus immaturus) and citrus Unshiu Peel (Citri unshiu pericarpium) induce P-glycoprotein and cytochrome P450 3A4 expression via upregulation of Pregnane X receptor. Front Pharmacol [Internet]. 2017 [cited 2019 Dec 20];8. Available from: https://www.frontiersin.org/articles/10.3389/fphar.2017.00084/full

Paine AJ. Hepatic cytochrome P-450. Essays Biochem [Internet]. 1981 [cited 2023 Jul 16];17:85–126. Available from: https://pubmed.ncbi.nlm.nih.gov/6795037/.

Pan MH, Chang WL, Lin-Shiau S-Y, Ho CT, Lin J. Induction of apoptosis by garcinol and curcumin through cytochrome c release and activation of caspases in human leukemia HL-60 cells. J Agric Food Chem. 2001;49(3):1464–74.

Pan Y, Abd-Rashid BA, Ismail Z, Ismail R, Mak JW, Pook PCK, et al. In vitro modulatory effects on three major human cytochrome P450 enzymes by multiple active constituents and extracts of Centella asiatica. J Ethnopharmacol. 2010;130(2):275–83.

Pang X, Tang C, Guo R, Chen X. Non-cytochrome P450 enzymes involved in the oxidative metabolism of xenobiotics: focus on the regulation of gene expression and enzyme activity. Pharmacol Ther. 2021:108020.

Paredes-Gonzalez X, Fuentes F, Jeffery S, Saw CLL, Shu L, Su ZY, et al. Induction of NRF2-mediated gene expression by dietary phytochemical flavones apigenin and luteolin. Biopharm Drug Dispos. 2015;36(7):440–51.

Patel J, Buddha B, Dey S, Pal D, Mitra AK. In vitro interaction of the HIV protease inhibitor ritonavir with herbal constituents: changes in P-gp and CYP3A4 activity. Am J Ther. 2004;11(4):262–77.

Patel M, Taskar KS, Zamek-Gliszczynski MJ. Importance of hepatic transporters in clinical disposition of drugs and their metabolites. J Clin Pharmacol. 2016;56:S23–39.

Perez J, Patil BS, Cadena A, Martinez E, Ahmad H. In vivo induction of phase II detoxifying enzymes, glutathione transferase and quinone reductase by citrus triterpenoids. 2010 [cited 2023 Jul 13];10(1). Available from: https://www.ncbi.nlm.nih.gov/pmc/articles/PMC2954937/#B21

Peterson S, Lampe JW, Bammler TK, Gross-Steinmeyer K, Eaton DL. Apiaceous vegetable constituents inhibit human cytochrome P-450 1A2 (hCYP1A2) activity and hCYP1A2-mediated mutagenicity of aflatoxin B. Food Chem Toxicol. 2006;44(9):1474–84.

Peterson S, Schwarz Y, Li SS, Li L, King IB, Chen C, et al. CYP1A2, GSTM1, and GSTT1 polymorphisms and diet effects on CYP1A2 activity in a crossover feeding trial. Cancer Epidemiol Biomark Prevent [Internet]. 2009 [cited 2020 Dec 7];18(11):3118–25. Available from: https://www.ncbi.nlm.nih.gov/pmc/articles/PMC2784188/.

Phang-Lyn S, Llerena VA. Biochemistry, biotransformation [Internet]. PubMed. Treasure Island: StatPearls Publishing; 2020. Available from: https://www.ncbi.nlm.nih.gov/books/NBK544353/.

Piver B, Berthou F, Dreano Y, Lucas D. Inhibition of CYP3A, CYP1A and CYP2E1 activities by resveratrol and other non volatile red wine components. Toxicol Lett [Internet]. 2001;125(1–3):83–9. Available from: https://pubmed.ncbi.nlm.nih.gov/11701226/

Platt KL, Edenharder R, Aderhold S, Muckel E, Glatt H. Fruits and vegetables protect against the genotoxicity of heterocyclic aromatic amines activated by human xenobiotic-metabolizing enzymes expressed in immortal mammalian cells. Mutat Res [Internet]. 2010;703(2):90–8. Available from: https://pubmed.ncbi.nlm.nih.gov/20713174/

Pool-Zobel B, Veeriah S, Frank-D B. Modulation of xenobiotic metabolising enzymes by anti-carcinogens—focus on glutathione S-transferases and their role as targets of dietary chemoprevention in colorectal carcinogenesis. Mutat Res Genet Toxicol Environ Mutagen. 2005;591(1–2):74–92.

Pouremamali F, Pouremamali A, Dadashpour M, Soozangar N, Jeddi F. An update of Nrf2 activators and inhibitors in cancer prevention/promotion. Cell Commun Signal. 2022;20(1):100.

Rahden-Staron I, Czeczot H, Szumilo M. Induction of rat liver cytochrome P450 isoenzymes CYP 1A and CYP 2B by different fungicides, nitrofurans, and quercetin. Mutat Res Genet Toxicol Environ Mutagen. 2001;498(1–2):57–66.

Reed L, Arlt VM, Phillips DH. The role of cytochrome P450 enzymes in carcinogen activation and detoxication: an in vivo–in vitro paradox. Carcinogenesis. 2018;39(7):851–9.

Rendic S, Guengerich FP. Survey of human oxidoreductases and cytochrome P450 enzymes involved in the metabolism of xenobiotic and natural chemicals. Chem Res Toxicol. 2014;28(1):38–42.

Rushmore TH, Morton MR, Pickett CB. The antioxidant responsive element. Activation by oxidative stress and identification of the DNA consensus sequence required for functional activity. J Biol Chem [Internet]. 1991;266(18):11632–9. Available from: https://pubmed.ncbi.nlm.nih.gov/1646813/

Sasaki K, Wada K, Tanaka Y, Yoshimura T, Matuoka K, Anno T. Thyme (Thymus vulgaris L.) leaves and its constituents increase the activities of xenobiotic-metabolizing enzymes in mouse liver. J Med Food [Internet]. 2005 [cited 2023 Jul 14];8(2):184–189. Available from: https://pubmed.ncbi.nlm.nih.gov/16117610/

Sasu A, Herman H, Mariasiu T, Rosu M, Balta C, Anghel N, et al. Protective effects of silymarin on epirubicin-induced mucosal barrier injury of the gastrointestinal tract. Drug Chem Toxicol [Internet]. 2015 [cited 2023 Jun 25];38(4):442–445. Available from: https://pubmed.ncbi.nlm.nih.gov/25609004/

Satoh H, Yamashita F, Tsujimoto M, Murakami H, Koyabu N, Ohtani H, et al. Citrus juices inhibit the function of human organic anion-transporting polypeptide OATP-B. Drug Metabol Dispos [Internet]. 2005 [cited 2023 Jul 12];33(4):518–23. Available from: https://pubmed.ncbi.nlm.nih.gov/15640378/.

Satoh T, Fujisawa H, Nakamura A, Takahashi N, Watanabe K. Inhibitory effects of eight green tea Catechins on cytochrome P450 1A2, 2C9, 2D6, and 3A4 activities. J Pharm Pharma Sci. [Internet]. 2016;19(2):188–97. Available from: https://www.ncbi.nlm.nih.gov/pubmed/27518169.

Sayago-Ayerdi S, García-Martínez DL, Ramírez-Castillo AC, Ramírez-Concepción HR, Viuda-Martos M. Tropical fruits and their co-products as bioactive compounds and their health effects: a review. Food Secur. 2021;10(8):1952.

Sevior DK, Pelkonen O, Ahokas JT. Hepatocytes: The powerhouse of biotransformation. Int J Biochem Cell Biol [Internet]. 2012 [cited 2019 Nov 25];44(2):257–6. Available from: https://www.sciencedirect.com/science/article/pii/S1357272511003098.

Shang A, Cao S, Xu X, Gan R, Tang G, Corke H, et al. Bioactive compounds and biological functions of garlic (Allium sativum L.). Foods [Internet]. 2019;8(7):246. Available from: https://www.mdpi.com/2304-8158/8/7/246/pdf

Shirasaka Y, Shichiri M, Mori T, Nakanishi T, Tamai I. Major active components in grapefruit, orange, and apple juices responsible for OATP2B1-mediated drug interactions. J Pharma Sci [Internet]. 2013;102(9):3418–26. Available from: https://pubmed.ncbi.nlm.nih.gov/23794501/

Siess MH, Vernevaut MF, Suschetet M. The effect of cholesterol and the nature of dietary fatty acids on the activities of some microsomal monooxygenases in the rat liver. Reprod Nutr Dev [Internet]. 1983 [cited 2023 Jul 2];23(2a):195–206. Available from: https://pubmed.ncbi.nlm.nih.gov/6405460/

Singh SK, Srivastava PN, Meena A, Luqman S. Dietary flavonoid narirutin as a prospective antagonist of oncogenic pri/ pre-microRNAs. Phytother Res. 2022;36(2):963–83.

Sivagami G, Karthikkumar V, Balasubramanian T, Nalini N. The modulatory influence of p-methoxycinnamic acid, an active rice bran phenolic acid, against 1,2-dimethylhydrazine-induced lipid peroxidation, antioxidant status and aberrant crypt foci in rat colon carcinogenesis. Chem Biol Interact. 2012;196(1–2):11–22.

Soundararajan P, Kim J. Anti-carcinogenic Glucosinolates in cruciferous vegetables and their antagonistic effects on prevention of cancers. Molecules. 2018;23(11):2983.

Srinivasan P, Suchalatha S, Babu PVA, Devi RS, Narayan S, Sabitha KE, et al. Chemopreventive and therapeutic modulation of green tea polyphenols on drug metabolizing enzymes in 4-Nitroquinoline 1-oxide induced oral cancer. Chem Biol Interact. 2008;172(3):224–34.

Surh YJ. Cancer chemoprevention with dietary phytochemicals. Nat Rev Cancer. 2003;3(10):768–80.

Svehlikova V. Interactions between sulforaphane and apigenin in the induction of UGT1A1 and GSTA1 in CaCo-2 cells. Carcinogenesis. 2004;25(9):1629–37.

Takanaga H, Ohnishi A, Yamada S, Matsuo H, Morimoto S, Shoyama Y, et al. Polymethoxylated flavones in orange juice are inhibitors of P-glycoprotein but not cytochrome P450 3A4. J Pharmacol Exp Ther [Internet]. 2000 [cited 2023 Jul 12];293(1):230–236. Available from: https://pubmed.ncbi.nlm.nih.gov/10734174/

Takano M, Yumoto R, Murakami T. Expression and function of efflux drug transporters in the intestine. Pharmacol Ther [Internet]. 2006 [cited 2020 Apr 28];109(1–2):137–136. Available from: https://www.ncbi.nlm.nih.gov/pubmed/16209890.

Takemura H, Itoh T, Yamamoto K, Sakakibara H, Shimoi K. Selective inhibition of methoxyflavonoids on human CYP1B1 activity. Bioorg Med Chem [Internet] 2010 [cited 2023 Jul 10];18(17):6310–6315. Available from: https://pubmed.ncbi.nlm.nih.gov/20696580/.

Tan XL, Shi M, Tang H, Han W, Spivack SD. Candidate dietary phytochemicals modulate expression of phase II enzymes GSTP1 and NQO1 in human lung cells. J Nutr [Internet] 2010 [cited 2020 Sep 12];140(8):1404–1410. Available from: https://www.ncbi.nlm.nih.gov/pmc/articles/PMC2903300/

Tanaka T, Kohno H, Shimada R, Kagami S, Yamaguchi F, Kataoka S, et al. Prevention of colonic aberrant crypt foci by dietary feeding of garcinol in male F344 rats. Carcinogenesis [Internet] 2000a [cited 2023 Jul 16];21(6):1183–1189. Available from: https://pubmed.ncbi.nlm.nih.gov/10837008/.

Tanaka T, Kohno H, Tsukio Y, Honjo S, Tanino M, Miyake M, et al. Citrus limonoids obacunone and limonin inhibit azoxymethane-induced colon carcinogenesis in rats. Biofactors. 2000b;13(1–4):213–8.

Teyssier C, Amiot MJ, Mondy N, Auger J, Kahane R, Siess MH. Effect of onion consumption by rats on hepatic drug-metabolizing enzymes. Food Chem Toxicol. 2001;39(10):981–7.

Thangaraj K, Natesan K, Palani M, Vaiyapuri M. Orientin, a flavanoid, mitigates 1, 2 dimethylhydrazine-induced colorectal lesions in Wistar rats fed a high-fat diet. Toxicol Rep. 2018;5:977–87.

Thapliyal R, Maru GB. Inhibition of cytochrome P450 isozymes by curcumins in vitro and in vivo. Food Chem Toxicol. 2001;39(6):541–7.

Ueng YF, Chang YL, Oda Y, Park SS, Liao JF, Lin MF, et al. In vitro and in vivo effects of naringin on cytochrome P450-dependent monooxygenase in mouse liver. Life Sci [Internet] 1999 [cited 2023 Jul 6];65(24):2591–2602. Available from: https://pubmed.ncbi.nlm.nih.gov/10619367/.

Valerio LG, Kepa JK, Pickwell GV, Quattrochi LC. Induction of human NAD(P)H:quinone oxidoreductase (NQO1) gene expression by the flavonol quercetin. Toxicol Lett. 2001;119(1):49–57.

Wadhwa K, Pahwa R, Kumar M, Kumar S, Sharma PC, Singh G, et al. Mechanistic Insights into the Pharmacological Significance of Silymarin. Molecules (Basel, Switzerland) [Internet]. 2022 [cited 2023 Mar 11];27(16):5327. Available from: https://pubmed.ncbi.nlm.nih.gov/36014565/

Walters DG. Cruciferous vegetable consumption alters the metabolism of the dietary carcinogen 2-amino-1-methyl-6-phenylimidazo[4,5-b]pyridine (PhIP) in humans. Carcinogenesis. 2004;25(9):1659–69.

Wark PA. Habitual consumption of fruits and vegetables: associations with human rectal glutathione S-transferase. Carcinogenesis. 2004;25(11):2135–42.

Warwick E, Cassidy A, Hanley B, Jouni ZE, Bao Y. Effect of phytochemicals on phase II enzyme expression in infant human primary skin fibroblast cells. Br J Nutr. 2012;108(12):2158–65.

White CM, Pasupuleti V, Roman YM, Li Y, Hernandez AV. Oral turmeric/curcumin effects on inflammatory markers in chronic inflammatory diseases: a systematic review and meta-analysis of randomized controlled trials. Pharmacol Res [Internet]. 2019;146:104280. Available from: https://www.sciencedirect.com/science/article/pii/S1043661818319236

Wilken R, Veena MS, Wang MB, Srivatsan ES. Curcumin: A review of anti-cancer properties and therapeutic activity in head and neck squamous cell carcinoma. Mol Cancer [Internet]. 2011 [cited 2019 Sep 13];10(1):12. Available from: https://www.ncbi.nlm.nih.gov/pmc/articles/PMC3055228/.

Wiśniewska A, Mazerska Z. Cytochrome P450 isoenzymes in metabolism of endo- and exogenic compounds. Postepy Biochem [Internet]. 2009;55(3):259–7. Available from: https://pubmed.ncbi.nlm.nih.gov/19928582/

Woo H, Lee JA, Park D, Jung E. Protective effect of mulberry (Morus alba L.) extract against benzo [A] pyrene induced skin damage through inhibition of aryl hydrocarbon receptor signaling. J Agric Food Chem. 2017;65(50):10925–32.

Wright KM, Magana AA, Laethem RM, Moseley CL, Banks TT, Maier CS, et al. Centella asiatica water extract shows low potential for cytochrome P450–mediated drug interactions. Drug Metab Dispos. 2020;48(10):1053–63.

Yaqoob A, Rehman Q, Rehman K, Akash MSH, Hussain I, Ahmad R. Chapter 3 - Role of drug-metabolizing enzymes in biotransformation of drugs [Internet]. Hamid Akash MS, Rehman K, editors. ScienceDirect. Academic; 2022 [cited 2023 Jul 15]. p. 73–108. Available from: https://www.sciencedirect.com/science/article/abs/pii/B9780323951203000130

Yoo JSH, Hong JY, Ning SM, Yang CS. Roles of dietary corn oil in the regulation of cytochromes P450 and glutathione S-transferases in rat liver. J Nutr. 1990;120(12):1718–26.

Yoxall V, Kentish P, Coldham N, Kuhnert N, Sauer MJ, Ioannides C. Modulation of hepatic cytochromes P450 and phase II enzymes by dietary doses of sulforaphane in rats: implications for its chemopreventive activity. Int J Cancer. 2005;117(3):356–62.

Zanger UM, Schwab M. Cytochrome P450 enzymes in drug metabolism: regulation of gene expression, enzyme activities, and impact of genetic variation. Pharmacol Ther [Internet]. 2013;138(1):103–4. Available from: https://www.sciencedirect.com/science/article/pii/S0163725813000065

Zapletal O, Procházková J, Dubec V, Hofmanová J, Kozubík A, Vondráček J. Butyrate interacts with benzo[a]pyrene to alter expression and activities of xenobiotic metabolizing enzymes involved in metabolism of carcinogens within colon epithelial cell models. Toxicology [Internet]. 2019 [cited 2023 Jul 13];412:1–1 Available from: https://pubmed.ncbi.nlm.nih.gov/30439556/

Zeng T, Zhang CL, Song F, Han X, Xie K. The modulatory effects of garlic oil on hepatic cytochrome P450s in mice. Hum Exp Toxicol. 2009;28(12):777–83.

Zeng W, Yang J, He Y, Zhu Z. Bioactive compounds in cruciferous sprouts and microgreens and the effects of sulfur nutrition. J Sci Food Agric [Internet]. 2023 May 30. [cited 2023 Jul 10]; Available from: https://pubmed.ncbi.nlm.nih.gov/37254614/

Zhao M, Ma J, Li M, Zhang Y, Jiang B, Zhao X, et al. Cytochrome P450 enzymes and drug metabolism in humans. Int J Mol Sci [Internet]. 2021;22(23):12808. Available from: https://pubmed.ncbi.nlm.nih.gov/34884615/

Zuiter AS. Proanthocyanidin: chemistry and biology: from phenolic compounds to Proanthocyanidins. Reference Module in Chemistry, Molecular Sciences and Chemical Engineering. 2014

Chapter 4
Nutraceuticals in the Prevention of Cancer

Priti Paul, Rahul Nair, Srushti Mahajan, Ujala Gupta, Mayur Aalhate, Indrani Maji, Rahul Kumar, Anshita Gupta Soni, and Pankaj Kumar Singh

Abbreviations

ALA	α-linolenic acid
ARA	Arachidonic acid
BC	Breast cancer
CRC	Colorectal cancer
DHA	Docosahexaenoic acid
EPA	Eicosapentaenoic acid
FAs	Fatty acids
IGF-1	Insulin-like growth factor 1
LC	Lung cancer
Mg	Magnesium
mTOR	Mammalian target of rapamycin
MUFAs	Monounsaturated fatty acids
NF	Nuclear factor
OVC	Ovarian cancer

Priti Paul and Rahul Nair contributed equally with all other contributors.

P. Paul · R. Nair · S. Mahajan · U. Gupta · M. Aalhate · I. Maji · P. K. Singh (✉)
Department of Pharmaceutics, National Institute of Pharmaceutical Education and Research, Hyderabad, India

R. Kumar
Department of Biological Sciences, National Institute of Pharmaceutical Education and Research, Hyderabad, India

A. G. Soni
Shri Rawatpura Sarkar Institute of Pharmacy, Kumhari, Durg, Chhattisgarh, India

S. M. Jafari et al. (eds.), *Nutraceuticals: A New Millennium Approach for Preventive Medicine*, Food Bioactive Ingredients,
https://doi.org/10.1007/978-3-031-76481-3_4

PA	Palmitic acid
PANC	Pancreatic cancer
PROC	Prostate cancer
PUFAs	Polyunsaturated fatty acids
SCFAs	Short-chain fatty acids
Se	Selenium
SFAs	Saturated fatty acids
TNF-α	Tumor necrosis factor-α
Zn	Zinc

1 Introduction

The field of nutrition study in the cancer process is quite vast, and it is becoming evident as research progresses that nutrition is a contributing factor in cancer. Research on nutrition and cancer prevention has sought to comprehend how particular foods and nutrients affect cancer-related pathways. The role of nutrition in the development of cancer was nevertheless supported in the latter part of the twentieth century by probing research and new epidemiological data (Martín Ortega and Segura Campos 2021). Doll and Peto forecasted in a 1981 paper titled "The Causes of Cancer" that nutrition was responsible for about 35% of cancer cases worldwide. Diet, nutrition, and bodily activity are key factors influencing the chances of developing cancer, owing to their links to obesity, a proven risk factor for many cancers (Mayne et al. 2016). Diet is a critical element influencing growth factor levels and other aspects of cell growth and can impact the genesis and progression of cancers in various ways. Several epidemiologic research studies have repeatedly associated a greater intake of plant-origin foods, including whole grains, vegetables, fruits, legumes, seeds, nuts, and tea, with a lower risk of acquiring different malignancies. Inadequate protein consumption is linked to lower IGF-1 (insulin-like growth factor 1) levels and lower cancer mortality, especially in the elderly. IGF-1 is required by normal cells for proliferative signaling. Yet, in the context of cancer, IGF-1 not only promotes cancer development via cell proliferation pathways but also inhibits cell cycle arrest and apoptosis (Levine et al. 2014). Inadequate vitamin D and calcium levels have also been associated with neoplastic colonocyte suppression (Aggarwal et al. 2016), (Madak Erdogan 2017). The association between diet and malignancy has now been identified at the atomic, cellular, and functional levels, leading to significant scientific and technological developments (Martín Ortega and Segura Campos 2021). The main objective of this segment is to give the audience a synopsis of how important micronutrients and macronutrients relate to preventing cancer.

2 Macronutrients

Macronutrients are the nutrients consumed by humans in larger quantities that supply energy. They comprise of carbohydrates, proteins, and fats. Elements including carbon, hydrogen, oxygen, nitrogen, sulfur, and phosphorus make up the macronutrients and form the structural bases of tissues. Macronutrients are necessary for substantial amounts to sustain body functioning and carry out daily tasks (Fellows 2009). The consumption of macronutrients influences the synthesis of numerous hormones such as insulin, glucagon, cortisol, growth hormone, IGF-1, leptin, and ghrelin. It is critical in regulating body functions and intracellular signaling mechanisms (Ranawana and Kaur 2013). The proportion of macronutrients in the consumed food can influence an individual's hormonal condition and the cell function of various tissues, boosting or inhibiting intracellular signal transduction pathways that govern cell division and death. Disproportionate macronutrient intake has been considered an essential axis in controlling the development of several illnesses, including malignancy. The mechanisms of different macronutrients in cancer prevention have been summarized in Fig. 4.1.

2.1 Proteins

Protein comes from the Greek word "proteios," which implies 'prime' or 'primary' implying that protein is a vital tissue component in humans and animals; hence, this phrase applies to nutrition (Reeds et al. 2000). A protein typically includes several

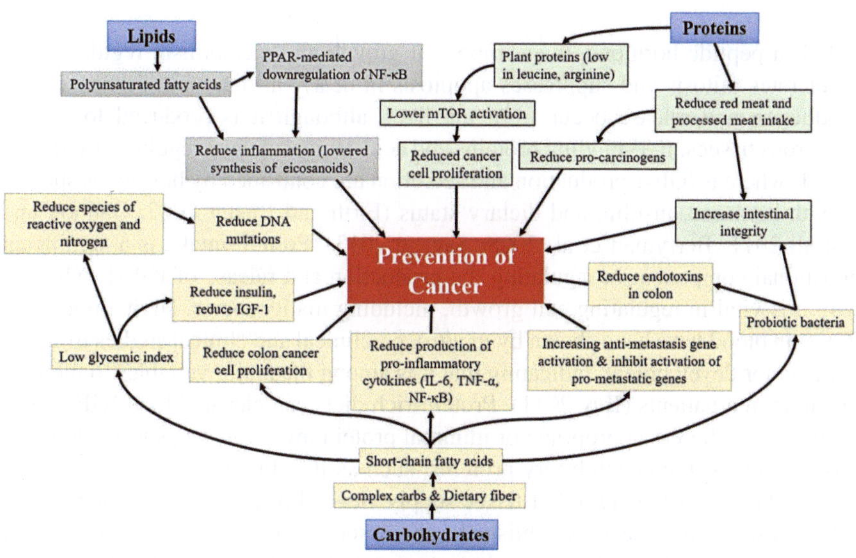

Fig. 4.1 Different mechanisms of macronutrients in the prevention of cancer

amino acids bonded by peptide bonds. Dietary protein provides little nutritional benefit unless it has been metabolized to amino acids, di or tripeptides, in the small intestines by peptidases and proteases. Thus, the nutritive worth of dietary protein is established by its amount, digestibility factors, and relative proportions of amino acids. These digestive products are either ingested by specialized cells that form the small intestine known as 'enterocytes' or utilized by bacteria. Amino acids that do not hydrolyze in the small gut reach circulation and are used for the biosynthesis of proteins in all body tissues (Wu 2016). Amino acids act as precursors in forming important biological substances like dopamine, serotonin, insulin, growth hormone, creatine, glutathione, albumin, immunoglobulins, DNA, and RNA (Wu 2009; Li et al. 2007).

Protein deficit causes metabolic diseases, kwashiorkor (induced by an extreme protein shortage), and marasmus (generated by a serious energy and protein shortage) in humans, notably in several developing-country youngsters (Leser 2013). Moderate insufficiency of dietary protein exists in the elderly, especially in developed nations, making them more vulnerable to metabolic and opportunistic disorders (Dasgupta et al. 2005). On the opposite end of the nutrition spectrum, excessive dietary amino acids, protein intake through meals, and heavy supplementation damage human health, particularly in individuals with renal or hepatic impairment. It is critical to comprehend how proteins regulate the generation of various hormones and how amino acids control processes within the cell. As a result, this segment outlines the key mechanisms that alter proteins and amino acids resulting from their metabolism in the gastrointestinal tract (Hoffer and Bistrian 2012).

2.1.1 Insulin-Like Growth Factor 1(IGF-1)

IGF-1, a peptide hormone, is an important growth and metabolism regulator that accelerates mitosis and suppresses apoptosis in nearly all mammal tissues. IGF-1 production and release occur from the liver, although it is produced locally in numerous tissues. IGF-binding proteins and acid-labile subunits regulate circulating IGF-1, whereas IGF-1 production and secretion are controlled by hormones such as growth hormone, insulin, and dietary status (Firth and Baxter 2002; Gannon and Nuttall 2011; Berryman et al. 2008; Savage 2013). Protein intake is a significant nutritional component in regulating the production and release of IGF-1, which is why it is vital in regulating cell growth, including malignancies. High amounts of IGF-1 in blood have been shown by *in vitro*, preclinical and clinical studies to accelerate tumor development, indicating that it is among the major variables to monitor in carcinoma patients (Key 2011). Protein-rich diets can elevate blood IGF-1 concentrations, whereas appropriate or minimal protein ingestion keeps levels low and raises levels of IGF-1 inhibitory proteins, such as IGF binding proteins and acid-labile subunit, decreasing IGF-1 tissue supply. Several observational epidemiological investigations have established an association between blood IGF-1 concentrations and the chances of breast cancer (BC) in females, prostate cancer (PROC) in males, and colorectal carcinoma (CRC) (Savage 2013; Yin et al. 2018).

2.1.2 Insulin

Pancreatic β-cells release insulin which is primarily responsible for glucose homeostasis. Protein ingestion, on the other hand, regulates the release of this hormone. Insulin is crucial in regulating protein and energy metabolism, boosting energy storage, and triggering muscle protein synthesis, among other things. Furthermore, insulin can increase cell growth. Insulin operates on various cells via insulin receptors (Ranawana and Kaur 2013; Taniguchi et al. 2006). Due to the ability of the insulin receptor to drive cell proliferation, pathological diseases like cancer and cardiovascular disorders can arise due to this functional mechanism in hyperinsulinemic circumstances. It has been discovered that insulin receptors are overexpressed in cancer cells, showing that elevated insulin levels can stimulate tumor mitogenesis (Vigneri et al. 2016). Proteins' capacity to increase the release of this hormone is determined by their amino acid profile, with animal sources having the highest potential to do so when opposed to plant-origin protein. Consumption of protein of plant origin has also been shown to lessen insulin resistance, lowering insulin levels and can be beneficial to cancer patients (Chen et al. 2018).

2.1.3 Mammalian Target of Rapamycin (mTOR) Signaling Cascade

A serine-threonine protein kinase, mTOR functions as a major controller of cell growth, replication, and viability in normal cells concerning dietary patterns, growth factors, and stress responses. It is critical to maintaining the equilibrium between cell growth and cell death based on cellular conditions and needs. Several dysregulations of this mechanism have been linked to various pathological conditions, including cancer (Hsieh et al. 2012). However, abnormally active mTOR in tumor cells provides signals stimulating tumor cells to proliferate, spread, and infect new healthy organs (Hsieh et al. 2012; Zou et al. 2020). mTORC1 and mTORC2 are physiologically and biochemically different complexes of mTOR. mTORC1 governs cell development and metabolism, whereas mTORC2 regulates the survival and proliferation of cells. Furthermore, several investigations have revealed that malignancies frequently overactivate the AKT/mTOR signaling cascade (Zou et al. 2020).

Different growth factors, like insulin and IGF-1, along with amino acids and glycogen, influence mTOR activation. Given that protein ingestion raises blood levels of insulin, IGF-1, and amino acids, it also serves as a major inducer of the mTOR protein. This pathway is heavily influenced by protein ingestion. Certain amino acids, such as leucine and arginine, present in greater concentrations in animal sources, have been identified as the primary activators. As a result, proteins derived from plants with lower levels of leucine and arginine are less efficient in mTOR pathway activation, protein production, and cell proliferation; this is a crucial aspect concerning cancer patients (McCarty 2011). Various newly discovered mTOR inhibitors (rapamycin, temsirolimus, and everolimus) have entered clinical studies in recent years. Also, it has been shown that certain medications work well when combined with mTOR inhibitors (Zou et al. 2020).

2.1.4 Protein Intake Recommendation

While establishing a protein recommendation, it is necessary to evaluate the person's present status, physical activity, and goals. As a result, the examination and advice of a skilled health expert are critical in managing disorders like cancer. According to epidemiological research, a recommended daily protein intake for adults with little physical activity will be 0.7–0.8 g/kg to minimize cancer risk. Protein consumption within this range is associated with a decreased risk of cancer death because of low plasma IGF-1 levels (Martín Ortega and Segura Campos 2021).

2.2 Carbohydrates

Carbohydrates are an essential macronutrient found in certain foods and drinks. They serve as the principal energy source, engage in blood glucose homeostasis and insulin metabolism, contribute to cholesterol and triglyceride metabolism, and aid in colon bacterial fermentation. Based primarily on their molecular makeup and level of polymerization, carbohydrates are further divided into simple and complex. Fewer than two sugar molecules make up simple carbohydrates, which are rapidly digested and turned into energy and cause a surge in blood glucose levels. Two or more sugar molecules make up complex carbohydrates that are digested and altered slowly compared to simple carbohydrates. Indigestible complex carbohydrates (also known as dietary fibers) contribute to bacterial growth and fermentation in the gut microflora (Maino Vieytes et al. 2019; Zeng 2014).

Carbohydrate digestion starts in the mouth, where they are broken down by salivary amylase into monosaccharides which are absorbed into the systemic circulation. It results in the elevation of blood glucose levels which stimulates the pancreatic β-cells to release insulin that directs the body's cells to take up sugar for storage or energy, whereas the surplus undergoes metabolism to synthesize cholesterol, fatty acids (FAs), and triglycerides. When blood glucose levels drop, the pancreatic α-cells generate glucagon, and the adrenals release cortisol. It encourages hepatic gluconeogenesis (glucose synthesis) and glycogenolysis (glycogen breakdown) to release stored glucose.

Based on the quantity and kind, ingested carbohydrate consumption has been suggested to influence cancer risk (Maino Vieytes et al. 2019). Carbohydrates have a wide range of impacts at the cellular, physiological, and ecological levels and their primary role as energy substrates. Among them are bacterial, epigenetic, endocrine, and systemic changes caused by their ingestion, which may impact the risk of developing cancer (Leturque et al. 2012). This section describes the biological pathways via which carbohydrates may be involved in cancer.

2.2.1 Glucose Metabolism and the Warburg Effect

The oxidative breakdown of the bonds of carbon in glucose, which is the main macronutrient, allows for the generation of energy in the form of ATP (mitochondrial oxidative phosphorylation). This process is essential to the continuing existence of all living things. After all of the glucose has been completely oxidized by mitochondrial respiration, the end product in mammals may be lactate or CO_2. In cancer cells and other normal proliferating cells, the glucose absorption rate increases drastically, and lactate is generated despite the availability of oxygen and perfectly functional mitochondria. This event, called the Warburg Effect, has been intensively studied. Yet, after careful examination, it is clear that its benefits for cell development and survival are not yet fully understood (Liberti and Locasale 2016; van der Heiden et al. 2009). Because carbohydrate is the principal macronutrient absorbed by tumor cells, this effect has raised various concerns concerning the quantity of carbohydrate cancer patients should ingest. Thus, different theories have been developed using very low carbohydrate or ketogenic diets to cause a cellular metabolic shift in which FAs and ketone bodies become the primary energy source (van der Heiden et al. 2009; Talib et al. 2021).

2.2.2 Insulin

As noted in the previous section on proteins, carbohydrates greatly impact insulin levels since they are an important factor influencing a patient's hormonal condition and hence in modulating cell signaling, including cancer cells. Carcinogenesis and the transition to aerobic glycolysis are thought to result from insulin receptor amplification in tumor cells caused by elevated blood sugar, insulin, and inflammatory cytokines (Makarem et al. 2018). Tumor suppressor gene mutations that affect the insulin receptor signaling cascade promote insulin receptor sensitivity and activity. It suggests that higher glucose and insulin concentrations in the tumor microhabitat may promote cell viability instead of cell death in DNA-damaged cells. Several investigations have discovered a link involving high blood sugar and hyperinsulinemia and tumor formation (Vigneri et al. 2016, 2020; Maino Vieytes et al. 2019). Carbohydrate restriction has been demonstrated to produce metabolic and hormonal changes. Restricting carbohydrate consumption can reduce mTOR system activation via another energy-detecting mechanism controlled by AMP-activated protein kinase (AMPK). AMPK activation and phosphorylation depend on the intracellular ratio of AMP to ATP, which is influenced by glucose and glycogen levels. Under carbohydrate restriction, insulin levels fall while glucagon levels rise (Weber et al. 2020).

2.2.3 Dietary Fiber, Gut Microflora, and Short-Chain Fatty Acids (SCFAs)

Dietary fibers comprise nondigestible plant substances that travel through the gut and resist enzymatic action (Zeng 2014). They can bind to nutrients forming new metabolites and regulating the disposition of nutrients. Dietary fibers comprise cellulose, lignin, pectins, oligosaccharides, and gums. Upon anaerobic fermentation in the GI tract, some nutritional fibers produce SCFAs, that have health benefits (Dhingra et al. 2012). It is helpful to modify the gut microflora by favoring the development of strains of useful bacteria such as the bifidobacterial. These bacteria reduce gut inflammation, elevate gastrointestinal transit rate and curb pathogenic strains' development (Vivarelli et al. 2019). Numerous investigations have demonstrated that consuming a diet rich in dietary fibers positively influences the GI microflora and substantially reduces the possibility of CRC. This phenomenon is the result of higher fecal matter production, reduced accumulation and exposure to cancer-causing substances, decrease in pathogenic bacteria promoting gut inflammation, lowering the levels of tumor necrosis factor-α (TNF-α), proinflammatory cytokines (Interleukin-6), high growth of beneficial bacteria having immunomodulatory and anticancer properties and increased synthesis of SCFAs (Maino Vieytes et al. 2019; Zeng 2014; Vivarelli et al. 2019; Kuo 2013).

Anaerobic fermentation of complex carbohydrates and undigested fiber leads to forming SCFAs (Matthews et al. 2012). A higher possibility of CRC is correlated with changes in the GI microflora due to decreased synthesis of SCFAs. Epithelial cells lining the GI lumen rapidly and efficiently take up the SCFAs. Three main SCFAs (acetate, propionate, and butyrate) are employed by colonic cells for energy production, in contrast to mutated CRC cells that utilize aerobic glycolysis for energy generation. CRC cells are more susceptible to SCFAs than normal colon cells, indicating that they play a significant part in cell homeostasis. Modifying SCFA concentrations in the gut via microflora alterations has been identified as a possible therapeutic or preventive method for CRC. Over the previous few decades, there has been an increase in fascination with figuring out how SCFAs work in the cells of CRC (Gomes et al. 2020).

Acetate accounts for approximately 60%–75% of the total SCFAs and is synthesized by several colonic microbes (van der Beek et al. 2017). Crypt cell multiplication is triggered by acetate which decreases the contraction rate of longitudinally oriented colonic smooth muscles. Also, it improves blood flow in the colon, augments ilial motility, lowers adipogenesis and modifies the immune system of the host by associating with the G protein-linked receptor in adipocytes and cells of the immune system (Brown et al. 2003; Hong et al. 2005). Acetate has also demonstrated the capacity to reduce nuclear factor (NF)-κB, TNF-α, and IL-6 that are induced by lipopolysaccharide, as well as to promote the formation of peripheral blood antibodies in a variety of tissues (Tedelind et al. 2007).

Similar to acetate, propionate has a concentration-dependent effect on the pace at which longitudinally oriented muscle contractions occur via enteric neurons in the terminal rat colon (Ono et al. 2004). In human and animal studies, propionate has demonstrated the ability to minimize food consumption and enhance satiety through leptin, a satiety hormone, and G protein-linked receptor activations (Xiong et al. 2004; Samuel et al. 2008). Propionate could also be anti-carcinogenic as it lowers the proliferation of human CRC cells and differentiation by histone protein hyperacetylation and activation of apoptosis (Jan et al. 2002; Hinnebusch et al. 2002). Additionally, propionate inhibits the tissue synthesis of proinflammatory cytokines (e.g., TNF-α, NF-κB) (Al-Lahham et al. 2010; Zapolska-Downar and Naruszewicz 2009).

Among SCFAs, butyrate is the most important to sustain colonic health and regulate cell proliferation and multiplication (Macfarlane and Macfarlane 2011). Butyrate oxidation has a role in the consumption of oxygen of over 70% in isolated colon cells. Butyric acid absorption and consumption by the colonic epithelium has been exhibited in an investigation of the concentration of SCFAs in arterial and portal circulation as well as in the colon's contents (Cummings et al. 1987). Butyrates display stronger anti-inflammatory characteristics in contrast to acetate and propionate. This phenomenon is likely arbitrated by suppressing the TNF-α generation, NF-κB stimulation, and expression of interleukins in cells of the immune system and colon (Bailón et al. 2010; Lührs et al. 2002).

SCFAs favor the proliferation of normal colonic cells while exhibiting growth inhibition of most CRC cells, producing cell cycle interruption and apoptosis induction by complicated cellular pathways. Substantial *in vitro* research has demonstrated that butyric acid suppresses histone deacetylacetylase enzymes and permits hyperacetylation of histone, which results in elevated gene transcription of cyclin D3 and p21/Cip1 genes (Waldecker et al. 2008). Furthermore, butyric acid limits cancer cell migration and invasion by increasing antimetastasis gene activation and preventing pro-metastatic gene activation (Waldecker et al. 2008).

2.2.4 Carbohydrate Intake Recommendation

It is well understood that excessive sugar consumption elevates the risk of various pathological conditions, including cancer. In contrast, a diet rich in complex carbohydrates can substantially decrease cancer risk. Thus, it is sensible to limit sugar consumption and increase the ingestion of vegetables, whole grains, and fruits to limit the possibility of malignancy and other ailments (Sieri et al. 2017).

The ketogenic diet has been extensively studied. It limits carbohydrate intake to the point where it causes the body to enter ketosis, a state brought on by increased fat oxidation and ketone body production that drives cells to change how they use energy. The ketogenic diet can create metabolically unfavorable conditions for cancer cells. Hence, it could be a potential supplement to a multifactorial cancer therapy tailored to each patient. Several preclinical and clinical investigations have yielded positive and neutral outcomes. The likelihood that it may function in some circumstances supports using the ketogenic diet in conjunction with established therapy (Leturque et al. 2012; Hagihara et al. 2020). Nonetheless, it is crucial to remember that every form of cancer is unique, and patients frequently have comorbidities, so a thorough evaluation of the patient is required. Regarding disease management, a diet rich in complex carbohydrates and fiber is unquestionably beneficial, as is reducing sugar consumption (Maino Vieytes et al. 2019; Sieri et al. 2017).

2.3 Lipids

Lipids are a class of substances that are immiscible in water but miscible in organic solvents and comprise sterols, FAs, sphingolipids, phospholipids, terpenes, and others (Fahy et al. 2011). Lipids carry out numerous biological functions, such as constructing cell membranes, energy storage, participating in signaling pathways, functioning as hormones and secondary messengers, maintenance of electrochemical gradients, and membrane anchoring. Lipid homeostasis is essential for health, and lipid deficiencies are important in the etiology of serious and deadly disorders (Watson 2006; Oresic et al. 2008). This section will examine FAs because they are the main lipids classified as macronutrients because of their higher proportion in food and energy supply.

Lipid breakdown commences in the buccal cavity when subjected to lingual lipases released by salivary glands to initiate the digestion of triacylglycerides. Digestion in the stomach proceeds due to the influence of both gastric and lingual enzymes. The peristalsis in the stomach contributes to the emulsifying process of lipid-soluble vitamins and dietary fat. Crude lipid emulsions in fine lipid droplets approach the duodenum and subsequently interact with pancreatic juice and bile, undergoing significant chemical and physical changes in the intestine emulsification proceeds with micellization and hydrolysis in readiness for absorption in the gut. Pancreatic juice and bile contain pancreatic bile salts, lipase, and colipase, which are crucial in ensuring the efficiency of breakdown and absorption of lipids. Free FAs from the gut lumen are absorbed by enterocytes which use them to produce neutral fats (Iqbal and Hussain 2009).

Dietary fats have long been thought to be involved in cancer genesis. Research has linked high fat intake to BC, CRC, PANC, and PROC. However, this discovery is no longer acknowledged because the carcinogenic properties of fat are not dependent on caloric quantity but only on the type and quantity of fat ingested. Thus, the following segment will focus on the importance of the important FAs types controlling the disorder.

Based on the double bond count and the degree of unsaturation in their structure, FAs can be classified into three major classes: saturated, monounsaturated, and polyunsaturated, which are further categorized as ω-3 FAs and ω-6 FAs. They play important parts in regulating cellular functions that could either cause or prevent cancer (Fahy et al. 2011).

2.3.1 Fatty Acids Supply for Membrane Synthesis in Cancer

Cell development and proliferation, including that of malignant cell growth, requires aggregation of substantial amounts of lipids for the production of cell organelles and cell membranes. These lipids may be acquired either exogenously (via diet) or synthesized endogenously via lipogenic pathways (Corn et al. 2020). As a result, the cell membrane and organelle composition is heavily influenced by the amount and type of fats ingested, as well as their synthesis endogenously, which is primarily mediated by an excess of carbohydrates, the metabolism of which improves the citrate availability and thus its production (Bojková et al. 2020). Recent investigations have demonstrated that tumors of several kinds absorb FAs from systemic circulation similarly to normal cells, which proves the significance of FAs.

Excessive lipid buildup or changes in saturated and unsaturated fatty acid levels might disturb homeostasis and increase cellular stress (Corn et al. 2020). Saturated fatty acids (SFAs) levels in tumor tissues are greater than in healthy tissues in malignancies of the breast, lung, prostate, endometrium, ovary, bladder, and colon (Forman and Mahabir 2010). Tumor cell membranes containing a larger proportion of SFAs have been associated with more aggressive tumors. It renders them highly resistant to apoptosis induction, whereas a greater rate of unsaturated FAs has been connected to a better response to tumor prevention and apoptosis. However, greater SFAs levels in the membrane diminish its fluidity, motility, and invasiveness while increasing the membrane's viability (Bojková et al. 2020; Chen and Huang 2019).

2.3.2 Palmitic Acid (PA)

PA is the most frequently researched SFAs in connection with cancer. Among SFAs, PA is found most abundantly in humans and is either procured from diet or generated from endogenous substances like FAs, carbohydrates, and amino acids. In membrane phospholipids and adipose triacylglycerol, PA makes about 20–30% of total FAs.

To comprehend the contribution of fat in cancer, there is a need to clarify the function of FAs in signal transduction pathways regulating cell growth. The research interest in this domain is growing with an emphasis on the role of PA in carcinogenesis. It is known that PA participates in intracellular signal transduction and plays a part in carcinogenesis and other conditions, including neurodegenerative, cardiovascular diseases, metabolic syndrome, and inflammation (Fatima et al. 2019). PA contributes to S-palmitoylation, a process in which PA-protein conjugation occurs via a thioester bond and catalyzes 23Asp-His-His-Cys (DHHC)-family palmitoyl S-acyltransferases. In contrast, PA elimination is caused by acyl-protein thioesterases. Palmitoylation controls the function of various proteins involved in homeostasis, such as the G protein-linked receptor. Palmitoylation's disruption of protein function is associated with metabolic and neuronal disorders and cancer (Fatima et al. 2019; Ko and Dixon 2018; Conibear and Davis 2010). Palmitoylation is vital in the functioning of both tumor suppressors and oncogenes (Lin et al. 2017a).

PA exhibited stimulative and suppressive properties on tumor proliferation in preclinical investigations. PA enhanced pancreatic cancer (PANC) cells (AsPC-1) invasiveness through the TLR-4/ROS/NF-B/matrix metalloproteinase-9 (MMP-9) signaling cascade and accelerated proliferation in many oral cancer cell lines expressing large amounts of clusters of differentiation (Binker-Cosen et al. 2017). Whereas PA inhibited growth, impeded metastasis and inhibited hepatocarcinoma cell proliferation *in vitro* and mouse xenograft models, suppressed mTOR, lowered membrane fluidity, and disrupted glucose metabolism (Lin et al. 2017b). Effects on PA in human studies showed inconsistency displaying a positive relationship between dietary PA and BC (Sczaniecka et al. 2012). However, another investigation described a detrimental relationship between PA consumption and the possibility of PANC (Nkondjock et al. 2005).

2.3.3 Monounsaturated Fatty Acids (MUFAs)

The chemical structure of MUFAs shows a single unsaturation. Among these, oleic acid is found most abundantly in the diet, comprising more than 20% of all FAs in several oils and fats, with the most abundant in olive oil. Preclinical findings on the impact of oleic acid on tumor formation and progression are inconsistent. Oleic acid increased the growth of BC cell lines i.e., MCF-7 cells (Gastón et al. 2017). But when human BC cell lines: BT-474 and SK-Br3 were subjected to oleic acid, HER-2/neu expression was suppressed (Menendez et al. 2005). Upon treatment with oleic acid, gastric cancer cell lines AGS and MKN-45 exhibited improved invasiveness through PI3K-Akt signaling cascade activation. As per a prospective cohort study, increased MUFAs consumption reduced the chances of digestive cancers (Sellem et al. 2019). In contrast, another investigation reported an elevated PROC risk with a higher MUFAs intake (Liss et al. 2019).

The comprehensive experimental data demonstrated a growth-promoting influence of oleic acid on tumor cells. However, a meta-analysis of case-control studies exhibited that olive oil lowered the chances of developing BC, tumors of the digestive system, and others. Olive oil's protective action against cancer could be attributed to compounds other than oleic acids, such as polyphenols, triterpenes, and other antioxidant substances (Psaltopoulou et al. 2011), (Sánchez-Quesada et al. 2013).

2.3.4 Polyunsaturated Fatty Acids (PUFAs)

The chemical structure of PUFAs shows the presence of multiple double bonds. Two main classes of biologically important PUFAs are ω-3 PUFAs, and ω-6 PUFAs, differentiated based on the double bond location. Linoleic acid (LA) and arachidonic acid (ARA) are examples of ω-6 PUFAs, while α-linolenic acid (ALA), docosahexaenoic acid (DHA), and eicosapentaenoic acid (EPA) are examples of ω-3 PUFAs (Michalak et al. 2016). LA and ALA are important as they are the precursors of ARA, EPA, and DHA, all of which contribute to body homeostasis (Saini and Keum 2018). PUFAs participate in regulating many intracellular pathways, such as the production of inflammatory mediators (eicosanoids), lipid biosynthesis, gluconeogenesis, and cell proliferation (Michalak et al. 2016; Jump 2011; Madsen et al. 2005; Ntambi and Bené 2001). Generally, ARA is a pro-inflammatory eicosanoid precursor. In contrast, EPA and DHA are anti-inflammatory eicosanoid precursors. Still, the interplay of ω-6 and ω-3 FAs and their lipid mediators in the setting of inflammation are complicated and remain unknown (Calder 2010; Innes and Calder 2018). Hence, sustaining a low ω-6/ω-3 ratio is important for decreasing inflammation, which contributes to several chronic disorders (DiNicolantonio and O'Keefe 2018). The advocated ratio of ω-6 to ω-3 is 4:1 (Simopoulos 2008). Thus, this ratio's differences can contribute to carcinogenesis (Lee et al. 2007; Stanford et al. 1995).

PUFAs regulate tumor cells through various processes, including regulation of tumor microhabitat, expression of genes and cell signaling systems, and alteration of the membrane phospholipid profile (Gu et al. 2015; Berquin et al. 2008).

2.3.4.1 ω-3 PUFAs

Epidemiological research has suggested that ω-3 PUFAs possess a cancer-protective role, whereas ω-6 PUFAs promote cancer. DHA uniquely influences membrane configuration and is regarded as the primary ω-3 PUFA in preventing cancer (Siddiqui et al. 2011). A study demonstrated that DHA obstructed the growth in MCF-7 cells through pAKT signaling (Huang et al. 2017). DHA resulted in apoptosis of PANC cells through suppression of the STAT3/ NF-κB -cyclin D1/survivin axis (Park et al. 2018). In mice with 4 T1 mammary tumor implants, ω-3 PUFAs-enriched diet reduced growth and development while augmenting apoptosis and tumor penetration by immune cells (Khadge et al. 2018). *In vitro* and *in vivo* investigations also revealed that DHA integration in cell membranes enhanced the absorption and activity of cytotoxic drugs (Siddiqui et al. 2011). Ingestion of long-chain ω-3 PUFAs was linked with a lowered probability of endometrial carcinoma, particularly in women with normal Quetelet's index (Brasky et al. 2015).

2.3.4.2 ω-6 PUFAs

In vitro, research produced conflicting outcomes on the influence of ω-6 PUFAs in carcinogenesis. LA and ARA lowered cell proliferation and survival in PROC cell lines: C4–2 cells and PC-3 cells (Bratton et al. 2019). Whereas, ALA increased MDA-MB-231 BC cell metastasis; the reported mechanism included overexpression of fascin, phospholipase D, and PI3K/Akt pathway activation (Serna-Marquez et al. 2017; Gonzalez-Reyes et al. 2018; Diaz-Aragon et al. 2019). *In vivo* studies showed growth-promoting effects of ω-6 PUFAs in tumors. ω-6 fats-enriched diet accelerated PANC proliferation in KRAS transgenic mice (Cheon et al. 2011). A systematic review of prospective cohort trials published in 2016 found no correlation between ω-6 PUFAs ingestion and the possibility of BC (Cao et al. 2016) However, a recent study found that consumption of a higher ω-3/ω-6 ratio is linked with a reduced possibility of BC in Asia than in the West (Nindrea et al. 2019).

2.3.5 Lipid Intake Recommendation

According to epidemiological, *in vivo*, and human data, a low-fat diet (<20% total energy) is the preferred choice to considerably lower cancer risk. The makeup of the ingested FAs is crucial, so intake of ω-3 fat sources such as salmon, tuna, walnuts, and ω-3 supplements should be prioritized. Olive oil consumption has demonstrated a lesser probability of all kinds of cancer. Saturated fat ingestion of red meat pork should be minimized (Bojková et al. 2020).

3 Micronutrients

Micronutrients are dietary components such as vitamins and trace elements that are impossible for humans or other vertebrates to synthesize. The body requires only micro- or milligram amounts of these substances to carry out diverse biochemical pathways and metabolic processes (Shergill-Bonner 2017). An inadequate supply of micronutrients leads to disease. A micronutrient's abundance or lack in the body is determined by the amount and type of food consumed and its use in biochemical reactions. When a diet is inadequate, macronutrients and micronutrients can be depleted or reduced, affecting various organic and metabolic functions, the immune system, and tolerance to treatment (Shenkin 2006; Johnson 2004). Cancer patients often neglect micronutrients due to dietary restrictions. The nutritional status of patients with cancer is highly correlated with their treatment success and healing process, with micronutrients playing an essential role described in Table 4.1. Many epidemiological studies have reported micronutrients'significance as supportive of nutrition treatment and that multivitamins and minerals can improve disease diagnosis and patient quality of life (Gröber et al. 2016). A recent study found that, based on the kind and stage of cancer, 30–90% of patients with cancer lack micronutrients. The presence of micronutrient deficiencies associated with cancer and treatment affects the progression of the illness and the efficacy of cytoreductive measures, along with the growing danger of complications (e.g., decreased immunity, depression, hindered injury recovery, and tiredness). To maintain immune stability, the patient must receive a sufficient supply of immune-improving micronutrients, such as vitamin D and selenium (Madiyal et al. 2016; Irimie et al. 2019).

Table 4.1 Micronutrients in the prevention of cancer

Micronutrient	Sources	Recommended dietary intake	Functions	Cancer type	References
Vit A	Yellow and orange fruits, green leafy vegetables, fish oils, eggs, dairy products	900 mcg for men and 700 mcg for women	Apoptosis, cell growth, differentiation	CRC, glioma, LC, cancer	Ilbert et al. (1996)
Vit B 1	Milk, fish, legumes, eggs, bread, and meat	1.2 mg for men and women, 1.1 mg daily	Embryonic development and carcinogenesis. The risk of cancer is reduced when doses are low, but it is increased when doses are high	BC, LC	Tsao et al. (2007)
Vit B 9	Green leafy vegetables, citrus fruits, asparagus whole grains, salmon poultry, shellfish, tuna, beans, and milk	400 mcg	DNA synthesis and methylation are key to maintaining DNA integrity and controlling gene expression.	BC, CRC, PANC and gastric cancer	Petrone et al. (2021)
Vit B12	Milk, fish, shellfish, egg and meat	2 to 5 mcg	DNA synthesis, cell division	Glioma, BC and gastric cancer	Kok et al. (2020), Galván-Portillo et al. 2010)
Vit C	Tomatoes, broccoli, citrus fruits, red peppers, strawberries, green peppers	Women-75 mg, Men 90 mg	An increase in apoptosis and autophagy. Inhibiting redox-sensitive transcription factors and molecules is one of its primary functions. Stresses that are metabolic and genotoxic to tumor cells can cause them to grow abnormally	Hematological malignancies and solid tumors	Zhou et al. (2015)

(continued)

Table 4.1 (continued)

Micronutrient	Sources	Recommended dietary intake	Functions	Cancer type	References
Vit D	Eggs, meat, fish, and dairy	15 mcg	Reduced danger of certain types of cancer	PANC, BC CRC, PROC	Grant (2020)
Vit E	Vegetables, cereal, meat, and eggs	15 mg for adults, and 5 mg for children	Targets oxidative stress and inflammatory markers to reduce unwanted side effects of cytotoxicity	CRC, PROC, BC	Barnett et al. (2002), Kline et al. (2004)
Selenium	Meat products, cereals, and seafood	55 mcg	Cancer risk is reduced, and antioxidants restore altered epigenetic events and genomic stability.	Skin, OVC, BC, CRC, PROC, LC leukemias, uterine, oropharyngeal and bladder cancers	Cai et al. (2016)
Zinc	Poultry, meat and seafood	15 mg	Apoptosis and DNA repair	PANC, head and neck squamous cell carcinoma and cervical cancer	Grattan and Freake (2012)
Magnesium	Legumes, leafy vegetables, nuts and grains	300 and 400 mg	DNA replication and repair, apoptosis, genomic stability, and chemical carcinogenesis prevention	PANC, CRC, LC	Dai et al. (2011)

3.1 Vitamin A (Vit A)

A lipid-soluble micronutrient, Vit A cannot be produced by the human body and must be consumed through diet. Fish oils, milk products, eggs, milk, and green leafy vegetables are rich in Vit A (Zinder et al. 2019). It is present in different forms, such as retinol, retinaldehyde, retinoic acid and carotenoids. It regulates methylation and cell growth, protecting against free radicals and DNA damage. Various biological processes are controlled by retinoid compounds, such as the development of an embryo, eyesight, differentiation, formation of bone, metabolic processes, hematogenesis, apoptosis, proliferation, and reproduction. Furthermore, Vit A suppresses tumorogenesis in the prostate, breast, skin, lung, mouth, and bladder.

There is a significant relationship between the progress of cancer and deficiency of Vit A has been proven by several epidemiological studies in recent years (Poulain et al. 2009). Ligand-activated transcription factors and nuclear retinoid receptors are also responsible for modulating epithelial differentiation and proliferation; retinoids reverse carcinogenesis through complex mechanisms (Smith and Saba 2005; Fontana and Rishi 2002; Maia et al. 2019). Increased Vit A consumption from fruits, vegetables, and fiber protects against pharyngeal and oral cancer, according to several studies (Chi et al. 2015; Kune et al. 1993). However, a greater intake of Vit A was linked in multiple studies to an augmented risk of cancer. Compared to most β-carotene supplements, a meta-analysis of many trials revealed a small increase in cancer incidence correlated with Vit A use (Huang et al. 2016). Eun et al. reported that Vit A prevents the proliferation of CRC cells without using retinoic acid receptors (Eun et al. 2007). Retinol consumption has a considerably reduced risk of melanoma than Vit A or β-carotene consumption, according to a meta-analysis study (Maalmi et al. 2018).

3.2 *Vitamin B1/Thiamine (Vit B1)*

It is regarded as a necessary aqueous-soluble vitamin that promotes glucose metabolism. It is an integral component of cellular metabolism, which is needed for the activity of the tricarboxylic acid cycle enzymes, and transketolase, α-ketoglutarate dehydrogenase, pyruvate dehydrogenase, which is involved in the pentose phosphate cycle. These enzymes catalyze amino acids. In addition to being naturally obtained in various foods, thiamine is also used to fortify a range of processed foods, including meat, milk bread, legumes, eggs, and fish. Early in the twenty-first century, there was a theory that excessive Vit B1 intake contributed to cancer development because it is an essential component of cellular metabolism (Liu et al. 2018; Wu and Lu 2012). According to research on BC cell lines, thiamine at higher concentrations promoted cell progress while inhibiting proliferation in PANC and neuroblastoma cancer cell lines. Patients with cancer are more likely to lack Vit B1 because chemotherapeutic treatments can deplete this vitamin, and tumor consumption depletes other tissues (Hanberry et al. 2014; Peterson et al. 2020). Epidemiological studies have shown that elevated dosages of Vit B1 for 24 hours reduced BC cell division and growth (Liu et al. 2018).

3.3 *Vitamin B9/Folate and Folic Acid (Vit B9)*

It is a B-complex vitamin that is water soluble and works in single-carbon transfer processes. Folic acid is a highly stable and oxidized form of folate. It is obtained from shellfish, salmon, whole grains, dark leafy vegetables, asparagus, poultry, beans, citrus fruits and dairy products. It is crucial for controlling gene expression by participating in DNA synthesis and methylation and also enhances immune function and maintains DNA integrity. Additionally, it plays a vital role in synthesizing proteins, nucleic acids, nerve cells, and blood cells. Numerous epidemiological studies found a link between folate inadequacy and malignancies, neural tube abnormalities, and atherosclerosis (Saghiri et al. 2017; Farhan Aslam et al. 2017), (Naderi and House 2018).

Folate deficiency can alter DNA methylation and disrupt genome integrity and DNA repair processes. Additionally, it alters S-adenosylmethionine levels, which decreases the action of important oncogenes and tumor suppressor genes (Verma 2016). According to epidemiological research, overconsumption of folate may help prevent ovarian cancer (OVC) (Zhang et al. 2012) and BC risks, especially for those who consume more alcohol (Chen et al. 2014), CRC (Dulai et al. 2016), and PROC. Several folic acid-conjugated nanoparticles have been shown to enhance antitumor activity in lung cancer (LC) patients (Zhao et al. 2017). A meta-analysis study that involved a large patient cohort of Chinese people and a patient cohort of Egyptians demonstrated that dietary folate and vitamin B6 exhibited protection against nasopharyngeal carcinoma (Galeone et al. 2015), (Zeng et al. 2016). Some controversial literature data suggest that folic acid contributes to cancer progression in some cases. Several meta-analyses showed that folic acid administration for up to 6 years did not shield people with previously resected adenomas from developing in the large intestine. Further research is required because there is insufficient proof that adenomas are linked to an enhanced possibility of developing cancer. Folic acid has been therapeutically used in many combinations to avoid CRC and chemotherapy side effects.

3.4 Vitamin B12/Cyanocobalamin (Vit B12)

Vit B12 is an aqueous-soluble molecule consisting of a corrin ring and a cobalt ion in the center. The primary sources of Vit B12 are shellfish, meat, milk, fish, and eggs (Watanabe 2007). The function of Vit B12 is to synthesize DNA, produce cellular energy, process fatty acids, catabolize amino acids, and methylate proteins to prevent the accumulation of homocysteine. A diet that contains few animal foods or has malabsorption issues frequently results in Vit B12 deficiencies. Those who consume less meat or have restricted nutritional patterns are also at risk of Vit B12 deficiency. Lack of Vit B12 causes weakness, fatigue, and anemia. Thus we consume foods containing Vit B12 every day or take supplements including Vit B12 (O'Leary and Samman 2010).

Several studies showed the connection between cancer and low plasma Vit B12 concentrations; also, some studies demonstrated a link between higher plasma cobalamin levels and an increased possibility of cancer (Wu et al. 2019; Collin 2013). According to some studies, patients with high cobalamin levels have a higher risk of hematological and solid tumors (Carmel et al. n.d.). According to the research by Campbel et al., Vit B12 helps prevent and treat unfavorable musculoskeletal outcomes associated with BC. In addition, Vit B12 supplementation is crucial for avoiding its deficiency in various cancer types, including gastric cancer, as its absorption can be hampered by surgical incisions of the stomach (Kim et al. 2011). According to Fanidi et al., high Vit B12 concentrations are linked with a greater threat of LC, and for every doubling of Vit B12, the risk increases by 15% (Fanidi et al. 2019). Maintaining enough blood concentrations of vitamins within an acceptable range is advised since either a vitamin deficit or excess might raise the chance of developing cancer. Therefore, it would be best to consume the daily required amount (2.4 mcg) to prevent illness.

3.5 Vitamin C/Ascorbic Acid (Vit C)

It is also called ascorbic acid, an important micronutrient that should be provided by the food or as a supplement. It comprises six-carbon lactone, naturally functional, effortlessly convertible, and a reduced form and oxidized forms like dehydroascorbic acid in ascorbate. Mammalian livers produce it, but not human beings, non-humans, or rodents (Granger and Eck 2018). When exposed to acidic environments and extreme temperatures, it easily oxidizes and degrades. In addition to citrus fruits and peppers, strawberries, tomatoes, broccoli, sprouts, and Indian gooseberries are all good sources of Vit C. It has been suggested that the daily required amount of Vit C is 90 mg for males and 75 mg for females (Doseděl et al. 2021; Li and Schellhorn 2007; Kaźmierczak-Barańska et al. 2020).

Oxidative processes and free radicals are responsible for initiating and promoting cancer. The antioxidant properties of ascorbic acid may inhibit some of these processes. Furthermore, ascorbic acid inhibits the development of carcinogenic nitrosamines by acting as a nitrite trap (Chambial et al. 2013; Pawlowska et al. 2019). As part of the immune system, it maintains the proper number of T lymphocytes and natural killer cells that fight tumors. It has been shown that Vit C significantly lowers the chances of cancer, diabetes, Alzheimer's, atherosclerosis, Parkinson's disease, and toxicity due to metal in humans (Johnson et al. 2003). As a preventive agent for cancer, Vit C stimulates the immune response, produces collagen, inhibits enzymes that cause metastasis, and stops the spread of cancer-causing viruses. Besides reducing chemotherapy harmfulness and avoiding oxidative stress, it enhances chemotherapy efficiency, neutralizes some carcinogens, and facilitates injury recovery in cancer patients (Li and Schellhorn 2007; Cimmino et al. 2018). In case-control studies, epidemiological research revealed a substantial inverse connection between the intake of Vit C and preventing PANC (Hua et al. 2016). Luo et al. dose-response study showed that increasing Vit C intake by 100 mg/day lowers the chance of developing LC by 7%. It suggests that Vit C may be effective in avoiding LC (Luo et al. 2014).

3.6 *Vitamin D (Vit D)*

Vit D contains two lipid-soluble compounds, cholecalciferol (D3) and ergocalciferol (D2), and their metabolites are regarded as essential for human health. Meat, eggs, dairy and fish are dietary sources of Vit D3, while ultraviolet-irradiated fungi and yeast are Vit D2 (Hohman et al. 2011). Humans produce Vit D3 when exposed to ultraviolet light, a precursor to endocrine hormones. It maintains a healthy skeleton of humans by sustaining an adequate supply of calcium and phosphorus (Wacker and Holick 2013; Severo et al. 2009).

Several types of cancer linked to Vit D deficiency have been recently recognized and are receiving considerable attention. Vit D has many biological functions contributing to cancer prevention that have only recently been realized. It has been found that the absence of Vit D rises the hazard of certain types of cancer. Some epidemiological studies indicate that a lack of Vit D contributes to yearly premature deaths from OVC, CRC, PROC, and BC (Garland et al. 2006; Holick 2014).

It has been demonstrated that 1,25-dihydroxy Vit D3, a Vit D metabolite, reduces the proliferation of tumor cells with Vit D receptors. It controls the entire process of tumorigenesis, from initiation to metastatic spread and interactions between cells and their microenvironment. It regulates several cell behaviors, including apoptosis, differentiation, proliferation, autophagy, and epithelial–epithelial-mesenchymal, as well as inflammation, antioxidants, angiogenesis and immune responses in the microenvironment. Researchers found a highly significant connection involving plasma Vit D concentration and a reduced risk of CRC in epidemiological studies. The greatest reductions in CRC incidents and mortality were seen in subjects with Vit D concentrations of more than 29 ng/mL. Another study by Dovnik et al. stated that optimal Vit D levels reduce cancer mortality and OVC risk. A study found that natural and synthetic Vit D compounds could induce apoptosis in VDR+ cells in oral precancerous lesions and oral squamous cell carcinomas (Chiang and Chen 2009).

3.7 *Vitamin E (Vit E)*

Vit E is a crucial lipid-soluble vitamin composed of four tocotrienols (α, β, γ, and δ) and four tocopherols (α, β, γ, and δ) (Khadangi and Azzi 2019). Vit E can be obtained from vegetables, meat, eggs, and cereals. The most prevalent and highly functional type of Vit E that is found naturally in the environment is tocopherol; in humans, it is the most prevalent type. Vit E intake recommended for adults is 15 milligrams, and for children, it is 5 milligrams (Sung et al. 2003). Vit E is a free-radical scavenger in cell membranes and inhibits DNA damage caused by carcinogens. Additionally, there is proof that a deficiency in Vit E might cause cancer, Parkinson's disease, cataracts, fibrocystic breast disease, heart disease, diabetes, stroke, epilepsy and, Alzheimer's disease (Pham and Plakogiannis 2005).

Chemotherapy may benefit from the use of Vit E. Research in earlier times has exhibited that Vit E in its different forms (γ and δ- tocotrienols) has anticancer effects. Anticancer activity of α- tocopherol has been studied extensively due to the contribution of free radical damage in carcinogenesis. There is a link between raised cancer risk and low Vit E or nutritional deficiencies. Some meta-analysis studies reported that supplementing α- tocopherol to populations with Vit E insufficiency has beneficial effects in lowering the cancer risk (Yang et al. 2020; Jiang 2017; Klein et al. 2003). The most effective anticancer effects belong to tocotrienols. They focus on mutual molecular pathways that prevent cell division, invasion, metastasis, autophagy, apoptosis, invasion, and metastasis. Protein kinase-C action, the growth of smooth muscle cells, and tumorgenesis are all suppressed by α -tocopherol. In cancer cell lines, Vit E activates the enzyme NADPH: quinone reductase and blocks the breakdown of ARA and prostaglandins (Constantinou et al. 2008; Tucker and Townsend 2005). Several epidemiological studies have shown no association with CRC or a reduced risk of CRC when Vit E intake is increased (Wu et al. 2015; Kristal et al. 2014; Malila et al. 2002). Another study showed that Vit E inhibited mouse colon carcinogenesis by lowering inflammation, trapping reactive nitrogen species, causing apoptosis and reducing free radicals (Ju et al. 2009).

3.8 Selenium (Se)

A non-metallic trace element known as selenium is an important nutrient for human health. Meat products, grains, and shellfish are the main dietary sources of Se; milk, vegetables, and fruit have comparatively low quantities of Se. The selenoproteins that execute the functions of Se, an essential trace element, include various oxidant resistance enzymes. For several enzymes, including glutathione peroxidase and the 5 deiodinases, Se is required. Inorganic types of Se consumed as dietary supplements include selenite and selenate (Rayman 2020).

Recent research findings have shown that Se plays a substantial defensive function in reducing the growth of cancerous tumors. Depleted plasma Se levels are linked to higher cancer-related illness and death. Se-metabolites inhibit cellular proliferation, enhance immune surveillance, suppress tumor neo-angiogenesis, enhance apoptosis, and alter carcinogen metabolism (Letavayová et al. 2006; Gromadzińska et al. 2008).

In numerous animal models featuring a dose-dependent response, sodium selenite and selenomethionine were reported to inhibit tumor development. In chemically stimulated carcinogenesis, selenite inhibits cancer cell proliferation more potently than selenomethionine. Selenomethionine treatment was observed to increase tissue selenium concentrations more than selenite supplementation (Combs et al. 2001; Tinggi 2008). The activation of apoptosis and its impact on androgen and estrogen receptor expression, suppression of protein kinase C activity and cell growth, cell-cycle arrest, and DNA-repair genes are some potential anticancer mechanisms of Se (Rayman 2005; Schrauzer 2000). Epidemiological data showed a strong connection between the deficiency of Se and cancer risk, and Se supplementation reduced cancer risk (Peters and Takata 2008; Lee et al. 2011). A recent case-control study (966 CRC cases and 966 matched controls) comprising 520,000 people in 10 Western European nations found indications that high levels of Se reduce the incidence of CRC (Hughes et al. 2015) Se may have a prognostic function found that Se may play a prognostic role in nasopharyngeal carcinoma (Chen et al. 2016), BC (Zhang et al. 2013), and renal cancer (Ha et al. 2014) according to several studies. Se successfully protects LC in persons with low selenium levels, according to a study by Fritz et al. (2011).

3.9 Zinc (Zn)

A trace element, Zn is vital for the normal functioning of human cells and involved in many cellular and structural proteins, enzymes and transcription factors. Major sources of Zn are seafood, poultry, and meats. Humans require trace amounts of Zn in their diet, approximately 15 mg of Zn daily. Zn deficiency is associated with growth retardation, hypogonadism, and brain dysfunctions. Besides improving cell-mediated immunity, Zn is also an anti-inflammatory and antioxidant (Santos et al. 2020; Prasad 2013).

Numerous studies have revealed the role of Zn in cancer. An increase in oxidative stress and chronic inflammation may promote the progression of many cancers. However, the exact mechanisms by which lack of Zn influences DNA integrity and cancer danger remain unclear (Ho 2004; Prasad et al. 2009). The control of oxidative stress and cellular signaling processes is likely crucial for DNA repair and apoptosis processes. The effects of Zn on apoptosis can be directly attributed to its action on the nucleus or its direct action on mitochondrial apoptosis (Lee et al. 2004).

Zn homeostasis dysfunction was reported in several epidemiological studies to be linked with PROC (To PK et al. 2020; Costello et al. 2004). Xie et al. found that cervical cancer patients had lower Zn levels than healthy individuals, based on a meta-analysis (Xie et al. 2018). In contrast to healthy controls, patients with head and neck squamous cell carcinoma had lower plasma Zn levels and a considerably greater copper: Zn ratio in the plasma, according to Abdulla et al. Additionally, it is understood that Zn plays a significant part in the initiation and advancement of PROC and may be used to prevent and treat the disease. In their study, Li et al. discovered that the consumption of Zn in the top category was associated with a lowered risk of PANC. It boosts immune system activity, inhibits free radicals, has a genetic influence, and minimizes oxidative DNA damage to prevent PANC (Li and Gai 2017).

3.10 Magnesium (Mg)

Mg, the fourth most prevalent cation in humans, contributes to over 300 enzymatic processes. Mg-rich foods contain grains, nuts, legumes, and green leafy vegetables. Adults should consume between 300 and 400 milligrams of Mg daily. It functions as a counter ion for the energetically dense ATP and nucleic acids, controls transmembrane transport, which is necessary for cell repair, RNA and DNA synthesis, the stability of cell function, and upholds the cell's antioxidant status. Additionally, it is crucial to the functioning of the gastrointestinal and reproductive systems. Numerous metabolic and inflammatory disorders, including cancer, metabolic syndrome, osteoporosis, diabetes mellitus, ischemic heart disease, and hypertension associated with Mg depletion. It is well established that consuming Mg can decrease the risk of developing CRC considerably (Wark et al. 2012; Volpe 2013). As a cofactor for DNA restoration and duplication, Mg is important for apoptosis (Sahmoun and Singh 2010), genomic stability, and the prevention of chemical carcinogenesis. The interplay between cancer and Mg status is complicated. It has been found that people and animals with Mg deficiency are more likely to develop tumors (Wolf et al. 2009).

An analysis by Dana et al. found that Mg consumption under 300 mg/day may reduce LC risk. LC incidence was not decreased by increasing Mg intake dosages to >300 mg/d. The most Mg-consuming individuals have a much minor threat of developing CRC later in life, according to several epidemiological studies (Qu et al. 2013; Chen et al. 2012). As a result of consuming less Mg, Dibaba et al. found a 24% rise in the danger of PANC (Dibaba et al. 2015) (Fig. 4.2).

4 Conclusion

Cancer is a multifaceted disease triggered via an interplay among environmental and genetic aspects. Scientists have reported that dietary intake and nutritional supplements are important factors in cancer development. Evidence suggests that high quality and quantity of dietary nutrients are linked to cancer occurrences and pathogenesis. Several dietary nutrients affect physiological, metabolic, and cellular signaling mechanisms associated with immune functions, thus influencing cancer growth and progression. Deficiencies of nutrients in the diet are thought to influence cancer development differently, but specific factors are still unknown. In the world, obesity and alcohol are the two most important nutrition factors contributing to cancer's burden. A healthy diet rich in macronutrients and micronutrients is crucial for preventing and treating cancer. Inadequate or excessive nutrients can contribute to cancer risk and progress. Sustaining normal blood concentrations of nutrients is a good approach for preventing and treating cancer; subsequently, this improves the cells' ability to function. Furthermore, physical activity, weight control, and proper nutrition may improve survival and decrease the probability of cancer recurrence.

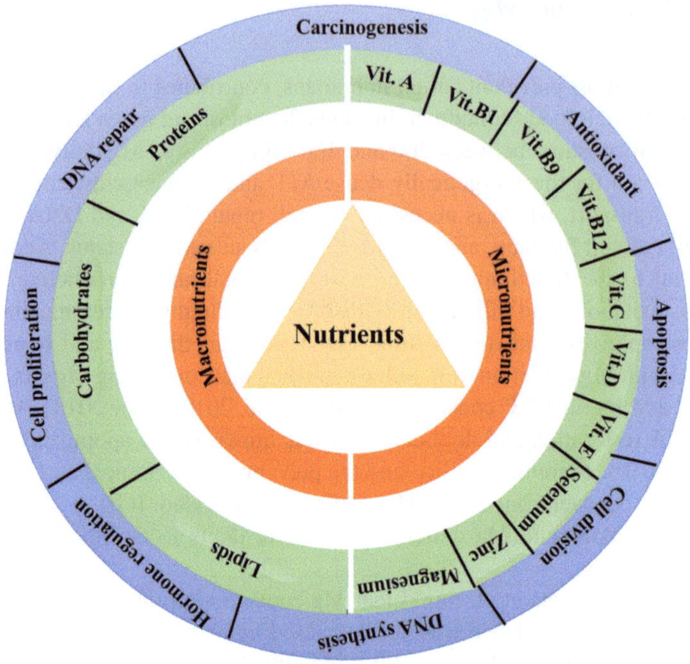

Fig. 4.2 Nutrients in the prevention of cancer

References

Aggarwal A, Höbaus J, Tennakoon S, Prinz-Wohlgenannt M, Graça J, Price SA, et al. Active vitamin D potentiates the anti-neoplastic effects of calcium in the colon: A cross talk through the calcium-sensing receptor. J Steroid Biochem Mol Biol. 2016;155:231–8.

Al-Lahham SH, Roelofsen H, Priebe M, Weening D, Dijkstra M, Hoek A, et al. Regulation of adipokine production in human adipose tissue by propionic acid. Eur J Clin Investig. 2010;40(5):401–7.

Bailón E, Cueto-Sola M, Utrilla P, Rodríguez-Cabezas ME, Garrido-Mesa N, Zarzuelo A, et al. Butyrate in vitro immune-modulatory effects might be mediated through a proliferation-related induction of apoptosis. Immunobiology. 2010;215(11):863–73.

Barnett KT, Fokum FD, Malafa MP. Vitamin E succinate inhibits colon cancer liver metastases. J Surg Res. 2002;106(2):292–8.

Berquin IM, Edwards IJ, Chen YQ. Multi-targeted therapy of cancer by omega-3 fatty acids. Cancer Lett. 2008;269(2):363–77.

Berryman DE, Christiansen JS, Johannsson G, Thorner MO, Kopchick JJ. Role of the GH/IGF-1 axis in lifespan and healthspan: lessons from animal models. Growth Hormone IGF Res. 2008;18:455–71.

Binker-Cosen MJ, Richards D, Oliver B, Gaisano HY, Binker MG, Cosen-Binker LI. Palmitic acid increases invasiveness of pancreatic cancer cells AsPC-1 through TLR4/ROS/NF-κB/MMP-9 signaling pathway. Biochem Biophys Res Commun. 2017;484(1):152–8.

Bojková B, Winklewski PJ, Wszedybyl-Winklewska M. Dietary fat and cancer—which is good, which is bad, and the body of evidence. Int J Mol Sci MDPI AG. 2020;21:1–56.

Brasky TM, Rodabough RJ, Liu J, Kurta ML, Wise LA, Orchard TS, et al. Long-chain ω-3 fatty acid intake and endometrial cancer risk in the Women's Health Initiative. Am J Clin Nutr. 2015;101(4):824–34.

Bratton BA, Maly IV, Hofmann WA. Effect of polyunsaturated fatty acids on proliferation and survival of prostate cancer cells. PLoS One. 2019;14(7):e0219822.

Brown AJ, Goldsworthy SM, Barnes AA, Eilert MM, Tcheang L, Daniels D, et al. The Orphan G protein-coupled receptors GPR41 and GPR43 are activated by propionate and other short chain carboxylic acids. J Biol Chem. 2003;278(13):11312–9.

Cai X, Wang C, Yu W, Fan W, Wang S, Shen N, et al. Selenium exposure and cancer risk: an updated meta-analysis and meta-regression. Sci Rep. 2016;6

Calder PC. Omega-3 fatty acids and inflammatory processes. Nutrients. 2010;2(3):355–74.

Cao Y, Hou L, Wang W. Dietary total fat and fatty acids intake, serum fatty acids and risk of breast cancer: A meta-analysis of prospective cohort studies. Int J Cancer. 2016;138(8):1894–904.

Carmel R, Vasireddy H, Aurangzeb I, George K. High serum cobalamin levels in the clinical setting ± clinical associations and holo-transcobalamin changes.

Chambial S, Dwivedi S, Shukla KK, John PJ, Sharma P. Vitamin C in disease prevention and cure: an overview. Indian J Clin Biochem. Springer. 2013;28:314–28.

Chen M, Huang J. The expanded role of fatty acid metabolism in cancer: new aspects and targets. Precis Clin Med. 2019;2(3):183–91.

Chen GC, Pang Z, Liu QF. Magnesium intake and risk of colorectal cancer: A meta-analysis of prospective studies. Eur J Clin Nutr. 2012;66:1182–6.

Chen P, Li C, Li X, Li J, Chu R, Wang H. Higher dietary folate intake reduces the breast cancer risk: a systematic review and meta-analysis. Br J Cancer. 2014;110(9):2327–38.

Chen F, Chen C, Qu Y, Xiang H, Ai Q, Yang F, et al. Selenium-binding protein 1 in head and neck cancer is low-expression and associates with the prognosis of nasopharyngeal carcinoma. Medicine (United States). 2016;95(35).

Chen Z, Zuurmond MG, van der Schaft N, Nano J, Wijnhoven HAH, Ikram MA, et al. Plant versus animal based diets and insulin resistance, prediabetes and type 2 diabetes: the Rotterdam Study. Eur J Epidemiol. 2018;33(9):883–93.

Cheon EC, Strouch MJ, Barron MR, Ding Y, Melstrom LG, Krantz SB, et al. Alteration of strain background and a high omega-6 fat diet induces earlier onset of pancreatic neoplasia in EL-Kras transgenic mice. Int J Cancer. 2011;128(12):2783–92.

Chi AC, Day TA, Neville BW. Oral cavity and oropharyngeal squamous cell carcinoma-an update. CA Cancer J Clin. 2015;65(5):401–21.

Chiang KC, Chen TC. Vitamin D for the prevention and treatment of pancreatic cancer. World J Gastroenterol. Baishideng Publishing Group Co. 2009;15:3349–54.

Cimmino L, Neel BG, Aifantis I. Vitamin C in stem cell reprogramming and cancer. Trends Cell Biol. Elsevier Ltd. 2018;28:698–708.

Collin SM. Folate and B12 in prostate cancer. In: Advances in clinical chemistry. Academic Press Inc.; 2013. p. 1–63.

Combs GF, Clark LC, Turnbull BW. An analysis of cancer prevention by selenium. Biofactors. 2001;14

Conibear E, Davis NG. Palmitoylation and depalmitoylation dynamics at a glance. J Cell Sci. 2010;123(Pt 23):4007–10.

Constantinou C, Papas A, Constantinou AI. Vitamin E and cancer: an insight into the anticancer activities of vitamin E isomers and analogs. Int J Cancer. 2008;123:739–52.

Corn KC, Windham MA, Rafat M. Lipids in the tumor microenvironment: from cancer progression to treatment. Prog Lipid Res. 2020;80:101055.

Costello LC, Feng P, Milon B, Tan M, Franklin RB. Role of zinc in the pathogenesis and treatment of prostate cancer: critical issues to resolve. Prostate Cancer Prostatic Dis. 2004;7(2):111–7.

Cummings JH, Pomare EW, Branch WJ, Naylor CP, Macfarlane GT. Short chain fatty acids in human large intestine, portal, hepatic and venous blood. Gut. 1987;28(10):1221–7.

Dai Q, Motley SS, Smith JA, Concepcion R, Barocas D, Byerly S, et al. Blood magnesium, and the interaction with calcium, on the risk of high-grade prostate cancer. PLoS One. 2011;6(4):e18237.

Dasgupta M, Sharkey JR, Wu G. Inadequate intakes of indispensable amino acids among homebound older adults. J Nutr Elder. 2005;24(3):85–99.

Dhingra D, Michael M, Rajput H, Patil RT. Dietary fibre in foods: A review. J Food Sci Technol. 2012;49:255–66.

Diaz-Aragon R, Ramirez-Ricardo J, Cortes-Reynosa P, Simoni-Nieves A, Gomez-Quiroz LE, Perez Salazar E. Role of phospholipase D in migration and invasion induced by linoleic acid in breast cancer cells. Mol Cell Biochem. 2019;457(1–2):119–32.

Dibaba D, Xun P, Yokota K, White E, He K. Magnesium intake and incidence of pancreatic cancer: the VITamins and lifestyle study. Br J Cancer. 2015;113(11):1615–21.

DiNicolantonio JJ, O'Keefe JH. Importance of maintaining a low omega–6/omega–3 ratio for reducing inflammation. Open Heart. 2018;5(2):e000946.

Doseděl M, Jirkovský E, Macáková K, Krčmová LK, Javorská L, Pourová J, et al. Vitamin C—sources, physiological role, kinetics, deficiency, use, toxicity, and determination. Nutrients. MDPI AG. 2021;13:1–36.

Dulai PS, Singh S, Marquez E, Khera R, Prokop LJ, Limburg PJ, et al. Chemoprevention of colorectal cancer in individuals with previous colorectal neoplasia: systematic review and network meta-analysis. BMJ (Online). BMJ Publishing Group. 2016;355

Eun YP, Wilder ET, Lane MA. Retinol inhibits the invasion of retinoic acid-resistant colon cancer cells in vitro and decreases matrix metalloproteinase mRNA, protein, and activity levels. Nutr Cancer. 2007;57(1):66–77.

Fahy E, Cotter D, Sud M, Subramaniam S. Lipid classification, structures and tools. Biochim Biophys Acta Mol Cell Biol Lipids. 2011;1811(11):637–47.

Fanidi A, Carreras-Torres R, Larose TL, Yuan JM, Stevens VL, Weinstein SJ, et al. Is high vitamin B12 status a cause of lung cancer? Int J Cancer. 2019;145(6):1499–503.

Farhan Aslam M, Majeed S, Aslam S, Irfan JA. Vitamins: Key role players in boosting up immune response-A mini review. Vitam Miner. 2017;06(01)

Fatima S, Hu X, Gong RH, Huang C, Chen M, Wong HLX, et al. Palmitic acid is an intracellular signaling molecule involved in disease development. Cell Mol Life Sci. 2019;76(13):2547–57.

Fellows PJ. Properties of food processing. In: Food processing technology. Elsevier; 2009. p. 11–95.

Firth SM, Baxter RC. Cellular actions of the insulin-like growth factor binding proteins. Endocr Rev. 2002;23:824–54.

Fontana JA, Rishi AK. SPOTLIGHT REVIEW classical and novel retinoids: their targets in cancer therapy. Leukemia [Internet]. 2002;16:463–72. Available from: www.nature.com/leu

Forman MR, Mahabir S. Saturated fatty acids and cancer. In: Bioactive compounds and cancer. Totowa: Humana Press; 2010. p. 213–33.

Fritz H, Kennedy D, Fergusson D, Fernandes R, Cooley K, Seely A, et al. Selenium and lung cancer: a systematic review and meta analysis. PLoS One. 2011;6

Galeone C, Edefonti V, Parpinel M, Leoncini E, Matsuo K, Talamini R, et al. Folate intake and the risk of oral cavity and pharyngeal cancer: a pooled analysis within the International Head and Neck Cancer Epidemiology Consortium. Int J Cancer. 2015;136(4):904–14.

Galván-Portillo MV, Oñate-Ocaña LF, Pérez-Pérez GI, Chen J, Herrera-Goepfert R, Chihu-Amparan L, et al. Dietary folate and vitamin B12 intake before diagnosis decreases gastric cancer mortality risk among susceptible MTHFR 677TT carriers. Nutrition. 2010 Feb;26(2):201–8.

Gannon MC, Nuttall FQ. Effect of a high-protein diet on ghrelin, growth hormone, and insulin-like growth factor-I and binding proteins 1 and 3 in subjects with type 2 diabetes mellitus. Metabolism. 2011;60(9):1300–11.

Garland CF, Garland FC, Gorham ED, Lipkin M, Newmark H, Mohr SB, et al. The role of vitamin D in cancer prevention. Am J Public Health [Internet]. 2006;96(2). Available from: http://www.noaa.gov

Gastón R, María Eugenia P, Das UN, Eynard AR. Polyunsaturated fatty acids differentially modulate cell proliferation and endocannabinoid system in two human cancer lines. Arch Med Res. 2017;48(1):46–54.
Gomes SD, Oliveira CS, Azevedo-Silva J, Casanova MR, Barreto J, Pereira H, et al. The role of diet related short-chain fatty acids in colorectal cancer metabolism and survival: prevention and therapeutic implications. Curr Med Chem. 2020;27(24):4087–108.
Gonzalez-Reyes C, Marcial-Medina C, Cervantes-Anaya N, Cortes-Reynosa P, Salazar EP. Migration and invasion induced by linoleic acid are mediated through fascin in MDA-MB-231 breast cancer cells. Mol Cell Biochem. 2018;443(1–2):1–10.
Granger M, Eck P. Dietary vitamin C in human health. In: Advances in food and nutrition research. Academic Press Inc.; 2018. p. 281–310.
Grant WB. Review of recent advances in understanding the role of vitamin D in reducing cancer risk: breast, colorectal, prostate, and overall cancer. Anticancer Res. 2020;40(1):491–9.
Grattan BJ, Freake HC. Zinc and cancer: implications for LIV-1 in breast cancer. Nutrients. 2012;4(7):648–75.
Gröber U, Holzhauer P, Kisters K, Holick MF, Adamietz IA. Micronutrients in oncological intervention. Nutrients. MDPI. 2016;8
Gromadzińska J, Reszka E, Bruzelius K, Wąsowicz W, Åkesson B. Selenium and cancer: Biomarkers of selenium status and molecular action of selenium supplements. Eur J Nutr D. Steinkopff-Verlag. 2008;47:29–50.
Gu Z, Shan K, Chen H, Chen YQ. N-3 polyunsaturated fatty acids and their role in cancer chemoprevention. Curr Pharmacol Rep. 2015;1(5):283–94.
Ha YS, Lee GT, Kim YH, Kwon SY, Choi SH, Kim TH, et al. Decreased selenium-binding protein 1 mRNA expression is associated with poor prognosis in renal cell carcinoma. World J Surg Oncol. 2014;12(1)
Hagihara K, Kajimoto K, Osaga S, Nagai N, Shimosegawa E, Nakata H, et al. Promising effect of a new ketogenic diet regimen in patients with advanced cancer. Nutrients. 2020;12(5)
Hanberry BS, Berger R, Zastre JA. High-dose vitamin B1 reduces proliferation in cancer cell lines analogous to dichloroacetate. Cancer Chemother Pharmacol. 2014;73(3):585–94.
Hinnebusch BF, Meng S, Wu JT, Archer SY, Hodin RA. The effects of short-chain fatty acids on human colon cancer cell phenotype are associated with histone hyperacetylation. J Nutr. 2002;132(5):1012–7.
Ho E. Zinc deficiency, DNA damage and cancer risk. J Nutr Biochem. 2004;15:572–8.
Hoffer LJ, Bistrian BR. Appropriate protein provision in critical illness: A systematic and narrative review. Am J Clin Nutr. 2012;96:591–600.
Hohman EE, Martin BR, Lachcik PJ, Gordon DT, Fleet JC, Weaver CM. Bioavailability and efficacy of vitamin D2 from UV-irradiated yeast in growing, vitamin D-deficient rats. J Agric Food Chem. 2011;59(6):2341–6.
Holick MF. Cancer, sunlight and vitamin D. J Clin Transl Endocrinol. Elsevier BV. 2014;1:179–86.
Hong YH, Nishimura Y, Hishikawa D, Tsuzuki H, Miyahara H, Gotoh C, et al. Acetate and propionate short chain fatty acids stimulate adipogenesis via GPCR43. Endocrinology. 2005;146(12):5092–9.
Hsieh AC, Liu Y, Edlind MP, Ingolia NT, Janes MR, Sher A, et al. The translational landscape of mTOR signalling steers cancer initiation and metastasis. Nature. 2012;485(7396):55–61.
Hua YF, Wang GQ, Jiang W, Huang J, Chen GC, de Lu C. Vitamin C intake and pancreatic cancer risk: A meta-analysis of published case-control and cohort studies. PLoS One. 2016;11(2)
Huang X, Gao Y, Zhi X, Ta N, Jiang H, Zheng J. Association between vitamin A, retinol and carotenoid intake and pancreatic cancer risk: evidence from epidemiologic studies. Sci Rep. 2016:6.
Huang LH, Chung HY, Su HM. Docosahexaenoic acid reduces sterol regulatory element binding protein-1 and fatty acid synthase expression and inhibits cell proliferation by inhibiting pAkt signaling in a human breast cancer MCF-7 cell line. BMC Cancer. 2017;17(1):890.

Hughes DJ, Fedirko V, Jenab M, Schomburg L, Méplan C, Freisling H, et al. Selenium status is associated with colorectal cancer risk in the European prospective investigation of cancer and nutrition cohort. Int J Cancer. 2015;136(5):1149–61.
Ilbert G, Menn SO, Ary G, Oodman EG, Hornquist ADT, Ohn J, et al. Effects of a combination of beta carotene and vitamin A on lung cancer and cardiovascular disease. 1996.
Innes JK, Calder PC. Omega-6 fatty acids and inflammation. Prostaglandins Leukot Essent Fatty Acids. 2018;132:41–8.
Iqbal J, Hussain MM. Intestinal lipid absorption. Am J Physiol Endocrinol Metab. 2009;296(6):E1183–94.
Irimie AI, Braicu C, Pasca S, Magdo L, Gulei D, Cojocneanu R, et al. Role of key micronutrients from nutrigenetic and nutrigenomic perspectives in cancer prevention. Medicina (Lithuania). MDPI AG. 2019;55
Jan G, Belzacq AS, Haouzi D, Rouault A, Métivier D, Kroemer G, et al. Propionibacteria induce apoptosis of colorectal carcinoma cells via short-chain fatty acids acting on mitochondria. Cell Death Differ. 2002;9(2):179–88.
Jiang Q. Natural forms of vitamin E as effective agents for cancer prevention and therapy. Adv Nutr. American Society for Nutrition. 2017;8:850–67.
Johnson IT. Micronutrients and cancer. Proc Nutr Soc. 2004 Nov;63(4):587–95.
Johnson LJ, Meacham SL, Kruskall LJ. The antioxidants-vitamin c, vitamin e, selenium, and carotenoids. J Agromedicine. 2003;9(1):65–82.
Ju J, Hao X, Lee MJ, Lambert JD, Lu G, Xiao H, et al. A γ-tocopherol-rich mixture of tocopherols inhibits colon inflammation and carcinogenesis in azoxymethane and dextran sulfate sodium-treated mice. Cancer Prev Res. 2009;2(2):143–52.
Jump DB. Fatty acid regulation of hepatic lipid metabolism. Curr Opin Clin Nutr Metab Care. 2011;14(2):115–20.
Kaźmierczak-Barańska J, Boguszewska K, Adamus-Grabicka A, Karwowski BT. Two faces of vitamin c—antioxidative and pro-oxidative agent. Nutrients. MDPI AG. 2020;12
Key TJ. Diet, insulin-like growth factor-1 and cancer risk. Proc Nutr Soc. 2011;70(3):385–8.
Khadangi F, Azzi A. Vitamin E—The next 100 years. IUBMB Life. Blackwell Publishing Ltd. 2019;71:411–5.
Khadge S, Thiele GM, Sharp JG, McGuire TR, Klassen LW, Black PN, et al. Long-chain omega-3 polyunsaturated fatty acids decrease mammary tumor growth, multiorgan metastasis and enhance survival. Clin Exp Metastasis. 2018;35(8):797–818.
Kim H, Hyung WJ, Song KJ, Choi SH, Kim CB, Noh SH. Oral vitamin B12 replacement: an effective treatment for vitamin B12 deficiency after total gastrectomy in gastric cancer patients. Ann Surg Oncol. 2011;18(13):3711–7.
Klein EA, Thompson IM, Lippman SM, Goodman PJ, Albanes D, Taylor PR, et al. SELECT: the selenium and vitamin E cancer prevention trial. 2003.
Kline K, Yu W, Sanders BG. Vitamin E and Breast Cancer 1 [Internet]. Vol. 134, International Research Conference on Food. 2004. Available from: https://academic.oup.com/jn/article-abstract/134/12/3458S/4688702
Ko P, Dixon SJ. Protein palmitoylation and cancer. EMBO Rep. 2018;19(10)
Kok DE, Steegenga WT, Smid EJ, Zoetendal EG, Ulrich CM, Kampman E. Bacterial folate biosynthesis and colorectal cancer risk: more than just a gut feeling. Vol. 60, Critical Reviews in Food Science and Nutrition. Taylor and Francis Inc.; 2020. p. 244–56.
Kristal AR, Darke AK, Morris JS, Tangen CM, Goodman PJ, Thompson IM, et al. Baseline selenium status and effects of selenium and vitamin E supplementation on prostate cancer risk. J Natl Cancer Inst. 2014;106(3)
Kune GA, Kune S, Field B, Watson LF, Cleland H, Merenstein D, et al. Oral and pharyngeal cancer, diet, smoking, alcohol, and serum vitamin A and ß-carotene levels: A case-control study in men. Nutr Cancer. 1993;20(1):61–70.
Kuo SM. The interplay between fiber and the intestinal microbiome in the inflammatory response. Adv Nutr. 2013;4:16–28.

Lee DH, Anderson KE, Harnack LJ, Folsom AR, Jacobs DR. Heme iron, zinc, alcohol consumption, and colon cancer: Iowa women's health study. J Natl Cancer Inst. 2004;96(5):403–7.
Lee J, Demissie K, Lu SE, Rhoads GG. Cancer incidence among Korean-American immigrants in the United States and native Koreans in South Korea. Cancer Control. 2007;14(1):78–85.
Lee EH, Myung SK, Jeon YJ, Kim Y, Chang YJ, Ju W, et al. Effects of selenium supplements on cancer prevention: meta-analysis of randomized controlled trials. Nutr Cancer. 2011;63(8):1185–95.
Leser S. The 2013 FAO report on dietary protein quality evaluation in human nutrition: recommendations and implications. Nutr Bull. 2013;38(4):421–8.
Letavayová L, Vlčková V, Brozmanová J. Selenium: From cancer prevention to DNA damage. Toxicology. 2006;227:1–14.
Leturque A, Brot-Laroche E, le Gall M. Carbohydrate intake. In: Progress in molecular biology and translational science. Elsevier B.V; 2012. p. 113–27.
Levine ME, Suarez JA, Brandhorst S, Balasubramanian P, Cheng CW, Madia F, et al. Low protein intake is associated with a major reduction in IGF-1, cancer, and overall mortality in the 65 and younger but not older population. Cell Metab. 2014;19(3):407–17.
Li L, Gai X. The association between dietary zinc intake and risk of pancreatic cancer: a meta-analysis. Biosci Rep. 2017;37(3)
Li Y, Schellhorn HE. Critical review new developments and novel therapeutic perspectives for vitamin C 1,2 [Internet]. J Nutr. 2007;137. Available from: https://academic.oup.com/jn/article-abstract/137/10/2171/4664436
Li P, Yin YL, Li D, Kim WS, Wu G. Amino acids and immune function. Br J Nutr. 2007;98:237–52.
Liberti M, Locasale JW. The Warburg effect: how does it benefit cancer cells? Trends Biochem Sci. 2016;41(3):211–8.
Lin DTS, Davis NG, Conibear E. Targeting the Ras palmitoylation/depalmitoylation cycle in cancer. Biochem Soc Trans. 2017a;45(4):913–21.
Lin L, Ding Y, Wang Y, Wang Z, Yin X, Yan G, et al. Functional lipidomics: palmitic acid impairs hepatocellular carcinoma development by modulating membrane fluidity and glucose metabolism. Hepatology. 2017b;66(2):432–48.
Liss MA, Al-Bayati O, Gelfond J, Goros M, Ullevig S, DiGiovanni J, et al. Higher baseline dietary fat and fatty acid intake is associated with increased risk of incident prostate cancer in the SABOR study. Prostate Cancer Prostatic Dis. 2019;22(2):244–51.
Liu X, Montissol S, Uber A, Ganley S, Grossestreuer AV, Berg K, et al. The effects of thiamine on breast cancer cells. Molecules. 2018;23(6)
Lührs H, Gerke T, Müller JG, Melcher R, Schauber J, Boxberge F, et al. Butyrate inhibits NF-kappaB activation in lamina propria macrophages of patients with ulcerative colitis. Scand J Gastroenterol. 2002;37(4):458–66.
Luo J, Shen L, Zheng D. Association between vitamin C intake and lung cancer: a dose-response meta-analysis. Sci Rep. 2014;4
Maalmi H, Walter V, Jansen L, Owen RW, Ulrich A, Schöttker B, et al. Dose-response relationship between serum retinol levels and survival in patients with colorectal cancer: results from the DACHS study. Nutrients. 2018;10(4)
Macfarlane GT, Macfarlane S. Fermentation in the human large intestine: its physiologic consequences and the potential contribution of prebiotics. J Clin Gastroenterol. 2011;45(Suppl):S120–7.
Madak Erdogan Z. Impact of diet and nutrition on cancer hallmarks. J Cancer Prev Curr Res. 2017;7(4)
Madiyal A, Shetty SR, Babu GS. Micronutrients and their role in oral cancer: A review. West Indian Med J. 2016;65(2):383–8.
Madsen L, Petersen RK, Kristiansen K. Regulation of adipocyte differentiation and function by polyunsaturated fatty acids. Biochim Biophys Acta. 2005;1740(2):266–86.
Maia SB, Souza ASR, Caminha MDFC, da Silva SL, de Cruz RSBLC, dos Santos CC, et al. Vitamin A and pregnancy: a narrative review. Nutrients. MDPI AG. 2019;11

Maino Vieytes CA, Taha HM, Burton-Obanla AA, Douglas KG, Arthur AE. Carbohydrate nutrition and the risk of cancer, vol. 8. Curr Nutr Rep. Current Science Inc.; 2019. p. 230–9.

Makarem N, Bandera E, Lin Y, Jacques PF, Hayes RB, Parekh N. Consumption of sugars, sugary foods, and sugary beverages in relation to adiposity-related cancer risk in the Framingham offspring cohort (1991–2013). Cancer Prev Res. 2018;11(6):347–58.

Malila N, Virtamo J, Virtanen M, Pietinen P, Albanes D, Teppo L. Original communication dietary and serum a-tocopherol, b-carotene and retinol, and risk for colorectal cancer in male smokers. Eur J Clin Nutr [Internet]. 2002;56:615–21. Available from: www.nature.com/ejcn

Martín Ortega AM, Segura Campos MR. Macronutrients and micronutrients in cancer prevention and treatment. In: Oncological functional nutrition: phytochemicals and medicinal plants. Elsevier; 2021. p. 99–124.

Matthews GM, Howarth GS, Butler RN. Short-chain fatty acids induce apoptosis in colon cancer cells associated with changes to intracellular redox state and glucose metabolism. Chemotherapy. 2012;58(2):102–9.

Mayne ST, Playdon MC, Rock CL. Diet, nutrition, and cancer: Past, present and future. Nat Rev Clin Oncol. Nature Publishing Group. 2016;13:504–15.

McCarty MF. MTORC1 activity as a determinant of cancer risk - rationalizing the cancer-preventive effects of adiponectin, metformin, rapamycin, and low-protein vegan diets. Med Hypotheses. 2011;77(4):642–8.

Menendez JA, Vellon L, Colomer R, Lupu R. Oleic acid, the main monounsaturated fatty acid of olive oil, suppresses Her-2/neu (erbB-2) expression and synergistically enhances the growth inhibitory effects of trastuzumab (Herceptin™) in breast cancer cells with Her-2/neu oncogene amplification. Ann Oncol. 2005;16(3):359–71.

Michalak A, Mosińska P, Fichna J. Polyunsaturated fatty acids and their derivatives: therapeutic value for inflammatory, functional gastrointestinal disorders, and colorectal cancer. Front Pharmacol. 2016;7

Naderi N, House JD. Recent developments in folate nutrition. In: Advances in food and nutrition research. Academic Press Inc.; 2018. p. 195–213.

Nindrea RD, Aryandono T, Lazuardi L, Dwiprahasto I. Association of dietary intake ratio of n-3/n-6 polyunsaturated fatty acids with breast cancer risk in Western and Asian countries: A meta-analysis. Asian Pac J Cancer Prev. 2019;20(5):1321–7.

Nkondjock A, Krewski D, Johnson KC, Ghadirian P. Canadian Cancer Registries Epidemiology Research Group. Specific fatty acid intake and the risk of pancreatic cancer in Canada. Br J Cancer. 2005;92(5):971–7.

Ntambi JM, Bené H. Polyunsaturated fatty acid regulation of gene expression. J Mol Neurosci. 2001;16(2–3):273–8.

O'Leary F, Samman S. Vitamin B12 in health and disease. Nutrients. MDPI AG. 2010;2:299–316.

Ono S, Karaki S, ichiro, Kuwahara A. Short-chain fatty acids decrease the frequency of spontaneous contractions of longitudinal muscle via enteric nerves in rat distal colon. Jpn J Physiol. 2004;54(5):483–93.

Oresic M, Hänninen VA, Vidal-Puig A. Lipidomics: a new window to biomedical frontiers. Trends Biotechnol. 2008;26(12):647–52.

Park M, Lim J, Kim H. Docoxahexaenoic acid induces apoptosis of pancreatic cancer cells by suppressing activation of STAT3 and NF-κB. Nutrients. 2018;10(11):1621.

Pawlowska E, Szczepanska J, Blasiak J. Pro- and antioxidant effects of vitamin C in cancer in correspondence to its dietary and pharmacological concentrations. Oxid Med Cell Longev. Hindawi Limited. 2019;2019

Peters U, Takata Y. Selenium and the prevention of prostate and colorectal cancer. Mol Nutr Food Res. 2008;52:1261–72.

Peterson CT, Rodionov DA, Peterson SN, Osterman AL. B vitamins and their role in immune regulation and cancer, vol. 12. Nutrients. MDPI AG; 2020. p. 1–23.

Petrone I, Bernardo PS, dos Santos EC, Abdelhay E. Mthfr c677t and a1298c polymorphisms in breast cancer, gliomas and gastric cancer: A review, vol. 12. Genes. MDPI AG; 2021.

Pham DQ, Plakogiannis R. Vitamin E supplementation in cardiovascular disease and cancer prevention: Part 1. Ann Pharmacother. 2005;39:1870–8.
Poulain S, Evenou F, Carré MC, Corbel S, Vignaud JM, Martinet N. Vitamin A/retinoids signalling in the human lung. Lung Cancer. 2009;66:1–7.
Prasad AS. Discovery of human zinc deficiency: Its impact on human health and disease. Adv Nutr. 2013;4:176–90.
Prasad AS, Beck FWJ, Snell DC, Kucuk O. Zinc in cancer prevention. Nutr Cancer. 2009;61:879–87.
Psaltopoulou T, Kosti RI, Haidopoulos D, Dimopoulos M, Panagiotakos DB. Olive oil intake is inversely related to cancer prevalence: a systematic review and a meta-analysis of 13800 patients and 23340 controls in 19 observational studies. Lipids Health Dis. 2011;10(1):127.
Qu X, Jin F, Hao Y, Zhu Z, Li H, Tang T, et al. Nonlinear association between magnesium intake and the risk of colorectal cancer. Eur J Gastroenterol Hepatol. 2013;25(3):309–18.
Ranawana V, Kaur B. Role of proteins in insulin secretion and glycemic control. In: Advances in food and nutrition research. Academic Press Inc.; 2013. p. 1–47.
Rayman MP. Selenium in cancer prevention: a review of the evidence and mechanism of action. Proc Nutr Soc. 2005;64(4):527–42.
Rayman MP. Selenium intake, status, and health: a complex relationship. Hormones. Springer. 2020;19:9–14.
Reeds PJ, Burrin DG, Stoll B, van Goudoever JB. Role of the gut in the amino acid economy of the host. In: Fürst P, Vevey S, editors. Peptides and amino acids in enteral nutrition, vol. 3. Karger AG; 2000.
Saghiri MA, Asatourian A, Ershadifar S, Moghadam MM, Sheibani N. Vitamins and regulation of angiogenesis: [A, B1, B2, B3, B6, B9, B12, C, D, E, K]. J Funct Foods. Elsevier Ltd. 2017;38:180–96.
Sahmoun AE, Singh BB. Does a higher ratio of serum calcium to magnesium increase the risk for postmenopausal breast cancer? Med Hypotheses. 2010;75(3):315–8.
Saini RK, Keum YS. Omega-3 and omega-6 polyunsaturated fatty acids: dietary sources, metabolism, and significance — A review. Life Sci. 2018;203:255–67.
Samuel BS, Shaito A, Motoike T, Rey FE, Backhed F, Manchester JK, et al. Effects of the gut microbiota on host adiposity are modulated by the short-chain fatty-acid binding G protein-coupled receptor, Gpr41. Proc Natl Acad Sci USA. 2008;105(43):16767–72.
Sánchez-Quesada C, López-Biedma A, Warleta F, Campos M, Beltrán G, Gaforio JJ. Bioactive properties of the Main triterpenes found in olives, virgin olive oil, and leaves of Olea europaea. J Agric Food Chem. 2013;61(50):12173–82.
Santos HO, Teixeira FJ, Schoenfeld BJ. Dietary vs. pharmacological doses of zinc: a clinical review, vol. 39. Clin Nutr. Churchill Livingstone; 2020. p. 1345–53.
Savage MO. Insulin-like growth factors, nutrition and growth. World Rev Nutr Diet. 2013;106:52–9.
Schrauzer GN. Anticarcinogenic effects of selenium, vol. 57. CMLS: Cell. Mol. Life Sci; 2000.
Sczaniecka AK, Brasky TM, Lampe JW, Patterson RE, White E. Dietary intake of specific fatty acids and breast cancer risk among postmenopausal women in the VITAL cohort. Nutr Cancer. 2012;64(8):1131–42.
Sellem L, Srour B, Guéraud F, Pierre F, Kesse-Guyot E, Fiolet T, et al. Saturated, mono- and polyunsaturated fatty acid intake and cancer risk: results from the French prospective cohort NutriNet-Santé. Eur J Nutr. 2019;58(4):1515–27.
Serna-Marquez N, Diaz-Aragon R, Reyes-Uribe E, Cortes-Reynosa P, Salazar EP. Linoleic acid induces migration and invasion through FFAR4- and PI3K−/Akt-dependent pathway in MDA-MB-231 breast cancer cells. Med Oncol. 2017;34(6):111.
Severo M, Lopes C, Lucas R, Barros H. Development of a tool for the assessment of calcium and vitamin D intakes in clinical settings. Osteoporos Int. 2009;20(2):231–7.
Shenkin A. The key role of micronutrients. Clin Nutr. 2006;2:1–13.
Shergill-Bonner R. Micronutrients. Paediatr Child Health (United Kingdom). Churchill Livingstone. 2017;27:357–62.

Siddiqui RA, Harvey KA, Xu Z, Bammerlin EM, Walker C, Altenburg JD. Docosahexaenoic acid: A natural powerful adjuvant that improves efficacy for anticancer treatment with no adverse effects. Biofactors. 2011;37(6):399–412.

Sieri S, Agnoli C, Pala V, Grioni S, Brighenti F, Pellegrini N, et al. Dietary glycemic index, glycemic load, and cancer risk: results from the EPIC-Italy study. Sci Rep. 2017;7(1)

Simopoulos AP. The importance of the Omega-6/Omega-3 fatty acid ratio in cardiovascular disease and other chronic diseases. Exp Biol Med. 2008;233(6):674–88.

Smith W, Saba N. Retinoids as chemoprevention for head and neck cancer: where do we go from here? Crit Rev Oncol Hematol. 2005;55:143–52.

Stanford JL, Herrinton LJ, Schwartz SM, Weiss NS. Breast cancer incidence in Asian migrants to the United States and their descendants. Epidemiology. 1995;6(2):181–3.

Sung L, Greenberg ML, Koren G, Tomlinson GA, Tong A, Malkin D, et al. Vitamin E: the evidence for multiple roles in cancer, vol. 46. Nutr Cancer. Lawrence Erlbaum Associates Inc.; 2003. p. 1–14.

Talib WH, Mahmod AI, Kamal A, Rashid HM, Alashqar AMD, Khater S, et al. Ketogenic diet in cancer prevention and therapy: molecular targets and therapeutic opportunities. Curr Issues Mol Biol. 2021;43(2):558–89.

Taniguchi CM, Emanuelli B, Kahn CR. Critical nodes in signalling pathways: insights into insulin action. Nat Rev Mol Cell Biol. 2006;7:85–96.

Tedelind S, Westberg F, Kjerrulf M, Vidal A. Anti-inflammatory properties of the short-chain fatty acids acetate and propionate: a study with relevance to inflammatory bowel disease. World J Gastroenterol. 2007;13(20):2826–32.

Tinggi U. Selenium: its role as antioxidant in human health. In: Environental Health and Preventive Medicine 2008. p. 102–108.

To PK, Do MH, Cho JH, Jung C. Growth modulatory role of zinc in prostate cancer and application to cancer therapeutics. Int J Mol Sci. 2020;21(8)

Tsao SM, Yin MC, Liu WH. Oxidant stress and B vitamins status in patients with non-small cell lung cancer. Nutr Cancer. 2007;59(1):8–13.

Tucker JM, Townsend DM. Alpha-tocopherol: roles in prevention and therapy of human disease. Biomed Pharmacother. 2005;59(7):380–7.

van der Beek CM, Dejong CHC, Troost FJ, Masclee AAM, Lenaerts K. Role of short-chain fatty acids in colonic inflammation, carcinogenesis, and mucosal protection and healing. Nutr Rev. 2017;75(4):286–305.

van der Heiden MG, Cantley LC, Thompson CB. Understanding the Warburg effect: the metabolic requirements of cell proliferation. Science. 2009;324(5930):1029–33.

Verma KA. Physico-chemical characteristics, analytics and metabolism of folate in plants. J Food Process Technol. 2016;07(01)

Vigneri R, Goldfine ID, Frittitta L. Insulin, insulin receptors, and cancer. J Endocrinol Investig. Springer International Publishing. 2016;39:1365–76.

Vigneri R, Sciacca L, Vigneri P. Rethinking the relationship between insulin and cancer. Trends Endocrinol Metabol. Elsevier Inc. 2020;31:551–60.

Vivarelli S, Salemi R, Candido S, Falzone L, Santagati M, Stefani S, et al. Gut microbiota and cancer: from pathogenesis to therapy. Cancers. MDPI AG. 2019;11

Volpe SL. Magnesium in disease prevention and overall health. Adv Nutr. 2013;4(3)

Wacker M, Holick MF. Sunlight and vitamin D: A global perspective for health. Dermato-Endocrinol. Landes Bioscience. 2013;5:51–108.

Waldecker M, Kautenburger T, Daumann H, Busch C, Schrenk D. Inhibition of histone-deacetylase activity by short-chain fatty acids and some polyphenol metabolites formed in the colon. J Nutr Biochem. 2008;19(9):587–93.

Wark PA, Lau R, Norat T, Kampman E. Magnesium intake and colorectal tumor risk: A case-control study and meta-analysis. Am J Clin Nutr. 2012;96(3):622–31.

Watanabe F. Vitamin B12 sources and bioavailability. Exp Biol Med. 2007;232:1266–74.

Watson AD. Thematic review series: systems biology approaches to metabolic and cardiovascular disorders. Lipidomics: a global approach to lipid analysis in biological systems. J Lipid Res. 2006;47(10):2101–11.

Weber DD, Aminzadeh-Gohari S, Tulipan J, Catalano L, Feichtinger RG, Kofler B. Ketogenic diet in the treatment of cancer—where do we stand? vol. 33. Mol Metabol. Elsevier GmbH; 2020. p. 102–21.

Wolf FI, Cittadini ARM, Maier JAM. Magnesium and tumors: Ally or foe? Cancer Treat Rev. 2009;35:378–82.

Wu G. Amino acids: metabolism, functions, and nutrition. Amino Acids. 2009;37:1–17.

Wu G. Dietary protein intake and human health. Food Funct. Royal Society of Chemistry. 2016;7:251–65.

Wu XY, Lu L. Vitamin B6 deficiency, Genome instability and cancer. Asian Pac J Cancer Prev. Asian Pacific Organization for Cancer Prevention. 2012;13:5333–8.

Wu QJ, Xiang YB, Yang G, Li HL, Lan Q, Gao YT, et al. Vitamin e intake and the lung cancer risk among female nonsmokers: A report from the Shanghai Women's Health Study. Int J Cancer. 2015;136(3):610–7.

Wu YH, Wu YC, Chu FY, Cheng SJ, Sun A, Chen HM. Significantly higher frequencies of hematinic deficiencies and hyperhomocysteinemia in oral precancer patients. J Formos Med Assoc. 2019;118(9):1299–307.

Xie Y, Wang J, Zhao X, Zhou X, Nie X, Li C, et al. Higher serum zinc levels may reduce the risk of cervical cancer in Asian women: a meta-analysis. J Int Med Res. 2018;46(12):4898–906.

Xiong Y, Miyamoto N, Shibata K, Valasek MA, Motoike T, Kedzierski RM, et al. Short-chain fatty acids stimulate leptin production in adipocytes through the G protein-coupled receptor GPR41. Proc Natl Acad Sci USA. 2004;101(4):1045–50.

Yang CS, Luo P, Zeng Z, Wang H, Malafa M, Suh N. Vitamin E and cancer prevention: studies with different forms of tocopherols and tocotrienols. Mol Carcinog. 2020;59(4):365–89.

Yin J, Ren W, Huang X, Li T, Yin Y. Protein restriction and cancer. Biochim Biophys Acta. Elsevier BV. 2018;1869:256–62.

Zapolska-Downar D, Naruszewicz M. Propionate reduces the cytokine-induced VCAM-1 and ICAM-1 expression by inhibiting nuclear factor-kappa B (NF-kappaB) activation. J Physiol Pharmacol. 2009;60(2):123–31.

Zeng H. Mechanisms linking dietary fiber, gut microbiota and colon cancer prevention. World J Gastrointest Oncol. 2014;6(2):41.

Zeng FF, Liu YT, Lin XL, Fan YY, Zhang XL, Xu CH, et al. Folate, Vitamin B6, Vitamin B12 and methionine intakes and risk for nasopharyngeal carcinoma in Chinese adults: a matched case-control study. Br J Nutr. 2016;115(1):121–8.

Zhang L, Liu W, Hao Q, Bao L, Wang K. Folate intake and methylenetetrahydrofolate reductase gene polymorphisms as predictive and prognostic biomarkers for ovarian cancer risk. Int J Mol Sci. 2012;13(4):4009–20.

Zhang S, Li F, Younes M, Liu H, Chen C, Yao Q. Reduced selenium-binding protein 1 in breast cancer correlates with poor survival and resistance to the anti-proliferative effects of selenium. PLoS One. 2013;8(5)

Zhao Q, Hu L, Wang Z, Li Z, Wang A, Liu J. Resveratrol-loaded folic acid-grafted dextran stearate submicron particles exhibits enhanced antitumor efficacy in non-small cell lung cancers. Mater Sci Eng C. 2017;(72):185–91.

Zhou S, Wang X, Tan Y, Qiu L, Fang H, Li W. Association between vitamin C intake and glioma risk: evidence from a meta-analysis. Neuroepidemiology. 2015;44(1):39–44.

Zinder R, Cooley R, Vlad LG, Molnar JA. Vitamin A and wound healing. Nutr Clin Pract. John Wiley and Sons Inc. 2019;34:839–49.

Zou Z, Tao T, Li H, Zhu X. MTOR signaling pathway and mTOR inhibitors in cancer: Progress and challenges. Cell Biosci. BioMed Central Ltd. 2020;10

Chapter 5
Nutraceuticals in Metabolic Syndrome and Diabetes

Ceren Hangul, Meltem Güngör, Zafer Cetin, Gokhan Gorgisen, and Eyup Ilker Saygili

1 Introduction

1.1 Diabetes: Types and Etiology

Diabetes mellitus is a chronic metabolic disorder characterized by high blood glucose level due to the defects in impaired insulin secretation and/or insulin responsiveness (Craig et al. 2009). The prevelance of diabetes mellitus is getting increased worldwide especially developing areas where adopted Western life styles habits. In diabetes mellitus, blood glucose level can not be regulated and the type and duration of

Authors Ceren Hangul and Meltem Güngör are the first co-authors contributed equally to this work.

C. Hangul
Department of Medical Biology, Faculty of Medicine, Akdeniz University, Antalya, Turkey
e-mail: cerenhangul@akdeniz.edu.tr

M. Güngör
Department of Medical Biochemistry, School of Medicine, SANKO University, Gaziantep, Turkey
e-mail: metem.gungor@sanko.edu.tr

Z. Cetin
Department of Medical Biology, School of Medicine, SANKO University, Gaziantep, Turkey

Department of Biological and Biomedical Sciences, Graduate Education Institute, SANKO University, Gaziantep, Turkey
e-mail: zafer.cetin@sanko.edu.tr

G. Gorgisen
Department of Medical Genetic, School of Medicine, Van Yüzüncü Yıl University, Van, Turkey
e-mail: gokhangorgisen@yyu.edu.tr

S. M. Jafari et al. (eds.), *Nutraceuticals: A New Millennium Approach for Preventive Medicine*, Food Bioactive Ingredients,
https://doi.org/10.1007/978-3-031-76481-3_5

diabetes affect the clinical phenotype of diabetes. (Mayer-Davis et al. 2017; Galtier 2010).

Treatment approaches are variable depends on the classification of diabetes. Although Diabetes mellitus can be classified into several types, each with distinct causes, characteristics, and treatment approaches. The main types of diabetes as proposed by American Diabetes Association is Type 1, Type 2, gestational diabetes mellitus and specific types of diabetes due to other causes (American Diabetes 2021).

Type 1 Diabetes also known as insulin-dependent diabetes mellitus (IDDM) or juvenile-onset diabetes it results from an autoimmune response in which the body's immune system mistakenly destroys the insulin-producing cells in the pancreas (Saberzadeh-Ardestani et al. 2018). Therefore, the biomarker of Type 1 diabetes is the presence of auto-antibodies against the pancreatic islet cells (Couper and Donaghue 2009; Hameed et al. 2011). Consequently, insulin therapy is the main therapeutic approach to regulate the blood glucose levels for individuals diagnosed with Type 1 diabetes during their life (Dabelea et al. 2014). Although underlying mechanism of autoimmune response has not been clarified yet, genetic and environmental factors have pivotal roles (Saberzadeh-Ardestani et al. 2018). In addition to auto-antibodies, human leukocyte antigen (HLA) region, play a significant role in susceptibility to Type 1 diabetes. Variations in DR and DQ genes can affect immune system regulation and increase the risk of developing the disease or leads to protection (Noble and Erlich 2012). During the autoimmune response, microenviroment of the pancreatic cells have been changed and especially cytokines such as interleukin-1 beta (IL-1β), interleukin-6 (IL-6), tumor necrosis factor-alpha (TNF-α), and interferon-gamma (IFN-γ) are released. These cytokines, particularly contribute to the destruction of beta cells and further inflammation within the pancreas (Lu et al. 2020). Reactive oxygen species (ROS) are another main reason that lead to development of Type 1 diabetes. Inflammation triggers to generate ROS and contribute to the damage of beta cells. Increased oxidative stress can impair the function and survival of these cells, exacerbating the progression of Type 1 diabetes (Yaribeygi et al. 2020).

Type 2 diabetes is the most common type of diabetes. Insulin resistance is a common pathological state in which insulin-target cells do not respond well to ordinary levels of circulating insulin and this leads to increase blood glucose level and triggers the development of Type 2 diabetes (Brunetti et al. 2014). Genetics, poor diet and obesity are the main factors in the development of Type 2 diabetes. At molecular level, chronic inflammation, ROS, celullar stress and lipid metabolism are the main inducers in Type 2 diabetes development (Galicia-Garcia et al. 2020).

E. I. Saygili (✉)
Department of Medical Biochemistry, School of Medicine, SANKO University, Gaziantep, Turkey

Department of Molecular Medicine, Graduate Education Institute, SANKO University, Gaziantep, Turkey
e-mail: isaygili@sanko.edu.tr; ilker.saygili@sanko.edu.tr

In Adipose tissue, particularly visceral adipose tissue, releases molecules known as adipokines, including adiponectin and resistin, which can influence insulin sensitivity and glucose metabolism. In individuals with obesity, dysregulation of adipokine production can contribute to insulin resistance (Kim et al. 2019). During the development of Type 2 diabetes, proinflammatory cytokines such as TNF-α and IL-6 and stress-induced MAPKs such as JNK and p38 MAPK inhibit the insulin signaling and this inhibition supresses the translocation of glucose transporters to the cellular membrane and impairs the glucose metabolism (Boura-Halfon and Zick 2009). Abnormal lipid metabolism, such as increased circulating free fatty acids and triglycerides, can also contribute to insulin resistance and impaired insulin signaling. Lipid accumulation in non-adipose tissues, such as the liver and muscle, further disrupts insulin action (Sears and Perry 2015).

Gestational diabetes occurs during pregnancy when hormonal changes impair insulin action, leading to high blood glucose levels (Ogurtsova et al. 2017). It usually develops around the 24th to 28th week of pregnancy and affects approximately 2–10% of pregnant women (Egan et al. 2017). The risk of the complications are increased by gestational diabetes for both the mother and the baby during pregnancy. However, with proper management, such as diet and exercise, blood glucose levels can be controlled, reducing the risk of complications (Padayachee and Coombes 2015).

There are other less common forms of diabetes is called as specific types. Latent Autoimmune Diabetes in Adults (LADA) has common features of both Type 1 and Type 2 diabetes. Initially, it may appear as Type 2 diabetes, but over time, individuals with LADA may require insulin therapy (Carlsson 2019). Maturity-Onset Diabetes of the Young (MODY) is a rare form of diabetes that is often hereditary. It typically develops in adolescence or early adulthood and mutations in the genes which are responsible for the insulin production are the main cause of MODY (Ellard et al. 2008).

1.2 Complications of Diabetes

Classical chronic complications such as diabetic kidney disease, retinopathy, peripheral neuropathy, peripheral vascular disease, stroke, and coronary heart disease are frequently seen in patients with diabetes. In addition to these, complications such as infections, functional dysability, cancer, liver diseases, and cognitive disability can be seen as emerging complications (Tomic et al. 2022) (Fig. 5.1).

1.2.1 Diabetic Retinopathy

Diabetes mellitus (DM) and its complications have become a public health issue over the world. Increased permeability of retinal capillaries, edema, loss of the blood-retina barrier and choroidal neovascularization stimulated by hyperglycemia play an important role in the development of diabetic retinopathy. In diabetic

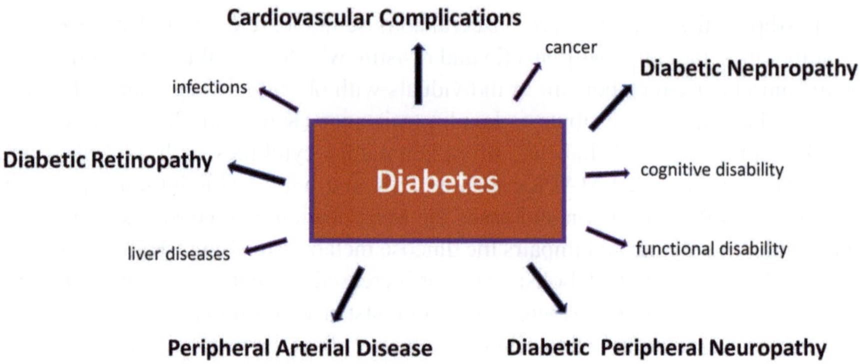

Fig. 5.1 Complications of diabetes

retinopathy patients, these changes manifest themselves clinically as microaneurysms, exudates and cotton wool spots and this stage of the disease is classified as nonproliferative retinopathy. If retinopathy cannot be detected at this stage and the necessary treatment approaches are not applied, the disease may progress to the stage of proliferative retinopathy with the development of neovascularization which manifest itself as a macular edema (Georgiou and Prokopiou 2023). inflammation, cell death, advanced glycosylation endproducts (AGEs), free radicals, polyol and hexamine pathways, activation of protein kinase C (PKC), reactive oxygen species (ROS), growth factors, chemokines, and proangiogenic Vascular Endothelial Growth Factor (VEGF) plays important roles in the pathophysiology of the DR (Saha et al. 2023; Invernizzi et al. 2023). Diabetic retinopathy is an important health problem seen in patients with diabetes, resulting in vision loss and blindness, and is seen in approximately 3–40% of patients with diabetes (Tan and Wong 2022).

1.2.2 Diabetic Nephrophathy

Diabetic nephropathy is a condition seen in patients with diabetes and causes proteinuric and nonproteinuric end stage renal disease and diabetic kidney disease (Pillai and Fulmali 2023). Chronic kidney disease, characterized by a decrease in estimated glomerular filtration rate (eGFR) and/or albuminuria, occurs in approximately 25–40% of diabetic patients. Studies conducted to elucidate the mechanisms involved in the emergence of diabetic nephropathy; polyol pathway, oxidative stress, chronical inflammation, fibrosis, advanced glycation end-products (AGEs), protein kinase C (PKC) pathway, transforming growth factor -β (TGF-β) pathway, renin-angiotensin-aldosterone system (RAAS), SGLT2 cotransporter expression, Janus kinase/signal transducers and activators of transcription (JAK-STAT) pathway and autophagy pathways show that they play an important role in the pathophysiology of diabetic nephropathy (Pelle et al. 2022). Disorders in these pathways cause histological disorders such as glomerular injury, tubulointerstitial inflammation and fibrosis, arteriolar hyalinosis, loss of the glycocalyx and loss of the endothelial surface layer in the kidneys (Tuttle et al. 2022).

1.2.3 Diabetic Peripheral Neuropathy

Diabetes can cause peripheral neuropathies such as DPN, autonomic neuropathy, radiculoplexus neuropathy, radiculopathy, and mononeuropathy, the most common of which is diabetic peripheral neuropathy (Elafros et al. 2022). Diabetic peripheral neuropathy characterized by axonal loss, neural demyelination and neuronal death is a condition that occurs with the destruction of peripheral nerves and causes clinical findings such as pain, loss of sensation, prickling, burning, and electric shock-like sensations in the hands and feet in patients and has an increased risk for foot ulcers and foot amputations (Eid et al. 2023; Pang et al. 2023). Although the incidence of peripheral neuropathy in patients with type 1 diabetes and type 2 diabetes varies, its incidence increases with age and disease duration. It has been shown that processes such as inflammatory process, blood–nerve barrier (BNB) disruption, reactive oxygen species (ROS), oxidative stress, lipid peroxidation, Advanced Glycation End Products, reduction in mitochondrial ATP production, autoimmunity, Polyol pathway, Wnt pathway, Mitogen activated protein kinase (MAPK) pathway and mTOR pathway play a role in the development of diabetic peripheral neuropathy (Galiero et al. 2023; Sanaye and Kavishwar 2023).

1.2.4 Cardiovascular Complications

Diabetes causes cardiovascular complications such as coronary artery disease, cardiac autonomic neuropathy and diabetic cardiomyopathy. These complications are the most important causes of morbidity and mortality in patients with diabetes (Pan et al. 2023). Diabetic cardiomyopathy is seen in approximately 12% of patients with diabetes (Hao and Liu 2023). While left ventricle hypertrophy, increased atrial filling pressure, myocardial stiffness, and left ventricle diastolic dysfunction with preserved ejection fraction (>50%) are observed in patients with early-stage diabetic cardiomyopathy, increased cardiac fibrosis, systolic and diastolic dysfunctions with reduced ejection fraction (<50%) are observed in advanced stages) findings are observed (Abdel Rhman and Owira 2022). Factors such as expansion of epicardial adipose tissue, cardiomyocyte apoptosis and necrosis, microvascular dysfunction, endothelial damage, insulin resistance, dysregulated Ca + methabolism, inflammation, fibrosis, overproduction of extracellular matrix proteins, non-enzymatic formation of AGEs, oxidative stress, Nf-KB signaling, RAAS, aldosterone activity, play an important role in the development of diabetic cardiomyopathy (Phang et al. 2023).

1.2.5 Peripheral Arterial Disease

Peripheral arterial disease is one of the atherosclerotic complications of diabetes and is characterized by narrowing and occlusion of the arteries due to fatty deposits. Peripheral arterial disease increases the risk for myocardial infarction and stroke. Factors such as inflammation, advanced glycation end-products (AGEs), oxidatively modified LDLs, NO signaling, vasoconscrtuction, insulin resistance, and dyslipidemia, thrombosis, oxidative stress, endothelial dysfunction, and mitochondrial dysfunction play an important role in the pathogenesis of peripheral arterial disease (Yadav et al. 2023). Clinical findings such as intermittent claudication, rest pain, ischaemia, ulcer and toe gangrene might be observed in diabetic patients with peripheral arterial disease (Zia Ur and Ram 2023). There are studies showing that the incidence of peripheral arterial disease in individuals with diabetes is between 20% and 50% (Liarakos et al. 2023).

1.2.6 Connection Between Oxidative Stress and Diabetes

It is generally agreed today that cellular oxidative stress mechanisms, such as reduction-oxidation (redox) imbalances in the cell, produce reactive free radicals that can damage cells and tissues and are the most important effects in the differentiation process between disease and homeostasis. Recent studies have explores more complete understanding interaction among oxidative stress and nutraceuticals in Metabolic Syndrome and Diabetes. Oxidative stress interferes with regulatory pathways involved in insulin resistance and T-cell dysfunction, auto-oxidation of glucose, increased the formation of advanced glycation end products (AGEs), which contributes to the etiology of diabetes and metabolic syndrome exacerbates the pathophysiology and consequences. But the precise mechanism is still unknown. Dyslipidemia, inflammation, and hyperglycemia are all associated with T2D and high OS. Excess ROS interacts with the cysteine residues (known as redox sensors) and causes their oxidation, functioning as a second messenger and controlling the function of key proteins such as Kelch-like ECH-associated protein 1 (Keap1), protein kinase C (PKC), and IB kinase (Fatima et al. 2023; Singh et al. 2009). Oxidative stress may occur a pivotal role in the development of complications in diabetes such as lens cataracts, nephropathy, and neuropathy (Bartlett and Eperjesi 2008). Millions of obese and diabetic people are largely due to own change in lifestyle to one of sedentary behavior and enormous industrialization, with access to foods and beverages high in energy, fat, sugar, or a mix of these (Nolan et al. 2011).

With this background, the central question that motivates this paper is to bring perspective to the comparison among nutraceuticals in diabetes and metabolic syndrome. In this chapter, we want to evaluate the molecular and systemic roles of nutraceuticals and how their effects on both oxidative stress and endothelial function could help prevent or delay the onset of vascular-related type 2 diabetes complications.

Endogenous antioxidants and nutraceuticals, which include the enzymes superoxide dismutase (SOD), catalase, and glutathione peroxidase, as well as the non-enzymatic antioxidants glutathione, thioredoxin, and peroxiredoxins, help maintain oxidative status by delaying, inhibiting, or preventing oxidation by scavenging free radicals (Halliwell 2011; Erejuwa et al. 2012).

Due to their antioxidant action, medicinal plants are extremely effective in the treatment of several diseases. All components of medicinal plants are useful in the discovery of new types of drugs and in the treatment of sickness. Due to their phytoconstituents, plants are a possible source of hypoglycemic medications (Rahimi et al. 2005).

Nutraceutials are substances able to slow or inhibit the oxidation of other molecules. Recently, the medical community has concentrated on using antioxidant treatments to treat many illnesses, particularly diabetes. Previous experimental investigations and clinical trials have revealed that antioxidants are effective at reducing complications from diabetes. The treatment approach makes use of synthetic antioxidants, combined drugs, and drugs with antioxidant activity. Generally speaking, medicinal herbs with antioxidant activity are taken into consideration for the treatment of type 2 diabetes (Kelly 1998).

Diets high in bioactive compounds, such as polyphenols may play a role in the prevention and treatment of several diseases thanks to its antioxidant, antiinflammatory, antimicrobial, hypoglycemic, and antineoplastic action. Polyphenols are an important group of bioactive substances that occur naturally and are divided into two major groups: non-flavonoids and flavonoids. All these biological and pharmacological activities are due to the peculiar chemical structure of flavonoid which is capable to have a great number of molecular targets. Regarding flavonoid effects the hypoglycemic one was examined (Gouveia et al. 2022).

Mediterranean diet is rich in polyphenols, which have been suggested as a possible treatment for Metabolic syndrome (Finicelli et al. 2019; Sohrab et al. 2018). Over 8000 different bioactive plant components are classified into two primary categories of polyphenols: non-flavonoids and flavonoids. The flavonoids include flavonols, anthocyanidins, anthocyanins, isoflavones, flavones, flavanols (or cathechins), flavanones, and flavanonols. The nonflavonoids include phenolic acids, stilbenes, and lignans (Lecour and Lamont 2011).

Benzo-pyrone compounds called flavonoids have two phenolic rings and a pyran ring. Variable pharmacological effects are produced by varied arrangements of hydroxyl (-OH), methoxy (-OCH3), and glycosidic side groups and conjugations among the rings A and B. Tannins and other composite species are produced when this nuclear structure is polymerized, aiding its antioxidant effect (Gonzales et al. 2015).

Gouveia et al. suggested that found evidence from the studies it included that flavonoids, when taken alone as supplements, can dramatically alter a number of metabolic parameters and hence lower the risk of disorders connected to the metabolic syndrome. It is pointed that only theaflavin and catechin, out of all the examined flavonoids, were unable to change metabolic parameters; however, only one study used these substances (Gouveia et al. 2022). Catechins effect diabetes via decreasing adipocyte differentiation and proliferation, lipogenesis and lipid absorption, lowering plasma triglyceride, free fatty acid, cholesterol, glucose, insulin and leptin levels, as well as increasing fatty acid β-oxidation and thermogenesis (Kawser Hossain et al. 2016).

The potential therapeutic uses of flavonoids for a variety of illnesses, including diabetes, cardiovascular disease, cancer, and Alzheimer's, have been examined. In the near future, further experimental and clinical research will be required to confirm the true therapeutic value of flavonoids. Results from both in vitro and in vivo studies strongly suggest that flavonoids increase glucose tolerance and improve insulin sensitivity, which may have a positive impact on insulin resistance (Zhou et al. 2023). Flavonoid-rich diets are advised for maintaining beneficial health and wellbeing and preventing diabetes, obesity, cardio-vascular illnesses, and neurodegenerative disorders due to their strong antioxidant and anti-inflammatory qualities. In addition to interacting with the insulin-related signaling system, flavonoids' biological anti-inflammatory and antioxidant properties also contribute to their potential to reduce IR. Numerous scientific studies have demonstrated that flavonoids may play a role in metabolic pathways that relate to glucose metabolism (Fogacci et al. n.d.).

A series of phosphorylation processes are triggered when the ligands for these receptors activate insulin resistance. And by recruiting and activating PI3K, which phosphorylates the serine/threonine residue of protein kinase B (Akt), irs1 promotes the PI3K-Akt pathway. Through AS160, Akt controls how GLUT4 is transported to the cell surface. When AGE binds to its receptor RAGE, a number of signaling cascades, including JNK, NF-B, and PKC activation, are activated, which reduces the effectiveness of the insulin signal (Khalid et al. 2021). Consuming a high-calorie diet may lead to an insulin and glucose imbalance and the emergence of diseases associated with lifestyle, such as diabetes and CVD. However, regular ingestion of polyphenols such as resveratrol may reduce plasma glucose levels linked to improved insulin activity and suppress cancer (Baur et al. 2006).

The capacity of metabolic and vascular stressors to induce cellular oxidative stress and impair mitochondrial function is one common cause of injury in diabetes (Vinik 1999; Russell et al. 2002). This theory has been validated by recent research that has measured the oxidative stress in sensory neurons both in vivo and in vitro as well as the antioxidant protection of the neurons. When 10–20 mm glucose is applied in vitro to dorsal root ganglion neurons, O2 and H2O2 are produced, which results in lipid oxidation and cell death. IGF-I prevents this glucose-induced mortality in part by reducing ROS generation (Russell et al. 1999).

Nonetheless; hyperglycaemia and hyperlipidaemia have been emphasized that to increase ROS production by oxidative stress, providing to the development of insulin resistance, altered endothelial dysfunction and energy metabolism, and the appearance of vascular complications (Oguntibeju 2019; Okon et al. 2007). An imbalance between antioxidant defense systems such as CAT and SOD enzymes and the production of ROS contributes to the probability of oxidative stress formation in diabetes mellitus and metabolic syndrome. In this sense, in various oxidative stress-related disorders, an increase in ROS production and/or a reduced ability to remove the excess ROS are connected with a decrease in plasma antioxidant enzyme activity, resulting in oxidative damage (Rizzo et al. 2012). The recommended mechanisms leading to insulin resistance include the accumulation of certain lipid mediators, abnormal features of mitochondrial function, an increase in stress-activated protein c-Jun-N-terminal-kinase (JNK), and inflammatory pathways (Masenga et al. 2023). During diabetes, the hexose monophosphate (HMP)/glycolysis pathways are redirected to provide energy where the amount of glucose entering the cells is minimal, generating oxidative stress (Abdel-Raheem et al. 2022). Oxidative stress is considered crosstalk between these mechanisms is associated to imbalance in glucose homeostasis and can lead to the diabetes-associated insulin-resistance status plays a vital role in the onset and change of this condition (Ashish et al. 2020).

2 Phytochemicals and Diabetes

2.1 Properties of Phytochemicals

Phytochemical is a combination of the Greek words ‘phyto’ meaning plant and ‘chemical’. Phytochemicals; They are non-nutrient bioactive components that are also involved in reducing the risk of chronic diseases, beyond providing the desired health benefits with nutrition (Liu 2004). These compounds are non-nutritive substances produced by plants through primary and secondary metabolism for survival. Phytochemicals that give color, odor and taste properties to plants; it also protects plants from herbivores, insects, etc. provides protection (Molyneux et al. 2007).

Phytochemicals with nutraceutical properties found in food. It has been shown to be of great importance due to its beneficial effects on human health (Thakur et al. 2020). It has been noted that dietary phytochemicals differ greatly in the composition of various fruits, vegetables, and grains and often have complementary mechanisms. Therefore, it is recommended to diversify the phytochemical sources throughout the day to obtain the highest health benefits (Bellik et al. 2012).

Isolated from fruit, vegetables, cereals; There are more than 5000 identified phytochemicals and much of it still remains a mystery (Liu 2003).

For thousands of years, plants have been used as a primary source of medicine (Veeresham 2012). Studies conducted in recent years; It shows that phytochemicals have positive effects on health by reducing the risk of chronic diseases by showing effects such as antioxidant, anti-inflammatory, antimicrobial, anticarcinogenic, immunomodulatory, cardioprotective (Demir and Akpınar 2020; Bahadoran et al. 2015).

In general terms; Phytochemicals are thought to have the following effects on health:

- Substrate for biochemical reactions
- Cofactors of enzymatic reactions
- Inhibitors of enzymatic reactions

- Absorbents/chelants that bind to and remove unwanted components in the gut
- Ligands that agonize or antagonize cell surface or intracellular receptors
- Cleaners of reactive or toxic chemicals
- Compounds that increase the absorption and/or stability of essential nutrients
- Selective growth factors for beneficial stomach and intestinal bacteria
- Fermentation substrates for beneficial mouth, stomach or intestinal bacteria
- Selective inhibitors of harmful intestinal bacteria (Dillard and German 2000).

Drugs of phytochemical origin show multiple therapeutic effects by inhibiting enzymes and hormones that activate many diseases, stimulating the DNA repair mechanism, promoting the production of protective enzymes, inducing antioxidant effect and strengthening immunity. The advantage of using medicinal plants for treatment over synthetic drugs is their less side effects, high potency, less patient toxicity, affordability and environmental friendliness. Secondary metabolites have anti-inflammatory, anti-cancer, antidiabetic, anti-obesity and immune-enhancing properties. With these features, it has become inevitable for pharmaceutical technologies to use many active phytochemicals directly or by modifying them in the development of active drug candidates (Pereira et al. 2012).

2.2 *Types of Phytochemicals and Food Sources*

Phytochemicals, which are bioactive compounds and have potential health benefits, are categorized according to the differences in their structure and functions (Figs. 5.2, 5.3, 5.4, and 5.5) (Cojocneanu Petric et al. 2015). In this subsection, the structures, classification and information on which foods contain phytochemicals are presented.

2.2.1 Polyphenols

Structurally all polyphenols contain more than one phenol unit and a varying number of hydroxyl groups, while most polyphenols have at least one carbohydrate (glycoside) attached to hydroxyl groups (Panickar 2013).

Polyphenols are divided into different groups according to the number and structure of phenol rings, as well as the number of structural elements attached to the polyphenols, and foods usually contain polyphenols in different chemical forms as complex mixtures (Lewandowska et al. 2016). Simple phenolic acids are basically classified into two groups as hydroxybenzoic acid and hydroxycinnamic acid. While hydroxybenzoic acids (p-hydroxybenzoic acid, protocatechuic acid, gallic acid) are not of high nutritional interest because they are found in a few edible plants, the other group (p-coumaric, caffeic and ferulic acids) is more common in foods (Gonzalez-Castejon and Rodriguez-Casado 2011).

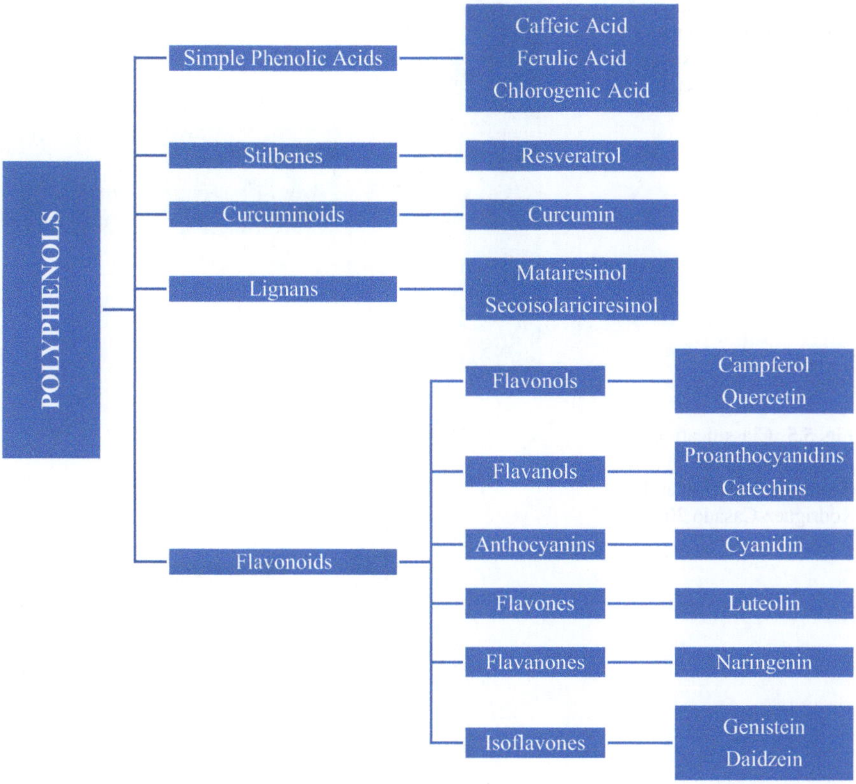

Fig. 5.2 Classification of polyphenols in the diet. (Gonzalez-Castejon and Rodriguez-Casado 2011)

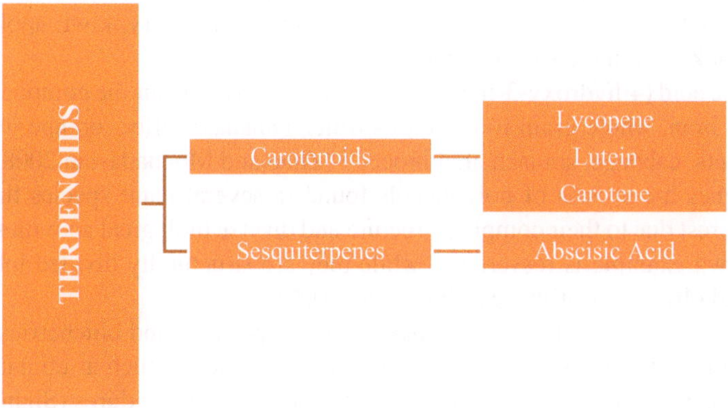

Fig. 5.3 Classification of terpenoids in the diet. (Gonzalez-Castejon and Rodriguez-Casado 2011)

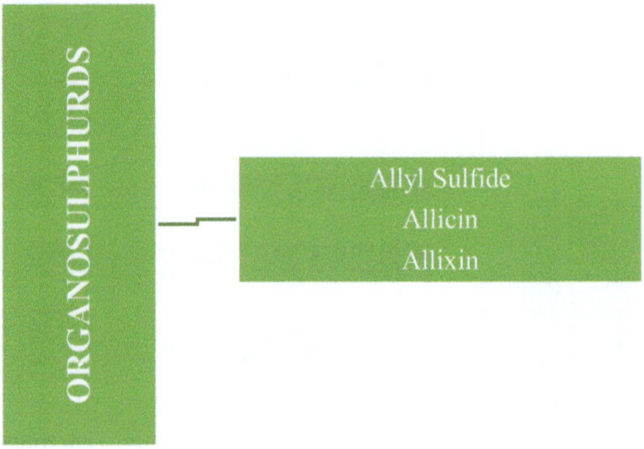

Fig. 5.4 Classification of organosulphurds in the diet. (Gonzalez-Castejon and Rodriguez-Casado 2011)

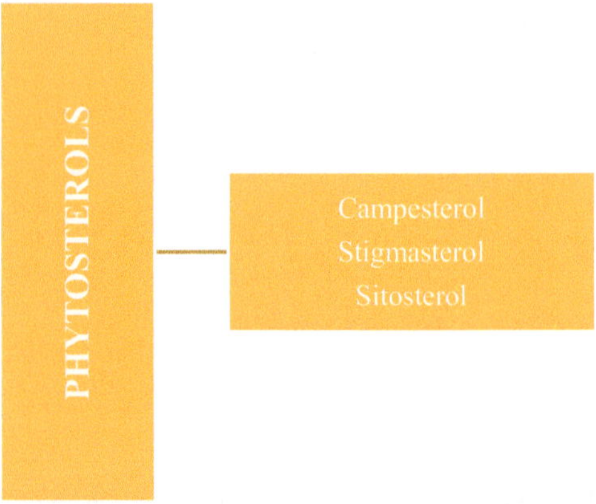

Fig. 5.5 Classification of phytosterols in the diet. (Gonzalez-Castejon and Rodriguez-Casado 2011)

Phenolic acids such as caffeic acid, ferulic acid and coumaric acid are found in high amounts in coffee, as well as in fruits such as blueberry, kiwi, apple, strawberry, blackberry and cherry (Manach et al. 2004).

Ferulic acid (4-hydroxy-3-methoxycinnamic acid) is a phenolic compound found in cereal bran, whole grain foods, citrus fruits, bananas, coffee, orange juice, eggplant, beets, cabbage, spinach, and broccoli (Zhao and Moghadasian 2008).

Stilbenes are a class of polyphenols found in several plant species that are of great interest due to their complex structure and diverse biological activities, and the best known example is resveratrol, while they are structurally divided into monomeric and oligomeric stilbenes (Shen et al. 2009).

While resveratrol is found in grapes, red wine, peanuts, and blueberries, its content in different sources varies widely depending on factors such as climate, fungal infections, exposure to ultraviolet rays, and winemaking procedures (Signorelli and Ghidoni 2005).

Curcuminoids are known as non-toxic yellow pigments naturally found in Curcuma (turmeric) and Zingiber (ginger) species (Aggarwal 2010). Curcumin is one of hundreds of compounds isolated from the spice turmeric and is currently being studied extensively for its potential therapeutic efficacy in the treatment of metabolic disorders associated with many diseases.

Lignans are a group of phytochemicals consisting of two phenylpropane units, found in plants in glycoside form. The richest dietary sources of lignans include flaxseed, secoisolariciresinol (>3.7 g/kg dry weight) and matairesinol, while this amount is up to 1000 times higher than the dietary sources of the same lignans found in other grains, some fruits and vegetables (Adlercreutz 2007). The lignans identified in rye as more diverse than flaxseed are pinoresinol, lariciresinol, isolariciresinol, and syringaresinol (Bondia-Pons et al. 2009). When the lignans in the structure of the plant are consumed, they are converted into enterolactone and enterodiol, two mammalian lignans with various biological activities by the intestinal microflora (Adlercreutz 2007).

Flavonoids are a large family that includes more than ten thousands of different phytochemicals with a common chemical structure, mainly consisting of 15 carbon skeletons containing two phenyl rings and a heterocyclic ring, defined in plants, especially fruits and vegetables (Kozlowska and Szostak-Wegierek 2014).

The first database for the flavonoid content of foods was established by the United States Department of Agriculture (USDA) in 2003, and today the content of 30 different flavonoids in approximately 2900 foods is obtained from this database (Bhagwat et al. 2013). Flavonoids are divided into different subclasses according to their chemical structure (see Fig. 5.2).

Flavonols are the most common flavonoids found in foods, and their main representatives are quercetin and campferol. It is usually found in relatively low concentrations of 15–30 mg/kg fresh weight, while its richest sources are onions, kale, leeks, broccoli, and blueberries (Manach et al. 2004).

Flavanols are found in various teas, fruits, cocoa and chocolates in monomer form (catechins) and polymer form (proanthocyanidins). The most common flavanols in fruit and cocoa are catechin and epicatechin; flavanols epicatechingallate (ECG), gallocatechin, epigallocatechin (EGC), and epigallocatechin gallate (EGCG) found in grapes, teas, and seeds of some leguminous plants (Kawser Hossain et al. 2016).

Catechins are the main polyphenols in green tea and make up about 35% of the dry weight (Manach et al. 2004). The anti-obesity potential of green tea catechins, particularly EGCG, has been demonstrated in cell culture, animal, and human studies.

Anthocyanins are compounds that dissolve in the epidermal cells of flowers and fruits and are responsible for the pink, red, blue or purple color and dark fruits such as strawberries, cherries, hawthorn, peaches, grapes, apples and plums, as well as red onions, red radishes, black beans, eggplant and red found in cabbage (Manach et al. 2004). Cyanidin has been shown to have the highest antioxidant activity (Wang et al. 1997).

Flavones are mostly found in glycoside form in fruits and vegetables, and the most important edible sources for which they have been analyzed are parsley and celery (Manach et al. 2004). Luteolin and apigenin are the best-known flavones found in many edible plants such as carrots, peppers, celery, parsley and spinach (Lopez-Lazaro 2009).

Since flavanones have a highly reactive structure, they are compounds that are exposed to hydroxylation, glycosylation and O-methylation reactions in plants. Edible plants such as tomatoes and mint are sources, and most importantly citrus fruits, which contain high concentrations of flavanones. It exists in aglycon form, including naringenin in grapefruit, hesperetin in orange, and eroodictiol in lemons (Manach et al. 2004; Tanwar and Modgil 2012).

Isoflavones are flavonoids that are structurally similar to estrogens, although they are not steroids, they have a hydroxyl group at 7 and 4′ positions similar to the estradiol molecule and have pseudohormonal properties such as binding to estrogen receptors. Isoflavones are mostly found in legumes, while soy and soy products are the main source in the human diet. Genistein, daidzein, and glycine are the most well-known isoflavones and exist in four forms: aglycone, 7-O-glycoside, 6'-O-acetyl-7-O-glycoside, and 6'-O-malonyl-7-O-glycoside (Manach et al. 2004).

Capsaicin is one of the naturally occurring capsaicinoids in capsicum and is responsible for the bitter taste. Since the worldwide consumption of spicy foods, especially hot red pepper, is quite common, studies to determine the metabolic effects of capsaicin and its derivatives are quite abundant (Reyes-Escogido Mde et al. 2011).

2.2.2 Terpenoids

Terpenoids (isoprenoids) are a large family representing more than 4000 primary and secondary metabolic compounds commonly found in plants. All terpenoids have an isoprene unit in their structure and are divided into different groups according to the number of carbon atoms.

Carotenoids are fat-soluble tetraterpene pigments responsible for the red, orange, yellow and green color produced by plants and some photosynthetic microorganisms. Although the main dietary sources are fruits and vegetables, they are also found in bread, eggs and oils. While 600 carotenoids have been identified in nature, only 50 are found in the human diet, and β carotene, α carotene, β cryptoxanthin, lycopene, lutein, and zeaxanthin have been identified in serum. Carotenoids, are divided into two as hydrocarbons (carotenes) and their oxygenated derivatives (xanthophylls); α carotene, β carotene, and β cryptoxanthin are known to be precursors for vitamin A biosynthesis (Eggersdorfer and Wyss 2018).

Lycopene is a carotenoid mostly found in fruits such as tomatoes, capsicum, watermelon and pink grapefruit, responsible for the red color whose bioavailability is increased by heat treatment. It is known that it exhibits various mechanisms of action, including inhibition of LDL cholesterol oxidation and lipid peroxidation, with its strong antioxidant activity (Costa-Rodrigues et al. 2018).

2.2.3 Organosulfides

Organosulfur compounds are bioactive compounds found in abundance in Allium plants such as garlic, onions, and leeks, which attract attention with their health benefits as well as giving the distinctive flavor and aroma of vegetables. The best known organosulfur compounds are allicin, allixin and allyl sulfides (Sahu 2002).

2.2.4 Phytosterols

Phytosterols are natural food components that are structurally similar to cholesterol. These phytochemicals also include phytostanols, which are formed by the natural hydrogenation of double bonds in phytosterols commonly found in nature, that is, they do not contain double bonds in the sterane ring. While it can be found in esterified and free alcohol forms, food sources are known to be unrefined vegetable oils, seeds, cereals, nuts and legumes. The most studied phytosterols are campesterol, brasicasterol, stigmasterol and sitosterol.

2.3 Phytochemıcal Index

Phytochemical index; it was introduced in 2002 by Mark F. McCarty with the proposal to determine the phytochemical content of the diet via an index. While suggesting this, McCarty stated that the concept of 'glycemic load', which characterizes another feature of the diet, has been used in epidemiological studies in recent years, and similarly, FI can help the researchers to characterize the phytochemical feature of the diet and to associate this index with a healthy one. He also stated that it can be used by nutritionists as a tool to determine the phytochemical content of individuals' diets and to measure their development (McCarty 2004).

While calculating the phytochemical index, fruits with high phytochemical content, all vegetables except potatoes, legumes, whole grains and foods containing whole grains, and oilseeds are taken as a basis. McCarty stated that although the phytochemical content of fruit and vegetable juices does not contain fiber or the phytochemical content is reduced by taking the crust, it is still rich in phytochemicals and that he can participate in the calculation. Similarly, he added that wine, cider and beer would also be used in the calculation, but hard drinks should not be included. He stated that soy protein should be added as it is a source of isoflavones, extra virgin olive oil can also be taken, but other cooking oils should not be taken because their phytochemical content per calorie is not rich enough. McCarty stated that FI has some shortcomings. Because the calculation is mainly based on calorie content, foods such as green tea, black tea, which have no calorie content but rich in phytochemicals, cannot be used in the index. At the same time, he stated that since phytochemicals in foods have different effects on health, they cannot be differentiated (McCarty 2004) The phytochemical index has been evaluated by very few studies. The Glucose-Lipid Study in Iran often used the phytochemical index as a parameter (Bahadoran et al. 2013a, b). However, since McCarty's FI calculation method could not include phytochemical rich foods such as spices and green tea in the calculation, the FI was calculated as the daily consumption amount of phytochemical rich foods in grams instead of the calorie amount/daily calorie amount from foods rich in phytochemical content. The total daily consumption was determined as the amount of food [FI = (foods rich in phytochemicals g/day/total food intake g/day) × 100]. Thus, foods that could not be included in the group could be taken. Bahadoran et al. while evaluating FI, the energy-adjusted phytochemical index, which determines the FI per 1000 calories for each individual, was calculated and evaluated according to the energy-adjusted FI. [(1000 kcal × FI)/daily calorie intake of the individual] (Bahadoran et al. 2013b).

2.4 Relation of Phytochemicals to Diabetes

A plethora of compounds can be classified as nutraceutical-phytochemicals in relation to the prevention and treatment of diabetes. In this section, diabetes and nutraceuticals will be discussed within the context of the phytochemicals summarised in the previous section.

2.4.1 Polyphenols

As previously stated, polyphenols are categorised into two main groups: flavonoids and non-flavonoids. Although flavonoids represent a large family, they share a common structural unit, the flavanic core. In contrast, the structures of non-flavonoids are heterogeneous and vary considerably, resulting in the existence of numerous subgroups (see Sect. 3). We will now examine the molecular mechanisms and clinical effects of these molecules on diabetes (Fig. 5.6).

Caffeic acid is a compound that has been shown to have a protective effect on diabetic complications. It was found to be protective in cardiac tissue of diabetic mice (Chao et al. 2009). Furthermore, in a human study, it was shown to improve wound healing via mutiple pathways as Nf-kB and NOS (Romana-Souza et al. 2018). In a rat model of diabetic retinopathy, caffeic acid was found to decrease oxidative stress in neurons and to increase neuronal survival (Fathalipour et al. 2019). In nephropathy, caffeic acid exerted a protective effect on the kidneys through the modulation of autophagy and microRNA levels. (Matboli et al. 2017; Salem et al. 2019).

In addition to caffeic acid, chlorogenic acid and ferulic acid, other phenolic acids have been shown to regulate molecular targets in diabetes, with beneficial effects on parameters such as B-cell protection, blood glucose level, hyperlipidaemia and diabetic complications (Li et al. 2022; Yan et al. 2020).

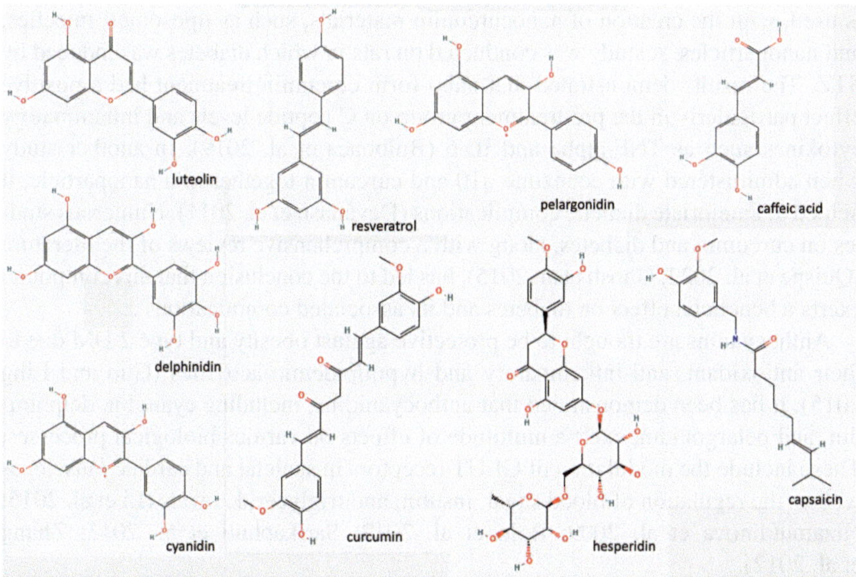

Fig. 5.6 The two-dimensional biochemical structures of polyphenols that have been demonstrated to be effective in the treatment of Type 2 DM. Obtained from PubChem (Luteolin 2024; Resveratrol 2024; Pelargonidin 2024; Caffeic 2024; Delphinidin 2024; Cyanidin 2024; Curcumin, 2024; Hesperidin 2024; Capsaicin, 2024)

As a highly efficacious compound, numerous studies have been conducted at the level of clinical trials on diabetes, investigating the potential of resveratrol as a therapeutic agent. Resveratrol has been demonstrated to possess protective properties against end-organ damage and to reduce complications through pleotropic effects on multiple tissues (Szkudelski and Szkudelska 2015) such as SIRT1 activation (Milne et al. 2007), attenuation of ROS related damage in the heart, kidneys and liver (Mohammadshahi et al. 2014; Schmatz et al. 2012). To summarize, the beneficial effects of resveratrol extend beyond glucose-related pathways, such as insulin sensitivity, GLUT4 translocation, and carbohydrate-related enzymes. Resveratrol also exerts anti-inflammatory (e.g., IL-6 and TNF-alpha regulation) and anti-oxidative effects. These properties make resveratrol one of the most beneficial molecules for the treatment and prevention of diabetes, including end-organ damage (Ozturk et al. 2017).

Curcumin is one of the most extensively researched biological compounds in the context of diabetes. Curcumin regulates energy metabolism, it reduces the level of intracellular lipids, suppresses angiogenesis necessary for growth in adipose tissue, regulates transcription factors that play a key role in adipogenesis and lipogenesis (Mohamed et al. 2014). Due to its low solubility and absorption, its bioavailability is limited at normal intake. In traditional use, it is taken by heating with black pepper and olive oil, which has been shown to significantly enhance its bioavailability. In order to achieve high bioavailability at the pharmaceutical level, nanotechnology is used, with the creation of nanocurcumin materials, such as liposomes, micelles, and nanoparticles. A study was conducted on rats in which diabetes was induced by STZ. The results demonstrated that nano-form curcumin treatment had a positive effect particularly in the pre-treatment group on C peptide levels and inflammatory cytokines such as TNF-alpha and IL-6 (Bulboaca et al. 2019). In another study when administered with coenzime q10 and curcumin together in a nanoparticle, it helped to ameloriate diabetic complications (Devadasu et al. 2011). Numerous studies on curcumin and diabetes, along with a comprehensive reviews of the literature (Quispe et al. 2022; Ghosh et al. 2015), has led to the conclusion that this compound exerts a beneficial effect on diabetes and its associated complications .

Anthocyanins are thought to be protective against obesity and type 2 DM due to their antioxidant, anti-inflammatory and hypolipidemic activities (Guo and Ling 2015). It has been demonstrated that anthocyanidins, including cyanidin, delphinidin, and pelargonidin, exert a multitude of effects on various biological processes. These include the modulation of GLUT receptors in skeletal and cardiac muscle, as well as the regulation of blood sugar, insulin, and triglyceride levels (Li et al. 2015; Nizamutdinova et al. 2009; Rojo et al. 2012; Sarikaphuti et al. 2013; Zhang et al. 2012).

Cyanidin and its derivatives may have a protective effect on diabetes through multiple molecular targets. Cyanidin can increase AMPKα phosphorilation (Tsuda et al. 2004). Cyanidin-3-rutinoside have ability to inhibit α-glucosidase (Adisakwattana et al. 2011). Cyanidin glycosides were found to be highly effective pancreatic α-amylase inhibitors, they can additionally block intestinal sucrose(Akkarachiyasit et al. 2010).

Flavone, flavanone, and flavonol derivatives have been directly linked to the protection against secondary neuronal changes in diabetic rats. In a systematic review, is reported flavonols have an effect on, Type 2 DM and obesity, can affect many biochemical signaling pathways, physiological and pathological processes, show anti-inflammatory, antioxidant and anti-proliferative properties, and improve endothelial function (Rangel-Huerta et al. 2015). Citrus flavonoids, a phytochemical in this subgroup, has been reported to have anti-diabetic, lipid-lowering, anti-atherogenic activities (Kawser Hossain et al. 2016). Luteolin has been demonstrated to improve cognitive decline (Liu et al. 2013), hesperidin has been shown to exert antidepressant effects (El-Marasy et al. 2014), an extract of green tea has been found to enhance spatio-temporal learning (Sharifzadeh et al. 2017), fisetin has been observed to protect peripheral nerves through the modulation of the NF-kB pathway (Sandireddy et al. 2016).

Some of the flavonol derivatives are more effective in regulating metabolism directly. High-cocoa polyphenol-rich chocolate has been tested directly in humans. In this study; HbA1C levels and trigliseryte levels decreased in addition to a decrease in blood pressure (Rostami et al. 2015). It was observed that another flavonol, quercetin, protected B-islets from damage via a reduction oxidative stress (Coskun et al. 2005).

Isoflavones are a family known with their strong radical-scavenging and antioxidative effects (Umeno et al. 2016). A meta-analysis revealed that isoflavones modulate the lipid profile and prevent cardiovascular effects (Chi et al. 2016). However, there was no discernible impact on the glucose profile, as evidenced by the HbA1c, insulin and glucose levels. (Baranska et al. 2021; Ye et al. 2015; Liu et al. 2010). Before reaching a definitive conclusion, it may be useful to keep in mind the possibility that factors related to dose and duration of use come to the fore in the meta-analysis and that individual variation may be missed in the response to nutraceuticals. Furthermore, clinical and *in vitro* studies provide evidence for the effect of isoflavones on diabetes. In a study involving approximately 600 individuals with type 2 diabetes and 600 controls, Nyugen et al. demonstrated that isoflavone intake was associated with a significantly lowered risk of developing type 2 diabetes (Nguyen et al. 2017). There are molecular evidence in mice studies for increased insulin sensivity, GLUT4 translocation, decreased IL-6 and TNF-alpha levels, increased glucose uptake and increased B-cell survival (Fu et al. 2012; Cao et al. 2013; Cheong et al. 2014). Depending on the estrogenic and antiestrogenic activities of isoflavones, it has been reported that they may be protective against obesity (Wang and Mazza 2002).

Capsaicin, when used topically at a concentration of 0.075% has been demonstrated to be an efficacious treatment for diabetic peripheral neuropathic pain (Care 1992).

2.4.2 Terpenoids

Caretinoids are natural antioxidant compounds. Although no definite conclusions can be drawn on its effects on diabetes, caretinoids were used both for the prevention and treatment of metabolic syndrome/diabetes. It has been reported that daily consumption of some terpenoids may be beneficial in the treatment of obesity-related metabolic disorders such as Type 2 DM, hyperlipidemia, insulin resistance, and metabolic syndrome (Goto et al. 2010). Dietary amounts and the caretionoid levels in plasma had been correlated with insulin resistance and plasma glucose concentrations. A study was conducted on 100 patients and 150 healthy controls, with the objective of examining serum carotenoids. The results demonstrated that retinol, α-carotene, and α-tocopherol exhibited similar levels, while β-carotene was significantly higher in the healthy group and retinol binding protein (RBP) was significantly higher in diabetics. (Abahusain et al. 1999). Furthermore, a reverse correlation was observed between fasting blood glucose and serum β-carotene when the entire group was examined (Abahusain et al. 1999). In a separate study, dietary intake of carotenoids was found to be associated with glucose metabolism. An inverse relationship between plasma glucose concentration and plasma α-tocopherol levels was observed (Ylonen et al. 2003). Plasma α-tocopherol levels were also found to be directly associated with 2 h plasma glucose levels (Ylonen et al. 2003). Additionally, plasma β-carotene levels exhibited an inverse direction with insulin resistance (Suzuki et al. 2002). In a single study, all serum carotenoids, including α-carotene, β-carotene, cryptoxanthin, lutein/zeaxanthin, and lycopene, demonstrated a linear inverse correlation with fasting serum insulin following the adjustment of multiple dependent variables (Ford et al. 1999).

Astaxanthin, diminished serine phosphorilation levels of IRS and its downstream PI3K-Akt pathway. It additionally modulated ezymes related to glucose metabolism. The aforementioned alterations have been demonstrated to enhance insulin sensitivity in comparison to mice that have not undergone treatment (Bhuvaneswari and Anuradha 2012). The glucose transporter 4 (GLUT4) was observed to be translocated to skeletal muscle (Arunkumar et al. 2012). In comparison to other carotenoids, astaxanthin is the most prominent due to its antioxidant properties (Naguib 2000). It has been demonstrated that astaxanthin can affect not all, but part of the ROS and inflammation-related molecules. COX2, ICAM-1 and iNOS were the downregulated cytokines, while NF-kB demonstrated no change in the level of expression (Park et al. 2015). Conversely, a different group of researchers observed a reduction in NF-kB (Bhuvaneswari et al. 2014). The previous group, in which diabetes-induced hepatotoxicity was prevented by astaxanthin and Corni Fructus, demonstrated a reduction in AGEs. Additionally, astaxanthin demonstrated no effect on hepatic and serum glucose levels, while Corni Fructus exhibited a reduction in both (Park et al. 2015). Another outstanding feature of astaxantin is its protective effect on diabetic nepropathy. Manabe and colleagues demonstrated that astaxanthin could prevent the production of ROS and pro-inflammatory cytokines that are associated with diabetes (Manabe et al. 2008). The administration of astaxanthin to rodents resulted in a reduction in the levels of 8-hydroxydeoxyguanosine, thereby preventing the development of kidney damage (Naito et al. 2004). The combination of astaxanthin with different antioxidant groups can result in divergent effects. For instance, the combination with α-tocopherol led to a reduction in oxidative stress in rats, while the combination with ascorbic acid led to an increase in oxidative stress. (Nakano et al. 2008).

Bixin was found to induce the activation of PPARγ and its target genes in adipocytes (Takahashi et al. 2009). In mice with a high-fat diet and obesity, the oral administration of bixin demonstrated hepatoprotective and hypoglycaemic effects (Pinzon-Garcia et al. 2018). The binding activity of bixin was found to be PPARγ-specific, as PPARα and PPARδ did not exhibit significant interaction. As an interesting data, despite bixin leading to hypoglycaemia, glucose levels were observed to be high in the latter phases, similar to those observed in metformin-treated animals. The destruction of Langerhans cells was also observed. (Keita et al. 2021). For this reason, it is important to assess the long-term side effects of the aforementioned treatment.

Crocin, the sole water-soluble carotenoid, was found to be effective in diabetic rats. Crocin demonstrated a significant effect on serum glucose, glycation end products, triglycerides, total cholesterol, LDL, HDL, HbA1c, microalbuminuria, and insulin resistance (Shirali et al. 2013). Furthermore, Crocin has the potential to protect renal tissue through its antioxidative effect (Rajaei et al. 2013).

Crocetin, via the insulin receptor substrate (IRS) pathway, has been demonstrated to block insulin resistance induced by palmitate in adipocytes. (Yang et al. 2010).

Lycopene has been demonstrated to regulate enzymes involved in the processing of ROS. In rats, administration of lycopene has been shown to increase the activity of catalase, superoxide dismutase and glutathione peroxidase enzymes, resulting in a reduction in lipid peroxidation of the serum compartment (Ali and Agha 2009). Another study depicted similar change in plasma while tissue lipid peroxidation remain unaltered (Duzguner et al. 2008). In rats with diabetes mellitus treated with lycopene, biochemical kidney parameters, including urea and creatinine, were found to decrease. This suggests that lycopene may play a protective role in the development of nephropathy (Li et al. 2014).

The research demonstrated that fucoxanthin and its converted form, fucoxanthinol, suppressed PPARγ and blocked adipocyte differentiation (Maeda et al. 2006). Fucoxanthin can promote expression of GLUT4 mRNA in skeletal muscles in mice. This compound also lead observable clinical improvements in body weight, in addition to hyperglycemia, hyperinsulinemia and hyperleptinemia (Maeda et al. 2009).

2.4.3 Organosulphides

The administration of allyl methyl sulfide has been demonstrated to be highly effective in the treatment of diabetes in rats subjected to streptozotocin (STZ) treatment. This compound has been shown to regulate blood glucose levels and to reduce the production of inflammatory cytokines, including TNF-alpha, IL-6 and NF-kB. Furthermore, it has been observed to improve liver function, particularly with a decline in oxidative stress (Sujithra et al. 2018). A recent study tested the in silico drug properties of organosulphides and one of the aniline compounds, which was found to be a drug-like ligand with no toxicity and to be more active than metformin, an antidiabetic agent (Rajalakshmi et al. 2021). Garlic as being an organosuphur containing food, suggested as a potential therapeutic option in diabetic retinopathy. Different types of garlic preparetes as extract, raw extract, aged extract, garlic oil, or combination with metformin have been applied on diabetes *in vitro* and *in vivo*. The results demonstrated that these substances elevated insulin levels and insulin sensitivity,and reduced blood glucose levels (Sanie-Jahromi et al. 2023).

2.4.4 Phytosterols

This group of compounds is related to glucose, lipid metabolism and insulin resistance. It has been reported that these compounds mechanically compete with cholesterol for the formation of micelle in the intestinal lumen, thus reducing cholesterol absorption and lowering serum total and LDL cholesterol levels, and may be protective against atherosclerosis (Marangoni and Poli 2010). In a specific study, it was revealed that phytosterols reduced LDL by an average of 13% and offered an alternative to statins (Malinowski and Gehret 2010). In contrast, a separate study of mice demonstrated that phytosterol was not effective in improving metabolic parameters, including insulin resistance, plasma lipid and glucose levels (Calpe-Berdiel et al. 2008). However, majority of the studies present data indicating that it may be effective in the treatment of diabetes.

3 Conclusion and Future Perspectives

There are various groups of phytochemicals, including flavonoids, polyphenol compounds and carotenoids, which are present in foods as nutraceuticals. Together with key mechanisms as antioxidant effect, reduction of mitochondrial damage, and additional mechanisms as vascular regeneration, anti-inflammatory effects; nutraceuticals can be used in diabetes for the (i) prevention of the disease, (ii) prevention of the complications (iii) treatment of the disease. Being natural compounds, nutraceuticals are recommended for use in diabetes with their potent properties in addition to lower side effects.

References

Abahusain MA, Wright J, Dickerson JW, de Vol EB. Retinol, alpha-tocopherol and carotenoids in diabetes. Eur J Clin Nutr. 1999;53(8):630–5.

Abdel Rhman M, Owira P. The role of microRNAs in the pathophysiology, diagnosis, and treatment of diabetic cardiomyopathy. J Pharm Pharmacol. 2022;74(12):1663–76.

Abdel-Raheem A, Hamed HI, Fahim E-S, Mohamed ASJ. Oxidative stress markers as early predictors of diabetes complications in Type 2 diabetic patients. IJoPP. 2022;66(2):111–9.

Adisakwattana S, Yibchok-Anun S, Charoenlertkul P, Wongsasiripat N. Cyanidin-3-rutinoside alleviates postprandial hyperglycemia and its synergism with acarbose by inhibition of intestinal alpha-glucosidase. J Clin Biochem Nutr. 2011;49(1):36–41.

Adlercreutz H. Lignans and human health. Crit Rev Clin Lab Sci. 2007;44(5–6):483–525.

Aggarwal BB. Targeting inflammation-induced obesity and metabolic diseases by curcumin and other nutraceuticals. Annu Rev Nutr. 2010;30:173–99.

Akkarachiyasit S, Charoenlertkul P, Yibchok-Anun S, Adisakwattana S. Inhibitory activities of cyanidin and its glycosides and synergistic effect with acarbose against intestinal alpha-glucosidase and pancreatic alpha-amylase. Int J Mol Sci. 2010;11(9):3387–96.

Ali MM, Agha FG. Amelioration of streptozotocin-induced diabetes mellitus, oxidative stress and dyslipidemia in rats by tomato extract lycopene. Scand J Clin Lab Invest. 2009;69(3):371–9.

American Diabetes. A. 2. Classification and diagnosis of diabetes: standards of medical care in diabetes-2021. Diabetes Care. 2021;44(Suppl 1):S15–33.

Arunkumar E, Bhuvaneswari S, Anuradha CV. An intervention study in obese mice with astaxanthin, a marine carotenoid—effects on insulin signaling and pro-inflammatory cytokines. Food Funct. 2012;3(2):120–6.

Ashish A, Shah A, Pandey SJ. Interaction between oxidative stress and diabetes: a mini-review. JDMDC. 2020;7(2):58–61.

Bahadoran Z, Karimi Z, Houshiar-rad A, Mirzayi HR, Rashidkhani B. Dietary phytochemical index and the risk of breast cancer: a case control study in a population of Iranian women. Asian Pac J Cancer Prev. 2013a;14(5):2747–51.

Bahadoran Z, Golzarand M, Mirmiran P, Saadati N, Azizi F. The association of dietary phytochemical index and cardiometabolic risk factors in adults: Tehran lipid and glucose study. J Hum Nutr Diet. 2013b;26(Suppl 1):145–53.

Bahadoran Z, Mirmiran P, Tohidi M, Azizi F. Dietary phytochemical index and the risk of insulin resistance and beta-cell dysfunction: a prospective approach in Tehran lipid and glucose study. Int J Food Sci Nutr. 2015;66(8):950–5.

Baranska A, Blaszczuk A, Polz-Dacewicz M, Kanadys W, Malm M, Janiszewska M, et al. Effects of soy isoflavones on glycemic control and lipid profile in patients with type 2 diabetes: a systematic review and meta-analysis of randomized controlled trials. Nutrients. 2021;13(6)

Bartlett HE, Eperjesi F. Nutritional supplementation for type 2 diabetes: a systematic review. Ophthalmic Physiol Opt. 2008;28(6):503–23.

Baur JA, Pearson KJ, Price NL, Jamieson HA, Lerin C, Kalra A, et al. Resveratrol improves health and survival of mice on a high-calorie diet. Nature. 2006;444(7117):337–42.

Bellik Y, Boukraa L, Alzahrani HA, Bakhotmah BA, Abdellah F, Hammoudi SM, et al. Molecular mechanism underlying anti-inflammatory and anti-allergic activities of phytochemicals: an update. Molecules. 2012;18(1):322–53.

Bhagwat S, Haytowitz DB, Wasswa-Kintu SI, Holden JM. USDA develops a database for flavonoids to assess dietary intakes. JPFS. 2013;2:81–6.

Bhuvaneswari S, Anuradha CV. Astaxanthin prevents loss of insulin signaling and improves glucose metabolism in liver of insulin resistant mice. Can J Physiol Pharmacol. 2012;90(11):1544–52.

Bhuvaneswari S, Yogalakshmi B, Sreeja S, Anuradha CV. Astaxanthin reduces hepatic endoplasmic reticulum stress and nuclear factor-kappaB-mediated inflammation in high fructose and high fat diet-fed mice. Cell Stress Chaperones. 2014;19(2):183–91.

Bondia-Pons I, Aura A-M, Vuorela S, Kolehmainen M, Mykkänen H, Poutanen KJ. Rye phenolics in nutrition and health. J. Cereal Sci. 2009;49(3):323–36.

Boura-Halfon S, Zick Y. Phosphorylation of IRS proteins, insulin action, and insulin resistance. Am J Physiol Endocrinol Metab. 2009;296(4):E581–91.

Brunetti A, Chiefari E, Foti D. Recent advances in the molecular genetics of type 2 diabetes mellitus. World J Diabetes. 2014;5(2):128–40.

Bulboaca AE, Boarescu PM, Bolboaca SD, Blidaru M, Festila D, Dogaru G, et al. Comparative effect of curcumin versus liposomal curcumin on systemic pro-inflammatory cytokines profile, MCP-1 and RANTES in experimental diabetes mellitus. Int J Nanomedicine. 2019;14:8961–72.

Caffeic A. National Center for Biotechnology Information. "PubChem Compound Summary for CID 689043, Caffeic Acid" PubChem, https://pubchem.ncbi.nlm.nih.gov/compound/Caffeic-Acid. Accessed 24 Apr 2024.

Calpe-Berdiel L, Escola-Gil JC, Rotllan N, Blanco-Vaca F. Phytosterols do not change susceptibility to obesity, insulin resistance, and diabetes induced by a high-fat diet in mice. Metabolism. 2008;57(11):1497–501.

Cao YK, Zhang SF, Zou SE, Xia X. Daidzein improves insulin resistance in ovariectomized rats. Climacteric. 2013;16(1):111–6.

Capsaicin. National Center for Biotechnology Information. "PubChem Compound Summary for CID 1548943, Capsaicin" PubChem, https://pubchem.ncbi.nlm.nih.gov/compound/Capsaicin. Accessed 24 Apr 2024.

Care CS. Effect of treatment with capsaicin on daily activities of patients with painful diabetic neuropathy. GJD. 1992;15(2):159–65.

Carlsson S. Etiology and pathogenesis of latent autoimmune diabetes in adults (LADA) compared to type 2 diabetes. Front Physiol. 2019;10:320.

Chao PC, Hsu CC, Yin MC. Anti-inflammatory and anti-coagulatory activities of caffeic acid and ellagic acid in cardiac tissue of diabetic mice. Nutr Metab (Lond). 2009;6:33.

Cheong SH, Furuhashi K, Ito K, Nagaoka M, Yonezawa T, Miura Y, et al. Antihyperglycemic effect of equol, a daidzein derivative, in cultured L6 myocytes and ob/ob mice. Mol Nutr Food Res. 2014;58(2):267–77.

Chi XX, Zhang T, Zhang DJ, Yu W, Wang QY, Zhen JL. Effects of isoflavones on lipid and apolipoprotein levels in patients with type 2 diabetes in heilongjiang province in China. J Clin Biochem Nutr. 2016;59(2):134–8.

Cojocneanu Petric R, Braicu C, Raduly L, Zanoaga O, Dragos N, Monroig P, et al. Phytochemicals modulate carcinogenic signaling pathways in breast and hormone-related cancers. Onco Targets Ther. 2015;8:2053–66.

Coskun O, Kanter M, Korkmaz A, Oter S. Quercetin, a flavonoid antioxidant, prevents and protects streptozotocin-induced oxidative stress and beta-cell damage in rat pancreas. Pharmacol Res. 2005;51(2):117–23.

Costa-Rodrigues J, Pinho O, Monteiro P. Can lycopene be considered an effective protection against cardiovascular disease? JFC. 2018;245:1148–53.

Couper J, Donaghue KC. Phases of diabetes in children and adolescents. Pediatr Diabetes. 2009;10(Suppl 12):13–6.

Craig ME, Hattersley A, Donaghue KC. Definition, epidemiology and classification of diabetes in children and adolescents. Pediatr Diabetes. 2009;10(Suppl 12):3–12.

Curcumin. National Center for Biotechnology Information. "PubChem Compound Summary for CID 969516, Curcumin" PubChem, https://pubchem.ncbi.nlm.nih.gov/compound/Curcumin. Accessed 24 Apr 2024.

Cyanidin. National Center for Biotechnology Information. "PubChem Compound Summary for CID 128861, Cyanidin" PubChem, https://pubchem.ncbi.nlm.nih.gov/compound/Cyanidin. Accessed 24 Apr 2024.

Dabelea D, Mayer-Davis EJ, Saydah S, Imperatore G, Linder B, Divers J, et al. Prevalence of type 1 and type 2 diabetes among children and adolescents from 2001 to 2009. JAMA. 2014;311(17):1778–86.

Delphinidin. National Center for Biotechnology Information. "PubChem Compound Summary for CID 68245, Delphinidin" PubChem, https://pubchem.ncbi.nlm.nih.gov/compound/Delphinidin. Accessed 24 Apr 2024.

Demir T, Akpınar Ö. Biological activities of phytochemicals in plants. JTJoA-FS, Technol. 2020;8(8):1734–46.

Devadasu VR, Wadsworth RM, Kumar MN. Protective effects of nanoparticulate coenzyme Q10 and curcumin on inflammatory markers and lipid metabolism in streptozotocin-induced diabetic rats: a possible remedy to diabetic complications. Drug Deliv Transl Res. 2011;1(6):448–55.

Dillard CJ, German JB. Phytochemicals: nutraceuticals and human health. Journal of Science of Food and Agriculture. 2000;80(12):1744–56.

Duzguner V, Kucukgul A, Erdogan S, Celik S, Sahin KJ. Effect of lycopene administration on plasma glucose, oxidative stress and body weight in streptozotocin diabetic rats. JoAAR. 2008;33(1):17–20.

Egan AM, Vellinga A, Harreiter J, Simmons D, Desoye G, Corcoy R, et al. Epidemiology of gestational diabetes mellitus according to IADPSG/WHO 2013 criteria among obese pregnant women in Europe. Diabetologia. 2017;60(10):1913–21.

Eggersdorfer M, Wyss A. Carotenoids in human nutrition and health. Arch Biochem Biophys. 2018;652:18–26.

Eid SA, Rumora AE, Beirowski B, Bennett DL, Hur J, Savelieff MG, et al. New perspectives in diabetic neuropathy. Neuron. 2023;

Elafros MA, Andersen H, Bennett DL, Savelieff MG, Viswanathan V, Callaghan BC, et al. Towards prevention of diabetic peripheral neuropathy: clinical presentation, pathogenesis, and new treatments. Lancet Neurol. 2022;21(10):922–36.

Ellard S, Bellanne-Chantelot C, Hattersley AT. European molecular genetics quality network Mg. Best practice guidelines for the molecular genetic diagnosis of maturity-onset diabetes of the young. Diabetologia. 2008;51(4):546–53.

El-Marasy SA, Abdallah HM, El-Shenawy SM, El-Khatib AS, El-Shabrawy OA, Kenawy SA. Antidepressant effect of hesperidin in diabetic rats. Can J Physiol Pharmacol. 2014;92(11):945–52.

Erejuwa OO, Sulaiman SA, Ab Wahab MS. Honey: a novel antioxidant. Molecules. 2012;17(4):4400–23.

Fathalipour M, Eghtedari M, Borges F, Silva T, Moosavi F, Firuzi O, et al. Caffeic acid alkyl amide derivatives ameliorate oxidative stress and modulate ERK1/2 and AKT signaling pathways in a rat model of diabetic retinopathy. Chem Biodivers. 2019;16(12):e1900405.

Fatima MT, Bhat AA, Nisar S, Fakhro KA, Al-Shabeeb Akil AS. The role of dietary antioxidants in type 2 diabetes and neurodegenerative disorders: an assessment of the benefit profile. Heliyon. 2023;9(1):e12698.

Finicelli M, Squillaro T, Di Cristo F, Di Salle A, Melone MAB, Galderisi U, et al. Metabolic syndrome, Mediterranean diet, and polyphenols: evidence and perspectives. J Cell Physiol. 2019;234(5):5807–26.

Fogacci F, Di Micoli V, Veronesi M, Cicero AFJ. Comparative effect of a nutraceutical compound based on a flavonoid complex from bergamot on plasma lipids, glucose metabolism and liver enzymes: a 3-arm, double-blind, placebo-controlled, randomized, clinical trial. AoMS. n.d.;19:1180.

Ford ES, Will JC, Bowman BA, Narayan KM. Diabetes mellitus and serum carotenoids: findings from the third National Health and Nutrition Examination Survey. Am J Epidemiol. 1999;149(2):168–76.

Fu Z, Gilbert ER, Pfeiffer L, Zhang Y, Fu Y, Liu D. Genistein ameliorates hyperglycemia in a mouse model of nongenetic type 2 diabetes. Appl Physiol Nutr Metab. 2012;37(3):480–8.

Galicia-Garcia U, Benito-Vicente A, Jebari S, Larrea-Sebal A, Siddiqi H, Uribe KB, et al. Pathophysiology of type 2 diabetes mellitus. Int J Mol Sci. 2020;21(17)

Galiero R, Caturano A, Vetrano E, Beccia D, Brin C, Alfano M, et al. Peripheral neuropathy in diabetes mellitus: Pathogenetic mechanisms and diagnostic options. Int J Mol Sci. 2023;24(4)

Galtier F. Definition, epidemiology, risk factors. Diabetes Metab. 2010;36(6 Pt 2):628–51.

Georgiou M, Prokopiou E. Diabetic retinopathy and the role of Omega-3 PUFAs: a narrative review. Exp Eye Res. 2023;231:109494.

Ghosh S, Banerjee S, Sil PC. The beneficial role of curcumin on inflammation, diabetes and neurodegenerative disease: a recent update. Food Chem Toxicol. 2015;83:111–24.

Gonzales GB, Smagghe G, Grootaert C, Zotti M, Raes K, Van Camp J. Flavonoid interactions during digestion, absorption, distribution and metabolism: a sequential structure-activity/property relationship-based approach in the study of bioavailability and bioactivity. Drug Metab Rev. 2015;47(2):175–90.

Gonzalez-Castejon M, Rodriguez-Casado A. Dietary phytochemicals and their potential effects on obesity: a review. Pharmacol Res. 2011;64(5):438–55.

Goto T, Takahashi N, Hirai S, Kawada T. Various terpenoids derived from herbal and dietary plants function as PPAR modulators and regulate carbohydrate and lipid metabolism. PPAR Res. 2010;2010:483958.

Gouveia H, Urquiza-Martinez MV, Manhaes-de-Castro R, Costa-de-Santana BJR, Villarreal JP, Mercado-Camargo R, et al. Effects of the treatment with flavonoids on metabolic syndrome components in humans: a systematic review focusing on mechanisms of action. Int J Mol Sci. 2022;23(15)

Guo H, Ling W. The update of anthocyanins on obesity and type 2 diabetes: experimental evidence and clinical perspectives. Rev Endocr Metab Disord. 2015;16(1):1–13.

Halliwell B. Free radicals and antioxidants—quo vadis? Trends Pharmacol Sci. 2011;32(3):125–30.

Hameed S, Ellard S, Woodhead HJ, Neville KA, Walker JL, Craig ME, et al. Persistently autoantibody negative (PAN) type 1 diabetes mellitus in children. Pediatr Diabetes. 2011;12(3 Pt 1):142–9.

Hao J, Liu Y. Epigenetics of methylation modifications in diabetic cardiomyopathy. Front Endocrinol (Lausanne). 2023;14:1119765.

Hesperidin. National Center for Biotechnology Information. "PubChem Compound Summary for CID 10621, Hesperidin" PubChem, https://pubchem.ncbi.nlm.nih.gov/compound/Hesperidin. Accessed 24 Apr 2024.

Invernizzi A, Chhablani J, Viola F, Gabrielle PH, Zarranz-Ventura J, Staurenghi G. Diabetic retinopathy in the pediatric population: pathophysiology, screening, current and future treatments. Pharmacol Res. 2023;188:106670.

Kawser Hossain M, Abdal Dayem A, Han J, Yin Y, Kim K, Kumar Saha S, et al. Molecular mechanisms of the anti-obesity and anti-diabetic properties of flavonoids. Int J Mol Sci. 2016;17(4):569.

Keita H, Dos Santos CBR, Ramos MM, Padilha EC, Serafim RB, Castro AN, et al. Assessment of the hypoglycemic effect of Bixin in alloxan-induced diabetic rats: in vivo and in silico studies. J Biomol Struct Dyn. 2021;39(3):1017–28.

Kelly FJ. Use of antioxidants in the prevention and treatment of disease. J Int Fed Clin Chem. 1998;10(1):21–3.

Khalid M, Alkaabi J, Khan MAB, Adem A. Insulin signal transduction perturbations in insulin resistance. Int J Mol Sci. 2021;22(16)

Kim WK, Bae KH, Lee SC, Oh KJ. The latest insights into adipokines in diabetes. J Clin Med. 2019;8(11)

Kozlowska A, Szostak-Wegierek D. Flavonoids—food sources and health benefits. Rocz Panstw Zakl Hig. 2014;65(2):79–85.

Lecour S, Lamont KT. Natural polyphenols and cardioprotection. Mini Rev Med Chem. 2011;11(14):1191–9.

Lewandowska H, Kalinowska M, Lewandowski W, Stepkowski TM, Brzoska K. The role of natural polyphenols in cell signaling and cytoprotection against cancer development. J Nutr Biochem. 2016;32:1–19.

Li W, Wang G, Lu X, Jiang Y, Xu L, Zhao X. Lycopene ameliorates renal function in rats with streptozotocin-induced diabetes. Int J Clin Exp Pathol. 2014;7(8):5008–15.

Li D, Zhang Y, Liu Y, Sun R, Xia M. Purified anthocyanin supplementation reduces dyslipidemia, enhances antioxidant capacity, and prevents insulin resistance in diabetic patients. J Nutr. 2015;145(4):742–8.

Li X, Wu J, Xu F, Chu C, Li X, Shi X, et al. Use of Ferulic acid in the management of diabetes mellitus and its complications. Molecules. 2022:27(18).

Liarakos AL, Tentolouris A, Kokkinos A, Eleftheriadou I, Tentolouris N. Impact of Glucagon-like peptide 1 receptor agonists on peripheral arterial disease in people with diabetes mellitus: a narrative review. J Diabetes Complicat. 2023;37(2):108390.

Liu RH. Health benefits of fruit and vegetables are from additive and synergistic combinations of phytochemicals. Am J Clin Nutr. 2003;78(3 Suppl):517S–20S.

Liu RH. Potential synergy of phytochemicals in cancer prevention: mechanism of action. J Nutr. 2004;134(12 Suppl):3479S–85S.

Liu ZM, Chen YM, Ho SC, Ho YP, Woo J. Effects of soy protein and isoflavones on glycemic control and insulin sensitivity: a 6-mo double-blind, randomized, placebo-controlled trial in postmenopausal Chinese women with prediabetes or untreated early diabetes. Am J Clin Nutr. 2010;91(5):1394–401.

Liu Y, Tian X, Gou L, Sun L, Ling X, Yin X. Luteolin attenuates diabetes-associated cognitive decline in rats. Brain Res Bull. 2013;94:23–9.

Lopez-Lazaro M. Distribution and biological activities of the flavonoid luteolin. Mini Rev Med Chem. 2009;9(1):31–59.

Lu J, Liu J, Li L, Lan Y, Liang Y. Cytokines in type 1 diabetes: mechanisms of action and immunotherapeutic targets. Clin Transl Immunology. 2020;9(3):e1122.

Luteolin. National Center for Biotechnology Information. “PubChem Compound Summary for CID 5280445, Luteolin” PubChem, https://pubchem.ncbi.nlm.nih.gov/compound/Luteolin. Accessed 24 Apr 2024.

Maeda H, Hosokawa M, Sashima T, Takahashi N, Kawada T, Miyashita K. Fucoxanthin and its metabolite, fucoxanthinol, suppress adipocyte differentiation in 3T3-L1 cells. Int J Mol Med. 2006;18(1):147–52.

Maeda H, Hosokawa M, Sashima T, Murakami-Funayama K, Miyashita K. Anti-obesity and anti-diabetic effects of fucoxanthin on diet-induced obesity conditions in a murine model. Mol Med Rep. 2009;2(6):897–902.

Malinowski JM, Gehret MM. Phytosterols for dyslipidemia. Am J Health Syst Pharm. 2010;67(14):1165–73.

Manabe E, Handa O, Naito Y, Mizushima K, Akagiri S, Adachi S, et al. Astaxanthin protects mesangial cells from hyperglycemia-induced oxidative signaling. J Cell Biochem. 2008;103(6):1925–37.

Manach C, Scalbert A, Morand C, Remesy C, Jimenez L. Polyphenols: food sources and bioavailability. Am J Clin Nutr. 2004;79(5):727–47.

Marangoni F, Poli A. Phytosterols and cardiovascular health. Pharmacol Res. 2010;61(3):193–9.

Masenga SK, Kabwe LS, Chakulya M, Kirabo A. Mechanisms of oxidative stress in metabolic syndrome. Int J Mol Sci. 2023;24(9)

Matboli M, Eissa S, Ibrahim D, Hegazy MGA, Imam SS, Habib EK. Caffeic acid attenuates diabetic kidney disease via modulation of autophagy in a high-fat diet/streptozotocin- induced diabetic rat. Sci Rep. 2017;7(1):2263.

Mayer-Davis EJ, Lawrence JM, Dabelea D, Divers J, Isom S, Dolan L, et al. Incidence trends of type 1 and type 2 diabetes among youths, 2002–2012. N Engl J Med. 2017;376(15):1419–29.

McCarty MF. Proposal for a dietary "phytochemical index". JMh. 2004;63(5):813–7.

Milne JC, Lambert PD, Schenk S, Carney DP, Smith JJ, Gagne DJ, et al. Small molecule activators of SIRT1 as therapeutics for the treatment of type 2 diabetes. Nature. 2007;450(7170):712–6.

Mohamed GA, Ibrahim SR, Elkhayat ES. El dine RSJBofop. Cairo University Natural anti-obesity agents. 2014;52(2):269–84.

Mohammadshahi M, Haidari F, Soufi FG. Chronic resveratrol administration improves diabetic cardiomyopathy in part by reducing oxidative stress. Cardiol J. 2014;21(1):39–46.

Molyneux RJ, Lee ST, Gardner DR, Panter KE, James LF. Phytochemicals: the good, the bad and the ugly? Phytochemistry. 2007;68(22–24):2973–85.

Naguib YM. Antioxidant activities of astaxanthin and related carotenoids. J Agric Food Chem. 2000;48(4):1150–4.

Naito Y, Uchiyama K, Aoi W, Hasegawa G, Nakamura N, Yoshida N, et al. Prevention of diabetic nephropathy by treatment with astaxanthin in diabetic db/db mice. Biofactors. 2004;20(1):49–59.

Nakano M, Onodera A, Saito E, Tanabe M, Yajima K, Takahashi J, et al. Effect of astaxanthin in combination with alpha-tocopherol or ascorbic acid against oxidative damage in diabetic ODS rats. J Nutr Sci Vitaminol (Tokyo). 2008;54(4):329–34.

Nguyen CT, Pham NM, Do VV, Binns CW, Hoang VM, Dang DA, et al. Soyfood and isoflavone intake and risk of type 2 diabetes in Vietnamese adults. Eur J Clin Nutr. 2017;71(10):1186–92.

Nizamutdinova IT, Jin YC, Chung JI, Shin SC, Lee SJ, Seo HG, et al. The anti-diabetic effect of anthocyanins in streptozotocin-induced diabetic rats through glucose transporter 4 regulation and prevention of insulin resistance and pancreatic apoptosis. Mol Nutr Food Res. 2009;53(11):1419–29.

Noble JA, Erlich HA. Genetics of type 1 diabetes. Cold Spring Harb Perspect Med. 2012;2(1):a007732.

Nolan CJ, Damm P, Prentki M. Type 2 diabetes across generations: from pathophysiology to prevention and management. Lancet. 2011;378(9786):169–81.

Oguntibeju OO. Type 2 diabetes mellitus, oxidative stress and inflammation: examining the links. Int J Physiol Pathophysiol Pharmacol. 2019;11(3):45–63.

Ogurtsova K, da Rocha Fernandes JD, Huang Y, Linnenkamp U, Guariguata L, Cho NH, et al. IDF diabetes atlas: global estimates for the prevalence of diabetes for 2015 and 2040. Diabetes Res Clin Pract. 2017;128:40–50.

Okon EB, Chung AW, Zhang H, Laher I, van Breemen C. Hyperglycemia and hyperlipidemia are associated with endothelial dysfunction during the development of type 2 diabetes. Can J Physiol Pharmacol. 2007;85(5):562–7.

Ozturk E, Arslan AKK, Yerer MB, Bishayee A. Resveratrol and diabetes: a critical review of clinical studies. Biomed Pharmacother. 2017;95:230–4.

Padayachee C, Coombes JS. Exercise guidelines for gestational diabetes mellitus. World J Diabetes. 2015;6(8):1033–44.

Pan KL, Hsu YC, Chang ST, Chung CM, Lin CL. The role of cardiac fibrosis in diabetic cardiomyopathy: from pathophysiology to clinical diagnostic tools. Int J Mol Sci. 2023;24(10)

Pang B, Zhang LL, Li B, Sun FX, Wang ZD. BMP5 ameliorates diabetic peripheral neuropathy by augmenting mitochondrial function and inhibiting apoptosis in Schwann cells. Biochem Biophys Res Commun. 2023;643:69–76.

Panickar KS. Effects of dietary polyphenols on neuroregulatory factors and pathways that mediate food intake and energy regulation in obesity. Mol Nutr Food Res. 2013;57(1):34–47.

Park CH, Xu FH, Roh SS, Song YO, Uebaba K, Noh JS, et al. Astaxanthin and Corni Fructus protect against diabetes-induced oxidative stress, inflammation, and advanced glycation end product in livers of streptozotocin-induced diabetic rats. J Med Food. 2015;18(3):337–44.

Pelargonidin. National Center for Biotechnology Information. "PubChem Compound Summary for CID 440832, Pelargonidin" PubChem, https://pubchem.ncbi.nlm.nih.gov/compound/Pelargonidin. Accessed 24 Apr 2024.

Pelle MC, Provenzano M, Busutti M, Porcu CV, Zaffina I, Stanga L, et al. Up-date on diabetic nephropathy. Life (Basel). 2022;12(8)

Pereira DM, Valentao P, Correia-da-Silva G, Teixeira N, Andrade PB. Plant secondary metabolites in cancer chemotherapy: where are we? Curr Pharm Biotechnol. 2012;13(5):632–50.

Phang RJ, Ritchie RH, Hausenloy DJ, Lees JG, Lim SY. Cellular interplay between cardiomyocytes and non-myocytes in diabetic cardiomyopathy. Cardiovasc Res. 2023;119(3):668–90.

Pillai A, Fulmali D. A narrative review of new treatment options for diabetic nephropathy. Cureus. 2023;15(1):e33235.

Pinzon-Garcia AD, Orellano LAA, de Lazari MGT, Campos PP, Cortes ME, Sinisterra RD. Evidence of hypoglycemic, lipid-lowering and hepatoprotective effects of the Bixin and Bixin: beta-CD inclusion compound in high-fat-fed obese mice. Biomed Pharmacother. 2018;106:363–72.

Quispe C, Herrera-Bravo J, Javed Z, Khan K, Raza S, Gulsunoglu-Konuskan Z, et al. Therapeutic applications of curcumin in diabetes: a review and perspective. Biomed Res Int. 2022;2022:1375892.

Rahimi R, Nikfar S, Larijani B, Abdollahi M. A review on the role of antioxidants in the management of diabetes and its complications. Biomed Pharmacother. 2005;59(7):365–73.

Rajaei Z, Hadjzadeh MA, Nemati H, Hosseini M, Ahmadi M, Shafiee S. Antihyperglycemic and antioxidant activity of crocin in streptozotocin-induced diabetic rats. J Med Food. 2013;16(3):206–10.

Rajalakshmi R, Lalitha P, Sharma SC, Rajiv A, Chithambharan A, Ponnusamy A. In silico studies: physicochemical properties, drug score, toxicity predictions and molecular docking of organosulphur compounds against diabetes mellitus. J Mol Recognit. 2021;34(11):e2925.

Rangel-Huerta OD, Pastor-Villaescusa B, Aguilera CM, Gil A. A systematic review of the efficacy of bioactive compounds in cardiovascular disease: phenolic compounds. Nutrients. 2015;7(7):5177–216.

Resveratrol. National Center for Biotechnology Information. "PubChem Compound Summary for CID 445154, Resveratrol" PubChem, https://pubchem.ncbi.nlm.nih.gov/compound/Resveratrol. Accessed 24 Apr 2024.

Reyes-Escogido Mde L, Gonzalez-Mondragon EG, Vazquez-Tzompantzi E. Chemical and pharmacological aspects of capsaicin. Molecules. 2011;16(2):1253–70.

Rizzo A, Roscino MT, Binetti F, Sciorsci RL. Roles of reactive oxygen species in female reproduction. Reprod Domest Anim. 2012;47(2):344–52.

Rojo LE, Ribnicky D, Logendra S, Poulev A, Rojas-Silva P, Kuhn P, et al. In vitro and in vivo anti-diabetic effects of anthocyanins from Maqui Berry (Aristotelia chilensis). Food Chem. 2012;131(2):387–96.

Romana-Souza B, Dos Santos JS, Monte-Alto-Costa A. Caffeic acid phenethyl ester promotes wound healing of mice pressure ulcers affecting NF-kappaB, NOS2 and NRF2 expression. Life Sci. 2018;207:158–65.

Rostami A, Khalili M, Haghighat N, Eghtesadi S, Shidfar F, Heidari I, et al. High-cocoa polyphenol-rich chocolate improves blood pressure in patients with diabetes and hypertension. ARYA Atheroscler. 2015;11(1):21–9.

Russell JW, Sullivan KA, Windebank AJ, Herrmann DN, Feldman EL. Neurons undergo apoptosis in animal and cell culture models of diabetes. Neurobiol Dis. 1999;6(5):347–63.

Russell JW, Golovoy D, Vincent AM, Mahendru P, Olzmann JA, Mentzer A, et al. High glucose-induced oxidative stress and mitochondrial dysfunction in neurons. FASEB J. 2002;16(13):1738–48.

Saberzadeh-Ardestani B, Karamzadeh R, Basiri M, Hajizadeh-Saffar E, Farhadi A, Shapiro AMJ, et al. Type 1 diabetes mellitus: cellular and molecular pathophysiology at a glance. Cell J. 2018;20(3):294–301.

Saha B, Roy A, Beltramo E, Sahoo OS. Stem cells and diabetic retinopathy: from models to treatment. Mol Biol Rep. 2023;50(5):4517–26.

Sahu SC. Dual role of organosulfur compounds in foods: a review. J Environ Sci Health C Environ Carcinog Ecotoxicol Rev. 2002;20(1):61–76.

Salem AM, Ragheb AS, Hegazy MGA, Matboli M, Eissa S. Caffeic acid modulates miR-636 expression in diabetic nephropathy rats. Indian J Clin Biochem. 2019;34(3):296–303.

Sanaye MM, Kavishwar SA. Diabetic neuropathy: review on molecular mechanisms. Curr Mol Med. 2023;23(2):97–110.

Sandireddy R, Yerra VG, Komirishetti P, Areti A, Kumar A. Fisetin imparts neuroprotection in experimental diabetic neuropathy by modulating Nrf2 and NF-kappaB pathways. Cell Mol Neurobiol. 2016;36(6):883–92.

Sanie-Jahromi F, Zia Z, Afarid MJ. A review on the effect of garlic on diabetes, BDNF, and VEGF as a potential treatment for diabetic retinopathy. Chin Med. 2023;18(1):18.

Sarikaphuti A, Nararatwanchai T, Hashiguchi T, Ito T, Thaworanunta S, Kikuchi K, et al. Preventive effects of Morus alba L. anthocyanins on diabetes in Zucker diabetic fatty rats. Exp Ther Med. 2013;6(3):689–95.

Schmatz R, Perreira LB, Stefanello N, Mazzanti C, Spanevello R, Gutierres J, et al. Effects of resveratrol on biomarkers of oxidative stress and on the activity of delta aminolevulinic acid dehydratase in liver and kidney of streptozotocin-induced diabetic rats. Biochimie. 2012;94(2):374–83.

Sears B, Perry M. The role of fatty acids in insulin resistance. Lipids Health Dis. 2015;14:121.

Sharifzadeh M, Ranjbar A, Hosseini A, Khanavi M. The effect of green tea extract on oxidative stress and spatial learning in streptozotocin-diabetic rats. Iran J Pharm Res. 2017;16(1):201–9.

Shen T, Wang XN, Lou HX. Natural stilbenes: an overview. Nat Prod Rep. 2009;26(7):916–35.

Shirali S, Zahra Bathaie S, Nakhjavani M. Effect of crocin on the insulin resistance and lipid profile of streptozotocin-induced diabetic rats. Phytother Res. 2013;27(7):1042–7.

Signorelli P, Ghidoni R. Resveratrol as an anticancer nutrient: molecular basis, open questions and promises. J Nutr Biochem. 2005;16(8):449–66.

Singh PP, Mahadi F, Roy A, Sharma P. Reactive oxygen species, reactive nitrogen species and antioxidants in etiopathogenesis of diabetes mellitus type-2. Indian J Clin Biochem. 2009;24(4):324–42.

Sohrab G, Ebrahimof S, Hosseinpour-Niazi S, Yuzbashian E, Mirmiran P, Azizi F. Association of dietary intakes of total polyphenol and its subclasses with the risk of metabolic syndrome: Tehran lipid and glucose study. Metab Syndr Relat Disord. 2018;16(6):274–81.

Sujithra K, Srinivasan S, Indumathi D, Vinothkumar V. Allyl methyl sulfide, an organosulfur compound alleviates hyperglycemia mediated hepatic oxidative stress and inflammation in streptozotocin—induced experimental rats. Biomed Pharmacother. 2018;107:292–302.

Suzuki K, Ito Y, Nakamura S, Ochiai J, Aoki K. Relationship between serum carotenoids and hyperglycemia: a population-based cross-sectional study. J Epidemiol. 2002;12(5):357–66.

Szkudelski T, Szkudelska K. Resveratrol and diabetes: from animal to human studies. Biochim Biophys Acta. 2015;1852(6):1145–54.

Takahashi N, Goto T, Taimatsu A, Egawa K, Katoh S, Kusudo T, et al. Bixin regulates mRNA expression involved in adipogenesis and enhances insulin sensitivity in 3T3-L1 adipocytes through PPARgamma activation. Biochem Biophys Res Commun. 2009;390(4):1372–6.

Tan TE, Wong TY. Diabetic retinopathy: looking forward to 2030. Front Endocrinol (Lausanne). 2022;13:1077669.

Tanwar B, Modgil RJSD. Flavonoids: dietary occurrence and health benefits. 2012;2(1):59–68.

Thakur M, Singh K, Khedkar R. Phytochemicals: extraction process, safety assessment, toxicological evaluations, and regulatory issues. Functional and preservative properties of phytochemicals: Elsevier; 2020. p. 341–61.

Tomic D, Shaw JE, Magliano DJ. The burden and risks of emerging complications of diabetes mellitus. Nat Rev Endocrinol. 2022;18(9):525–39.

Tsuda T, Ueno Y, Aoki H, Koda T, Horio F, Takahashi N, et al. Anthocyanin enhances adipocytokine secretion and adipocyte-specific gene expression in isolated rat adipocytes. Biochem Biophys Res Commun. 2004;316(1):149–57.

Tuttle KR, Agarwal R, Alpers CE, Bakris GL, Brosius FC, Kolkhof P, et al. Molecular mechanisms and therapeutic targets for diabetic kidney disease. Kidney Int. 2022;102(2):248–60.

Umeno A, Horie M, Murotomi K, Nakajima Y, Yoshida Y. Antioxidative and antidiabetic effects of natural polyphenols and isoflavones. Molecules. 2016;21(6)

Veeresham C. Natural products derived from plants as a source of drugs. J Adv Pharm Technol Res. 2012;3(4):200–1.

Vinik AI. Diabetic neuropathy: pathogenesis and therapy. Am J Med. 1999;107(2B):17S–26S.

Wang J, Mazza G. Inhibitory effects of anthocyanins and other phenolic compounds on nitric oxide production in LPS/IFN-gamma-activated RAW 264.7 macrophages. J Agric Food Chem. 2002;50(4):850–7.

Wang H, Cao G, Prior RL. Oxygen radical absorbing capacity of anthocyanins. JJoa, Chemistry F. 1997;45(2):304–9.

Yadav U, Kumar N, Sarvottam K. Role of obesity related inflammation in pathogenesis of peripheral artery disease in patients of type 2 diabetes mellitus. J Diabetes Metab Disord. 2023;22(1):175–88.

Yan Y, Zhou X, Guo K, Zhou F, Yang H. Use of chlorogenic acid against diabetes mellitus and its complications. J Immunol Res. 2020;2020:9680508.

Yang L, Qian Z, Ji H, Yang R, Wang Y, Xi L, et al. Inhibitory effect on protein kinase Ctheta by Crocetin attenuates palmitate-induced insulin insensitivity in 3T3-L1 adipocytes. Eur J Pharmacol. 2010;642(1–3):47–55.

Yaribeygi H, Sathyapalan T, Atkin SL, Sahebkar A. Molecular mechanisms linking oxidative stress and diabetes mellitus. Oxidative Med Cell Longev. 2020;2020:8609213.

Ye YB, Chen AL, Lu W, Zhuo SY, Liu J, Guan JH, et al. Daidzein and genistein fail to improve glycemic control and insulin sensitivity in Chinese women with impaired glucose regulation: a double-blind, randomized, placebo-controlled trial. Mol Nutr Food Res. 2015;59(2):240–9.

Ylonen K, Alfthan G, Groop L, Saloranta C, Aro A, Virtanen SM. Dietary intakes and plasma concentrations of carotenoids and tocopherols in relation to glucose metabolism in subjects at high risk of type 2 diabetes: the Botnia dietary study. Am J Clin Nutr. 2003;77(6):1434–41.

Zhang AJ, Rimando AM, Fish W, Mentreddy SR, Mathews ST. Serviceberry [Amelanchier alnifolia (Nutt.) Nutt. ex. M. Roem (Rosaceae)] leaf extract inhibits mammalian alpha-glucosidase activity and suppresses postprandial glycemic response in a mouse model of diet-induced obesity and hyperglycemia. J Ethnopharmacol. 2012;143(2):481–7.

Zhao Z, Moghadasian MH. Chemistry, natural sources, dietary intake and pharmacokinetic properties of ferulic acid: a review. Food Chem. 2008;109(4):691–702.

Zhou M, Konigsberg WH, Hao C, Pan Y, Sun J, Wang X. Bioactivity and mechanisms of flavonoids in decreasing insulin resistance. J Enzyme Inhib Med Chem. 2023;38(1):2199168.

Zia Ur R, Ram N. Peripheral arterial disease (PAD) in diabetics: diagnosis and management—a narrative review. J Pak Med Assoc. 2023;73(3):621–6.

Chapter 6
Nutraceuticals in the Prevention of Chronic Renal Disease

Siddharth Goswami, Rohit Dimri, and Pallavi Singh

1 Introduction

The WHO estimates that Chronic Renal Disease (CRD) caused 1.3 million deaths worldwide in 2016 (Remuzzi et al. 2017). The disease impacted 753 million individuals around the world in 2016, out of which there are 336 million men and 417 million women in total.(Bikbov et al. 2018; Tjempakasari et al. 2022) More than 1.2 million patients died from it in the year of 2015, compared to 409,000 in 1990 (Wang et al. 2016) (Naghavi et al. 2015). There are 550,000 deaths from hypertension, 418,000 from diabetes, and 238,000 from glomerulonephritis (Wang et al. 2016). According to the Centers for Disease Control and Prevention (CDC), CRD was responsible for over 50,000 deaths in America in 2016. A increasing worldwide health issue, the condition is thought to afflict between 10% and 15% of the world's population. Ongoing examination recommends that around 1 of every 10 of the populace might have CRD, yet it is less usual in youthful grown-ups, being available in 1 out of 50 individuals. In those matured more than 75 years, CRD is available in 1 out of 2 individuals. CRD is a kidney illness where a steady loss of kidney capability happens over months to years (National Institute of Diabetes and Digestive and Kidney Diseases 2017). CRD decreases the kidneys' capacity for function and is a chronic, progressive disorder. The basic functions of the kidneys include removing waste from the blood and maintaining the body's fluid and electrolyte balance. When the kidneys are damaged, they can no longer perform these functions effectively, causing a buildup of fluid and waste products in the body, which complicates things including high blood pressure, anaemia, nerve damage,

S. Goswami · R. Dimri · P. Singh (✉)
Department of Biotechnology, Graphic Era Deemed to Be University, Dehradun, Uttarakhand, India
e-mail: pallavisingh.bt@geu.ac.in

S. M. Jafari et al. (eds.), *Nutraceuticals: A New Millennium Approach for Preventive Medicine*, Food Bioactive Ingredients,
https://doi.org/10.1007/978-3-031-76481-3_6

and a greater hazard of heart disease and stroke. At first, no side effects are seen, yet later side effects might include leg enlarging, tiredness, heaving, loss of craving, and confusion (National Institute of Diabetes and Digestive and Kidney Diseases 2017). Difficulties can connect with hormonal brokenness of the kidneys and hypertension (frequently connected with the enactment of the renin-angiotensin framework), bone sickness, and anaemia (Liao et al. 2012) (KDIGO: Kidney Disease Improving Global Outcomes, 2009). Furthermore, CRD patients have uniquely expanded cardiovascular entanglements with expanded dangers of death and hospitalization (Go et al. 2004). Kidney issues can result from a number of diseases or disorders, including diabetes, hypertension, glomerulonephritis, and polycystic kidney disease (Wang et al. 2016). Disorders are often diagnosed using urine tests to measure albumin levels and blood tests to calculate glomerular filtration rate (eGFR) (National Institute of Diabetes and Digestive and Kidney Diseases, 2017). For this reason, a variety of severity-based classification systems are available. In certain situations, an ultrasound or a kidney biopsy may be performed to diagnose the underlying cause of the sickness (Ferri et al. 2017).

Pharmaceutical therapy is an imperative characteristic of managing CRD. It is important to understand that while pharmaceutical treatment can help manage the symptoms and complications of CRD, it cannot cure the disease. In some cases, CRD may progress to the point where a kidney transplant or dialysis is necessary. Therefore, it is important for individuals with CRD to closely monitor their condition and work closely with their healthcare team to ensure proper treatment and management. Several drug classes, each with a unique mechanism of action and effect on the disease, are frequently used to treat this condition. Angiotensin-converting enzyme (ACE) inhibitors and angiotensin receptor blockers control (ARBs) blood pressure and save the kidneys. Angiotensin II, a hormone that causes blood vessels to constrict and raises blood pressure, was inhibited by these medications (Zhang et al. 2021). Diuretics remove excess fluid from the body and relieve symptoms such as swelling and shortness of breath (Jia et al. 2021; Al-Sayed et al. 2020). CRD caused anaemia, an illness characterized by a drop in RBCs. Fe alternatives increased iron in the body causing anaemia (Fattinger et al. 2019). CRD can lead to a decrease in calcium and Vitamin D levels, which can cause bone problems such as osteoporosis. Calcium and Vitamin D supplements are used to prevent or treat these problems (Bawazeer et al. 2020). Erythropoiesis-stimulating agents (ESAs) are the medications that stimulate the synthesis of RBCs and are used to treat anaemia in people with CRD (Xia et al. 2021; Hu et al. 2020; Li et al. 2019b). It has been demonstrated that several medications can control CRD by reducing symptoms and decreasing the disease's course.

Although these medications can be effective in managing the symptoms of CRD, they can also cause various side effects as shown in Table 6.1.

Nutraceuticals has been emerged from "nutrition"and "pharmaceutical", referring the products made from natural sources such as food and herbs that give health and therapeutic advantages in addition to their nutritional worth. Nutraceuticals have gained popularity as an alternative to conventional medications in recent years for a number of causes.

Table 6.1 Side effects of Pharmaceuticals utilized in Chronic Renal Disease (CRD) on human body

S. No	Pharmaceuticals utilized in chronic renal disease (CRD)	Side effects	References
1.1.	Erythropoiesis-stimulating agents (ESAs)	Headache, dizziness, nausea, a higher chance of blood clots, and a higher chance of stroke	Nazarko et al. (2018)
		Increased risk of tumour growth	Venkatesh et al. (2017)
1.2.	Diuretics	Dehydration, low potassium levels, low magnesium levels, and decreased kidney function	Soto et al. (2018)
		Increased risk of developing gout, hypokalaemia, and hypomagnesemia	Jia et al. (2021)
1.3.	Angiotensin-converting enzyme (ACE) inhibitors	Headache, dizziness, cough, low blood pressure, and decreased kidney function	Ibrahim et al. (2019)
		Increased chance of getting a skin rash, decreased renal function, and an increase in potassium levels	Heerspink et al. (2017)
4.	Angiotensin receptor blockers (ARBs)	Headache, dizziness, low blood pressure, and increased risk of kidney dysfunction	Mousa et al. (2018)
		Higher hazard of developing an allergic reaction, emerging angioedema and hyperkalaemia	Cushman et al. (2019)

1. Nutraceuticals are often considered safer and have fewer side effects compared to pharmaceutical drugs, as they are made from natural ingredients and do not undergo the same level of chemical processing. This makes them more appealing to consumers who are concerned about the negative impact that pharmaceutical drugs may have on their health.
2. Nutraceuticals are more accessible and affordable for many people, as they are widely available and do not require a prescription from a doctor. This allows individuals to take control of their health and wellness, without relying on expensive medical treatments.
3. Nutraceuticals are becoming increasingly popular as people are becoming more aware of the importance of a balanced diet and natural remedies for maintaining good health. The trend towards a more holistic approach to health and wellness has also driven the demand for nutraceuticals.
4. The growing scientific evidence supporting the health benefits of nutraceuticals has also contributed to their increased popularity. Nutraceuticals have gained attention as an alternative to pharmaceuticals due to their perceived safety, accessibility, affordability, popularity, and scientific support. As the demand for natural and holistic health solutions continues to grow, nutraceuticals will likely show an increasingly important role in the field of health and wellness.

Nutraceuticals have been gaining attention as an alternative to pharmaceuticals in the therapy of CRD. Nutraceuticals are food-based products that contain nutrients that are derived from natural sources and promote health, including vitamins, minerals, and antioxidants. These products are marketed as dietary supplements, functional foods, and fortified foods and can provide the necessary nutrients to support the health of the kidneys and help manage the symptoms of CRD. Nutraceuticals have a beneficial effect on CRD, such as, omega-3 fatty acids possess inflammation restricting characteristics and can slow down the progression of CRD (Navarro-Gonzalez et al. 2020). Moreover, vitamin D has been demonstrated to have a protective impact on the kidneys, and it may enhance the results of CRD. (Chowdhury et al. 2020). In addition, antioxidants including vitamins C and E have a key role in decreasing oxidative stress, a major factor in CRD. (Chauhan et al. 2019).

The current chapter presents a systematic literature review on the use of nutraceuticals, such as antioxidants and anti-inflammatory substances, in the prevention of CRD. Inflammation as well as oxidative stress are significant contributors to the onset and progression of CRD. Furthermore, certain nutraceuticals, including omega-3 fatty acids and phytochemicals, have demonstrated a protective impact on the kidneys by lowering oxidative stress and enhancing kidney performance. Research has shown that several metabolic pathways altering CRD, including those involved in glucose metabolism, lipid metabolism, and nitrogen metabolism provide a review for the development of effective treatments for CRD have been discussed. Genome and Proteome changes resulting in alterations in the genetic expression involved in kidney function and the production of proteins that have a role in the disease are described. Micronutrients like vitamins, minerals and macronutrients like carbohydrates, proteins and omega-3 fatty acids etc. having efficiency in treatment are reviewed. Additionally, keto analogues with low-protein, high-fat diets, reported to improve kidney function in some patients with CRD, maintaining a healthy diet that is low in salt, fat, and sugar, and high in fibre and micronutrients have been proposed. Moreover, the correlation between ions uptake and progression of CRD indicated that levels of sodium and potassium (high and low respectively) could increase the hazard of developing CRD, while a balanced intake of these ions can help to reduce the risk describing the relation between diet pattern and mortality rate have been highlighted in this review analysis. This chapter not only enhances the understanding of reducing the role of pharmaceuticals while providing the efficiency of nutraceuticals over pharmaceuticals.

2 Nutraceuticals Alternative to Pharmaceuticals Involved to Treat CRD with Their Mechanism of Action

Nutraceuticals that contain specific nutrients such as iron, vitamin B12, folic acid, potassium, magnesium, vitamin C, cocoa, omega-3 fatty acids, and green tea have shown potential benefits in treating Chronic Renal Disease (CRD). These nutrients have been found to exert positive results on various physiological processes such as blood pressure regulation, glucose metabolism, endothelial function, inflammation, oxidative stress, and cardiovascular risk.

1. **Folic acid, iron, and vitamin B12:** In individuals with chronic renal disease, folic acid, iron, and vitamin B12 can enhance erythropoiesis and treat anaemia. Several studies have shown how supplementing with folic acid, iron, and vitamin B12 can improve haemoglobin levels in these individuals and lessen their requirement for erythropoiesis-stimulating medications (Zhu et al. 2018; Jafari et al. 2019).
2. **Potassium and Magnesium:** Potassium and magnesium are essential minerals that are involved in several physiological processes in the body, including blood pressure regulation and glucose metabolism. Patients with chronic renal disease often have reduced levels of potassium and magnesium, which can lead to hypertension and insulin resistance. Nutraceuticals containing potassium and magnesium can help to correct these deficiencies and improve blood pressure control and glycemic control (Raj et al. 2018).
3. **Vitamin C:** Patients with chronic renal disease frequently experience oxidative stress and inflammation, which can be mitigated by the potent antioxidant vitamin C. Many studies have demonstrated that vitamin C administration in these individuals can enhance endothelial function and lessen inflammation and oxidative stress (Saran et al. 2015; Bahrami et al. 2020).
4. **Cocoa:** Flavonoids, which are polyphenolic chemicals with antioxidant and anti-inflammatory effects, are present in cocoa. Several studies have demonstrated that consuming cocoa can increase insulin sensitivity, decrease inflammation and oxidative stress, and enhance endothelial function in people with chronic renal impairment (Taubert et al. 2007; Farhat et al. 2020).
5. **Omega-3 Fatty Acids:** Omega-3 polyunsaturated fatty acids are among those having anti-inflammatory properties. Several studies have demonstrated that individuals with chronic renal illness who take omega-3 fatty acid supplements experience decreased oxidative stress, decreased inflammation, and improved endothelial function (Zhang et al. 2018; Mollazadeh et al. 2019a).
6. **Green Tea:** Catechins, polyphenolic substances found in green tea, have anti-inflammatory and antioxidant activities. Green tea drinking has been linked to improved endothelial function, decreased oxidative stress and inflammation, and lowered cardiovascular risk in those with chronic renal illness, according to many studies (Yamabe et al. 2013; Li et al. 2021) (Table 6.2).

3 Metabolic Pathways Altered in Chronic Renal Disease

CRD is connected to a wide range of metabolic changes that might hasten the illness's development and raise the risk of consequences including cardiovascular disease, anaemia, and bone disease.

Metabolic pathways that is commonly altered in CRD is the urea cycle. The urea cycle is responsible for converting toxic ammonia, a by-product of protein metabolism, into less toxic urea, which can be safely excreted by the kidneys. Because the kidneys are less able to carry out this excretory function in chronic kidney disease,

Table 6.2 Nutraceuticals that can be used as alternatives to pharmaceuticals to treat chronic renal disease with their sources, applications, side effects, recommended use, and advantages

Pharmaceuticals utilized in chronic renal disease (CRD)	Nutraceuticals	Sources	Side effects	Use recommended or not	Advantages	References
Erythropoiesis-stimulating agents (ESAs)	Patients with chronic renal illness can treat anaemia by promoting erythropoiesis with dietary supplements containing iron, vitamin B12, and folic acid	Foods high in iron, such as spinach, beans, pork, fish, and chicken; foods high in vitamin B12, such as fish, pork, and dairy; foods high in folic acid, such as fruits, leafy green vegetables, and fortified cereals	Excessive intake of iron supplements can cause gastrointestinal disturbances, constipation, and nausea	Although most nutraceuticals are safe to take, it is advised to consult with a doctor before using any supplements	Nutraceuticals are a natural and safe alternative to ESAs and can provide the body with other essential nutrients	Feng et al. (2018)
Diuretics	In individuals with chronic renal disease, potassium and magnesium-containing nutraceuticals can assist to minimise fluid retention and lower blood pressure	Foods rich in potassium, such as bananas, avocados, spinach, and sweet potatoes; foods rich in magnesium, such as dark chocolate, almonds, cashews, and spinach	Excessive intake of potassium supplements can cause hyperkalemia, which can lead to irregular heartbeat and muscle weakness	Nutraceuticals are generally safe to use, but it is recommended to consult a healthcare professional before taking any supplements	Nutraceuticals can help to reduce fluid retention and lower blood pressure without the side effects associated with diuretics	Kim et al. (2018) and Awotidebe et al. (2016)

Angiotensin-converting enzyme (ACE) inhibitors	Patients with chronic kidney disease may benefit from taking nutraceuticals that contain nitrates, vitamin C, and cocoa because they can reduce blood pressure and enhance endothelial function	Foods rich in nitrates, such as beetroot, spinach, and arugula; foods high in vitamin C, such as oranges, kiwi, and strawberries; Cocoa and dark chocolate	Excessive intake of nitrates can cause headaches and dizziness; excessive intake of cocoa and dark chocolate can cause gastrointestinal disturbances and weight gain	Nutraceuticals are generally safe to take, however it is always best to contact with a doctor before taking any supplements	Nutraceuticals can provide additional health benefits, such as improved cognitive function and reduced risk of cardiovascular disease	Al Sunni et al. (2014) and Xiaomin Zhang et al. (2012)
Angiotensin receptor blockers (ARBs)	Nutraceuticals containing omega-3 fatty acids, potassium, and green tea can help to lower blood pressure and reduce inflammation in patients with chronic kidney disease	Foods rich in omega-3 fatty acids, such as fatty fish, flaxseeds, and walnuts; foods high in potassium, such as bananas, sweet potatoes, and avocados; green tea	Excessive intake of omega-3 fatty acids can cause gastrointestinal disturbances and increased bleeding risk; excessive intake of potassium can cause hyperkalemia, which can lead to irregular heartbeat and muscle weakness	Nutraceuticals are typically safe to use, although it is best to consult a healthcare practitioner before taking any supplements	Nutraceuticals can provide additional health benefits, such as improved cardiovascular health and reduced risk of cognitive decline	Bjørke-Monsen et al. (2016)

harmful uremic solutes build up in the blood. The urea cycle is thus disturbed, and the concentrations of urea cycle intermediates including citrulline, arginine, and ornithine are changed. In chronic kidney illness, important urea cycle enzymes including arginase and carbamoyl phosphate synthetize also exhibit decreased activity (Krishnan et al. 2017). This disruption of the urea cycle can lead to a number of complications, including hyperammonemia, which can cause neurological symptoms, as well as alterations in protein metabolism and impaired nitrogen balance.

Insulin resistance is among the metabolic changes that are most noticeable in CRD. This happens as a result of diminished insulin sensitivity and decreased glucose metabolism, which results in hyperglycaemia (Schrauben et al. 2019). Several mechanisms are thought to contribute to insulin resistance in CRD, including oxidative stress, inflammation, and fibrosis. These processes can lead to dysfunction of the insulin signalling pathway and impaired glucose uptake by cells, ultimately resulting in insulin resistance. Insulin resistance can also develop as a result of additional variables that are frequently linked to CRD, such as obesity and hypertension. Insulin resistance can result from both hypertension and obesity, both of which can impede insulin signalling and promote chronic inflammation (Hall et al. 2015). Also, it might raise the risk of cardiovascular disease, a leading cause of death among CRD patients (Chueakula et al. 2018).

Dyslipidemia is a common metabolic abnormality observed in individuals with chronic kidney disease (CRD). It is characterized by low levels of HDL and elevated levels of LDL and triglycerides in the bloodstream, leading to abnormal blood lipid levels. Lipid metabolism is a vital function of the kidneys, and in CRD, this function may be impaired, leading to the development of dyslipidemia. Dyslipidemia is a major risk factor for atherosclerosis, which is a complication of CRD and increases the risk of cardiovascular disease. Moreover, dyslipidemia can further exacerbate renal damage by promoting the accumulation of lipids in the kidneys, which can cause inflammation, fibrosis, and worsen renal function (Khandelwal et al. 2016; Hager et al. 2017). Dyslipidemia is a significant metabolic alteration associated with CRD that can have detrimental effects on both renal and cardiovascular health.

Many amino acids, including methionine, cysteine, and tryptophan, exhibit decreased metabolism, which is another alteration of amino acid metabolism in CRD (Kalantar-Zadeh and Fouque 2017). Increased inflammation and oxidative stress may result from this, which might worsen kidney damage. The synthesis of erythropoietin, a hormone that promotes the creation of red blood cells, can be impacted by impaired amino acid metabolism, which can also contribute to the emergence of anaemia, a typical consequence of CRD (Griffin and Bradshaw 2017).

Inflammation and oxidative stress are common occurrences in chronic kidney disease (CRD) and can lead to changes in several metabolic pathways. High levels of oxidative stress can cause damage to DNA, lipid peroxidation, mitochondrial dysfunction, and exacerbate kidney damage (Wu et al. 2021; Valentijn et al. 2021). Furthermore, two inflammatory cytokines, interleukin-6 (IL-6) and tumor necrosis factor-alpha (TNF-α), are often elevated in CRD and contribute to systemic inflammation and metabolic abnormalities. These changes can adversely affect the metabolism of amino acids, lipids, and glucose, leading to anemia, dyslipidemia, and insulin resistance (Lee et al. 2015; Khanaghaei et al. 2022). Oxidative stress and inflammation are one of key factors in the development and progression of CRD, leading to alterations in several metabolic pathways and complications such as anemia, dyslipidemia, and insulin resistance.

Alterations in several metabolic pathways, such as oxidative stress, inflammation, mineral metabolism, amino acid metabolism, glucose, and lipid metabolism, are linked to CRD. The risk of consequences including cardiovascular disease, anaemia, and bone disease might rise as a result of these changes, which can also speed up the advancement of kidney impairment. To create CRD therapies that work and stop the illness from becoming worse, it is essential to comprehend these metabolic alterations (Fig. 6.1).

Pathways altered are primarily due to the reduced function of the kidneys, which are responsible for filtering waste products and maintaining the balance of various substances in the blood.

Protein Metabolism A critical function of the kidneys is the elimination of waste products from protein metabolisms, such as urea and creatinine. The kidney's capacity to filter out these waste products is impaired in chronic renal disease, resulting in their accumulation in the blood and the development of uraemia. Additionally, protein intake may need to be restricted in chronic renal disease to prevent the build-up of waste products (Garibotto et al. 2010).

Carbohydrate Metabolism Insulin resistance, which raises blood sugar levels and raises the chance of developing diabetes, is a potential side effect of chronic renal illness. It is not entirely known how renal illness produces insulin resistance, although it is believed to be connected to the buildup of harmful compounds in the blood. A reduction in the body's capacity to utilise glucose for energy as a result of

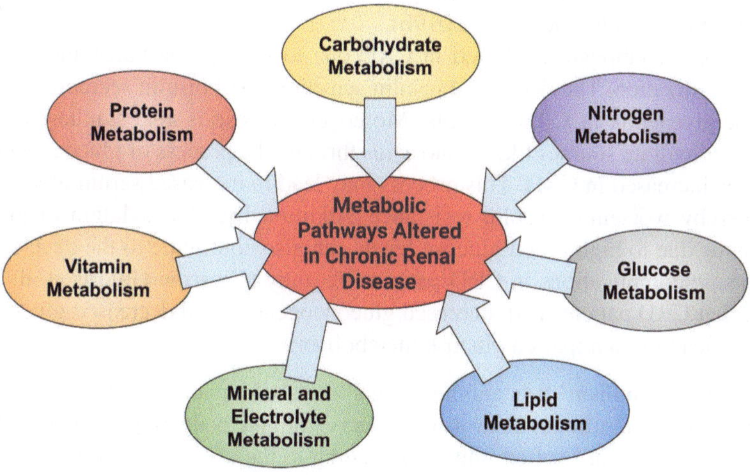

Fig. 6.1 The metabolic pathways that are altered in CRD

insulin resistance might increase the synthesis and buildup of ketone bodies in the body (De et al. 2013).

Lipid Metabolism The kidneys eliminate triglycerides and other fats from the blood. Chronic renal failure impairs the kidney's ability to eliminate these lipids, causing them to accumulate in the blood and increase the risk of cardiovascular disease. The kidneys' inability to adequately remove extra cholesterol from the body may be to blame for the decline in HDL cholesterol levels (Baek et al. 2022).

Mineral and Electrolyte Metabolism The kidneys help regulate the balance of minerals and electrolytes in the body, such as potassium, sodium, and calcium. In chronic renal disease, the kidneys' ability to regulate these substances is impaired, leading to imbalances in their levels. High levels of potassium, for example, can result from impaired kidney function and can lead to muscle weakness or heart rhythm disturbances, while low levels of calcium can lead to bone disease (Kistler et al. 2021).

Vitamin Metabolism Some vitamins, such as vitamin D and vitamin B12, are processed by the kidneys. In chronic renal disease, the kidney's ability to produce active vitamin D is reduced, leading to decreased calcium absorption and bone disease. The kidneys' ability to absorb and metabolize vitamin B12 is also impaired, leading to anaemia and neurological symptoms.

Glucose Metabolism It is the method used to break down glucose in order to create energy. Patients with CRD frequently have impaired glucose metabolism, which can result in side effects include insulin resistance and poor glucose tolerance. Several research have examined how dietary therapies affect individuals with CRD's glucose metabolism. A low-carb diet improved glucose metabolism and decreased insulin resistance in CRD patients, according to a randomised controlled experiment (Tay et al. 2015). Another investigation revealed that omega-3 fatty acid supplementation can enhance glucose metabolism by lowering oxidative stress and inflammation (Mollazadeh et al. 2019b).

Glucose metabolism is altered in CRD due to insulin resistance and impaired glucose utilization. The decreased insulin sensitivity and insulin secretion can lead to hyperglycaemia in CRD patients. Moreover, glucose is also synthesized from non-carbohydrate sources like amino acids through the process of gluconeogenesis, which is increased in CRD. This process may lead to increased serum glucose levels, thereby worsening insulin resistance. Additionally, glycosylation of proteins and lipids due to high blood glucose levels can also contribute to the development of diabetic complications in CRD patients. A study (Cigarran-Guldris et al. 2019) found that CRD patients had increased gluconeogenesis and decreased glucose utilization, leading to impaired glucose metabolism.

Nitrogen Metabolism It is another crucial metabolic pathway that is responsible for the removal of nitrogenous waste products from the body. In patients with CRD, nitrogen metabolism is often impaired, leading to the accumulation of toxic waste products like urea and creatinine. Low-protein diets and keto acid supplementation have been shown to improve nitrogen metabolism and reduce the accumulation of toxic waste products in patients with CRD (Klahr et al. 1982; Shao et al. 2019).

Nitrogen metabolism is also altered in CRD due to reduced glomerular filtration rate (GFR) and impaired renal function. Urea, the main nitrogenous waste product, is usually excreted by the kidneys. However, in CRD, the accumulation of urea and other nitrogenous compounds can lead to various complications like uremic syndrome and chronic inflammation. Moreover, CRD patients have decreased protein intake due to dietary restrictions and proteinuria, which can lead to decreased amino acid availability for protein synthesis and increased protein catabolism. This process can further worsen nitrogen metabolism and lead to muscle wasting and sarcopenia. According to one study, CRD patients have impaired nitrogen metabolism as a result of reduced protein synthesis and increased protein catabolism (Filiopoulos et al. 2016).

4 Genome and Proteome Change in Chronic Renal Diseases

A variety of inherited and environmental factors impact chronic renal disorders. Alterations in the affected person's DNA and proteome define these disorders at the molecular level. Affected people's proteomes and genomes can alter due to chronic renal disorders. The proteome is the whole collection of proteins that are expressed by an organism or cell at any one moment. In contrast, the genome refers to the entire set of genetic material, including all of a person's genes.

4.1 Chronic Renal Diseases and Genome Changes

A person's entire genetic makeup, including DNA and genes, is referred to as their genome. In chronic renal disorders such chronic kidney disease, mutations or changes in the expression of genes that control kidney function can have an impact on the genome (CKD) (Table 6.3).

Genetic anomalies or changes in the expression of genes involved in metabolism, immune response, and kidney function can be responsible for chronic renal diseases. For example, the formation and progression of chronic kidney disease (CKD) can be attributed to genes that produce RAAS (renin-angiotensin-aldosterone system) proteins. The RAAS system's dysregulation can lead to hypertension and ultimately renal failure, as it is essential for maintaining fluid balance and managing blood pressure (Gupta et al. 2020). In addition, the altered expression of several genes linked to fibrosis and inflammation, including matrix metalloproteinases (MMPs), transforming growth factor-beta (TGF-β), and interleukin-6 (IL-6), has been linked to CRD. Variations in the angiotensinogen (AGT) and renin (REN) genes, which regulate inflammation and blood pressure, have also been associated with an increased risk of developing CRD (Chen et al. 2018; Arablou et al. 2021).

Table 6.3 Genes associated with chronic kidney diseases: The table provides a summary of several genes that are associated with chronic kidney diseases, their functions, and their corresponding associations

Gene	Function or abbreviation	Association with chronic renal diseases	Reference
ACE	Angiotensin-converting enzyme	Increasing activity has been associated to hypertension and the advancement of chronic kidney disease.	Anguiano et al. (2015) and Sanchis-Gomar et al. (2020)
APOL1	Apolipoprotein L1	APOL1 gene variants, specifically G1 and G2, have been linked to an increased risk of chronic kidney disease (CKD).	Friedman and Pollak (2016) and Hung et al. (2022)
COL4A3, COL4A4, COL4A5	Collagen type IV alpha chains	Alport syndrome is a hereditary kidney disease brought on by mutations in these genes.	Alge et al. (2023) and Warady et al. (2020)
MYH9	Non-muscle myosin heavy chain IIA	Associated with different types of chronic kidney disease in addition to focal segmental glomerulosclerosis	Furlano et al. (2019)
NPHS1, NPHS2	Nephrin, podocin	Mutations in these genes cause congenital nephrotic syndrome, a rare genetic kidney disease	Guo et al. (2019) and Ramanathan et al. (2017)
PKD1, PKD2	Polycystin 1, polycystin 2	Mutations in these genes cause autosomal dominant polycystic kidney disease, a genetic kidney illness characterised by the formation of many cysts in the kidneys.	Heyer et al. (2016)
TGFβ1	Transforming growth factor beta 1	Contributes to the fibrosis of the kidneys in chronic kidney disease	Mai et al. (2020)
TNFα	Tumor necrosis factor alpha	a component of the inflammatory response brought on by chronic renal disease	Lee et al. (2015) and Pavkov et al. (2015)
VEGFA	Vascular endothelial growth factor A	Plays a part in the angiogenesis and renal fibrosis development of chronic kidney disease.	Stevens and Oltean (2018) and Anderson et al. (2018)

Moreover, genetic polymorphisms in the vitamin D receptor gene (VDR) and apolipoprotein L1 (APOL1) have been linked to an increased risk of renal disease, particularly in African Americans (Kumar et al. 2019). Genetic changes and variations in genes involved in metabolism, immune response, and kidney function can be the underlying cause of CRD. Gene expressions such as RAAS proteins, MMPs, TGF-β, IL-6, AGT, REN, VDR, and APOL1 can significantly increase the risk of developing CRD.

Environment-related factors including diet, toxin exposure, and stress can have an influence on these changes in gene expression patterns. For instance, it has been demonstrated that eating a lot of salt causes the RAAS system's genes to be expressed more often, which can lead to hypertension and kidney damage. Besides injuring the kidneys and causing sickness, toxins like heavy metals and pesticides can also alter gene expression patterns (Rust and Ekmekcioglu 2017).

Mutations in the genes that code for the proteins involved in the metabolism of poisons and medications are another example of genetic abnormalities that cause chronic kidney disorders. Kidney disease can result from these mutations because they can alter how the body removes and processes toxic chemicals. For instance, mutations in the CYP1A1 gene, which encodes a drug-metabolizing enzyme, have been associated with an increased risk of developing chronic kidney disease (CKD) in individuals exposed to environmental contaminants including cigarette smoke and industrial pollutants (Siddarth et al. 2022).

4.2 Chronic Renal Diseases and the Proteome

All the proteins that are expressed by a multi-cellular organism, tissues, and organs collectively make up their proteome. By altering the amounts of protein production, modifications, and destruction, among other things, CRD may modify the proteome.

At the proteome level, chronic renal disorders (CRD) may affect the expression of proteins involved in fibrosis, inflammation, and renal function. CKD is associated with increased generation of pro-inflammatory cytokines such as monocyte chemoattractant protein-1 (MCP-1), tumor necrosis factor-alpha (TNF-α), and interleukin-6 (IL-6). These cytokines contribute to the development of chronic inflammation in the kidney, leading to fibrosis and tissue damage (Romanova et al. 2020). Chronic renal disorders can also affect the production of extracellular matrix (ECM) proteins, which are critical for maintaining the structural integrity of the kidney. For example, in CKD, reduced production of ECM proteins such as collagen and laminin may lead to the collapse of renal structure and subsequent fibrosis. Conversely, there is an increase in the production of fibrosis-associated proteins such as transforming growth factor-beta (TGF-β) and fibronectin, which promote the deposition of ECM proteins and the onset of fibrosis (Gerrits et al. 2022; Zhong et al. 2017).

According to studies, CRD can cause proteins involved in renal function, inflammation, and fibrosis to express themselves at different levels. Furthermore, post-translational changes of proteins such glycosylation, phosphorylation, and acetylation may be abnormally high in patients with CRD.

Overall, the modifications to the genome and proteome that take place in chronic renal disorders reflect the intricate interplay between hereditary and environmental variables that contribute to the onset and progression of these ailments. In order to create more successful therapies for individuals with chronic renal illnesses, it might be helpful to understand these changes in order to find new therapeutic targets. New diagnostic and treatment strategies for CRD may be developed with the use of this understanding. To identify those who are at a high risk of getting CRD, for example, genetic testing can be utilized, enabling early intervention and prevention. Moreover, targeted treatments that control.

5 Role of Micronutrients in Combating CRD

Micronutrients are critical nutrients that the body needs in minute quantities to function normally. They consist of vitamins, minerals, and trace elements that are essential for supporting the body's physiological and metabolic activities. Micronutrient deficiencies are common in CRD patients, according to several studies, and they may accelerate the disease's development.

One such micronutrient that has been well studied in connection to CRD is vitamin D. In order to preserve bone health, vitamin D is crucial, and a lack of it has been related to the development of secondary hyperparathyroidism, a frequent consequence of CRD (Jean et al. 2017). Inflammation, immunological response, and blood pressure regulation are all critical elements in the onset and progression of CRD, and vitamin D is thought to modulate these processes. According to studies, vitamin D supplements can help CRD patients with proteinuria (excess protein in the urine) and renal function (Kim and Kim 2014).

Omega-3 fatty acids are another micronutrient that has been investigated in relation to CRD. Fish and other plant-based sources include omega-3 fatty acids, which are polyunsaturated fatty acids. They are well-recognized for having anti-inflammatory and anti-thrombotic qualities, and they could also help lower blood pressure and improve lipid profiles. According to studies, people with CRD who take omega-3 fatty acid supplements experience less proteinuria and a slower loss of renal function (Saglimbene et al. 2020; Hirahashi 2017).

Another category of micronutrients that has been investigated in regard to CRD is antioxidants. Antioxidants are chemicals that shield cells from oxidative stress, which is thought to contribute to the onset of CRD. Antioxidants including vitamin C (Hongsawong et al. 2021), vitamin E (Daenen et al. 2019), and beta-carotene may lessen oxidative stress and enhance renal function in people with CRD, according to studies. Antioxidant vitamin C helps lessen oxidative stress on the kidneys. According to studies, vitamin C supplements may benefit CRD patients with proteinuria (excess protein in the urine) and renal function (Hongsawong et al. 2021).

Selenium, magnesium, and zinc have all been investigated in connection to CRD (Fukasawa et al. 2023; Filler et al. 2021). Although magnesium is crucial for maintaining healthy bones and controlling blood pressure, zinc has a role in immunological health and wound healing. Antioxidant selenium has been found to lessen oxidative stress and inflammation. Research has revealed that these individuals with CRD frequently have mineral deficiencies, and supplementation may enhance kidney function and lessen proteinuria.

While medicinal therapies continue to be the predominant CRD management method, it is important to recognize the impact that diet and micronutrients have in halting or delaying the disease's progression. It has been demonstrated that micronutrients including vitamin D, omega-3 fatty acids, antioxidants, and minerals can enhance kidney function, lessen proteinuria, and control oxidative stress and inflammation. Optimizing nutrient intake may be a beneficial adjunct therapy for people with CRD, according to existing data, while further study is required to understand the function of micronutrients in CRD management fully (Table 6.4).

Table 6.4 Role of micronutrients in combating chronic renal disease (CRD)

Micronutrient	Role in combating CRD	Natural sources	Reference
Vitamin D	Maintains bones strong and the body's calcium and phosphorus levels stable, all of which are necessary for healthy kidney function. An increased risk of cardiovascular disease and the development of CKD is associated with low vitamin D levels	Fatty fish (including salmon, tuna, and mackerel), mushrooms, egg yolks, cheese, and meals supplemented with vitamins and minerals (such as milk and cereal)	Jean et al. (2017) and Kim and Kim (2014)
Vitamin C	Has antioxidant properties, protects against inflammation, and aids in the absorption of iron, which can be important for those with CRD who may experience frequent infections or anemia	Citrus fruits (such as oranges and grapefruits), strawberries, kiwi, broccoli, Brussels sprouts, and bell peppers	Hongsawong et al. (2021)
Vitamin B6	Helps regulate homocysteine levels in the blood, which can contribute to kidney damage if levels are too high	Chickpeas, salmon, chicken	Chen et al. (2016)
Vitamin E	Has antioxidant properties and can protect against inflammation and oxidative stress	Nuts, seeds, spinach	Daenen et al. (2019)
Zinc	Helps support immune function and wound healing, zinc deficiency include ESA hyporesponsive anaemia, nutritional challenges, or cardiovascular problems	Oysters, beef, pork, chicken, beans, and nuts	Fukasawa et al. (2023)
Iron	Required for red blood cell formation and oxygen transport. Iron deficiency anemia is common in CKD patients and can worsen kidney function. Supplementation may improve anemia and slow the progression of CKD	Red meat, chicken, fish, beans, tofu, and fortified cereals	Batchelor et al. (2020)
Omega-3 fatty acids	Reduced proteinuria in patients with CKD, and also help to lower blood pressure, which is another common complication of CKD	Fatty fish (such as salmon, tuna, and mackerel), flaxseed, chia seeds, walnuts, and soybeans	Saglimbene et al. (2020) and Hirahashi (2017)

6 Role of Macronutrients in Controlling the Progression of Chronic Renal Disease

Macronutrients, which comprise carbohydrates, proteins, and fats, have been found to have a substantial impact on how the body responds to CRD. Carbohydrates are necessary for the body's energy demands, but they can also contribute to the onset of insulin resistance in CRD patients. Insulin resistance, which occurs when the body's cells cease reacting to the actions of insulin, is the cause of high blood sugar levels. The loss of renal function can be accelerated and kidney disease can get worse when blood sugar levels are high. As a result, patients with CRD must closely watch their carbohydrate consumption. Whole grains, legumes, and non-starchy vegetables, which have a low glycaemic index, can help manage blood sugar levels and avoid insulin resistance.

Protein is required for tissue maintenance and repair, however, in CRD, high protein consumption might hasten the loss of renal function (Isaka 2021). Waste products are produced as a result of protein metabolism and are typically eliminated by the kidneys. As a result of waste products building up in the circulation, damaged kidneys and other organs suffer severe damage. Those with CRD are thus advised to have a diet that contains very little or no protein. High-quality protein sources including eggs, fish, poultry, and lean meats should be used since they supply the body's tissues with crucial amino acids. Beans, lentils, and tofu are plant-based sources of protein that should only be used in moderation to preserve healthy renal function (Tallman et al. 2022; Jakše et al. 2019). A balanced diet must include fats since they are vital to the body's cellular and hormonal processes. Nevertheless, consuming too much fat can lead to the emergence of metabolic conditions such as obesity, dyslipidaemia, and insulin resistance. Several disorders can further impede renal function, and patients with CRD are more likely to develop them. As a result, a diet high in unsaturated fats, such as those found in fatty fish, nuts, seeds, and plant oils, and low in saturated fats is suggested.

Macronutrients are essential for controlling CRD. A balanced diet low in carbs, moderate in protein, and moderate in fat should be consumed by those with CRD. This dietary strategy can control blood sugar levels, manage metabolic diseases, and delay the onset of CRD (Table 6.5).

7 Role of Proteins and Keto Analogues in Control of CRD

Proteins

Low Protein Diet (LPD) reduces nitrogen side effects and decreases kidney activity by lowering intraglomerular pressure, which may protect the kidneys, especially in individuals with decreased nephron capital and renal capacity (Fouque et al. 2007; Bellizzi et al. 2013). LPD prompts great metabolic impacts that can protect kidney capability and uremic side effects as recorded underneath and portrayed in Fig. 6.2 (Kovesdy et al. 2016). Conversely, High Protein Diet (HPD) put an adverse consequence on CRD as displayed in Fig. 6.3.

LPD is effective in the treatment of metabolic acidosis in CRD.The quantity of sulfur-containing amino acids created during protein digestion, as well as the amount of predialysis serum bicarbonate fixed, were lower in individuals who consumed more protein (Scialla et al. 2013; Goraya et al. 2012). In CRD, ongoing metabolic acidosis results in protein digestion inhibition, muscle catabolism, wasting, and bothers decrease in kidney ability and uremic symptoms (Goraya et al. 2015). Indeed, LPD decreased metabolic acidosis in individuals

Table 6.5 Macronutrients and their roles in controlling chronic renal disease

Macronutrient	Role in controlling chronic renal disease	Natural sources	Reference
Protein	Restricting protein intake slows the progression of CKD by reducing the workload on the kidneys and decreasing proteinuria	Chicken, fish, eggs, tofu, lentils, beans, quinoa	Ko et al. (2017)
Carbohydrates	Consuming complex carbs rather than simple sweets can aid in blood glucose regulation and insulin sensitivity. Low-carbohydrate diets enhance glycemic control in individuals with diabetic nephropathy, a frequent consequence of CKD	Whole grains, fruits, vegetables	Mirabelli et al. (2020) and Zainordin et al. (2021)
Fat	Limiting saturated and trans fats can help prevent cardiovascular disease, which is a common complication of chronic renal disease. Consuming healthy fats, such as monounsaturated and polyunsaturated fats, may also improve lipid profiles and reduce inflammation	Nuts, seeds, avocado, olive oil, fatty fish	Mafra et al. (2021) and Sabatino et al. (2017)
Fiber	High-fiber diets improve lipid profiles and reduce inflammation, which are risk factors for CKD progression.	Whole grains, fruits, vegetables, nuts, seeds.	Ranganathan et al. (2022)
Sodium	Restricting sodium intake can help control blood pressure and fluid buildup, which are common complications of chronic renal disease	Fresh fruits and vegetables, whole grains, lean proteins	Nerbass et al. (2015)

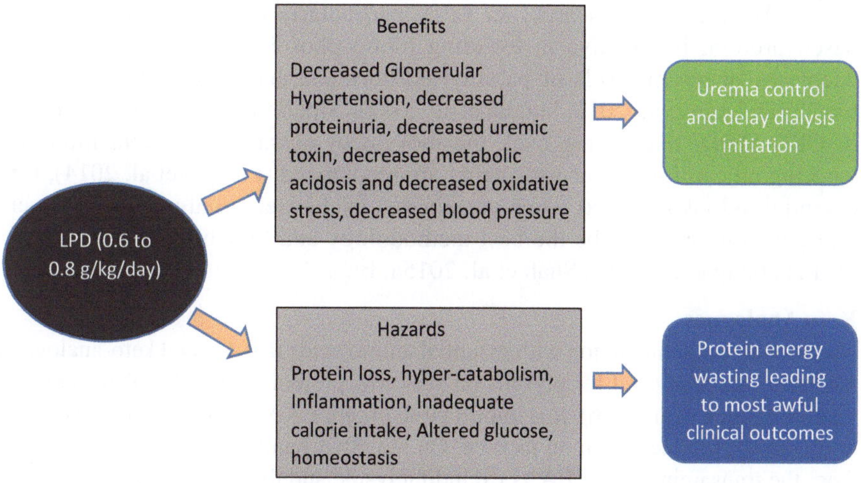

Fig. 6.2 Low protein diet in chronic renal disease

Fig. 6.3 High protein diet on kidney

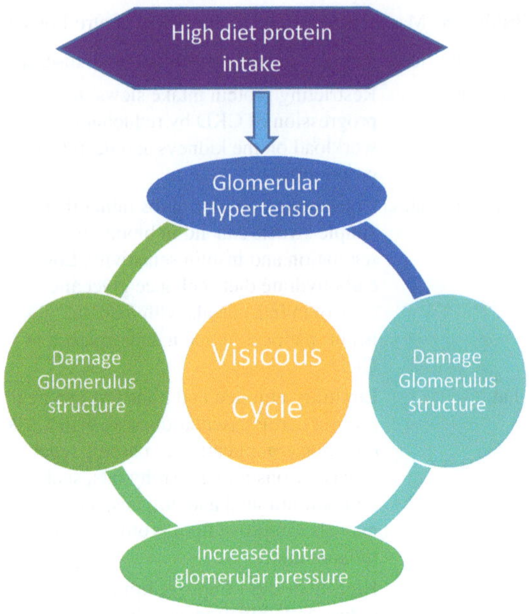

with advanced CRD. A study investigating the effects of an Improved Very Low Protein Diet showed that mean serum bicarbonate levels in the control (LPD) group remained below 19 mmol/L, while they rose to normal levels in the VLPD group (Garneata et al. 2016). LPD offers further developed control of ongoing kidney illness mineral and bone problems (CRD-Minerals and Bone Disorder (MBD)). The administration of hyperphosphatemia in CRD requires cautious dietary control, as dietary protein, particularly from creature sources, is a critical donor of phosphorus (St-Jules et al. 2016; Streja et al. 2013; Cupisti et al. 2013). According to research, an LPD, particularly one that contains plant-based protein, is effective in lowering blood phosphorus levels and causing decreases in serum levels of parathyroid hormone and fibroblast growth factor-23 (Di Iorio et al. 2012; Scialla et al. 2012). Albeit the advantages of an LPD in CRD-MBD, like easing back the movement of vascular calcification and working on cardiovascular well-being are recognized (Fouque et al. 2014), it is essential to look at that as a low-protein diet exclusively used to control hyperphosphatemia may not be the best methodology and may bring about adverse results (Jiang et al. 2016; Shah et al. 2015a, b).

Keto Analogues

Nutritional supplementation with essential amino acids (EAAs) and keto-analogues has been recommended for persons on very low protein diets (VLPD) for optimal health (Fouque et al. 2016). It is important to obtain EAAs from food sources as the body cannot produce them on its own. On the other hand, keto-analogues, derived from the transamination of EAAs excluding lysine and threonine, can be converted into individual EAAs without producing nitrogen waste (Shah et al. 2015a, b).

Protein-energy status can be maintained without the increase in nitrogen by-products with EAA and keto-analogue supplementation, phosphorus and acidic load can be reduced because purified EAA and keto-analogues contain no phosphorus, and protein degradation and synthesis can be improved. However, it has been argued that a long-term protein intake of less than 25% of recommended levels (as the minimum protein requirement to avoid negative nitrogen balance is less than 0.6 g/kg/day) may compromise overall health and well-being due to the risk of hypercatabolic conditions such as illness. This has sparked a rising tendency in the field of renal nutrition that contends that EAAs or keto-analogues should be added to even a protein intake of 0.6 g/kg/day (Kovesdy et al. 2013). It is recommended that at least half of the daily protein intake on both low protein diets and VLPD come from high biological value proteins. Although research on the use of keto-analogues in regulating CRD is still ongoing, the use of keto-analogues in conjunction with a VLPD has been demonstrated in certain trials to enhance metabolic control in instances of hyperphosphatemia and anaemia, two common consequences of CRD.

8 Diet Profile/Pattern of Patients with Acute Renal Disease and Chronic Renal Disease

Acute Renal Disease

A fast and temporary reduction in kidney function is a defining feature of acute renal disease, also known as acute kidney injury (AKI). If left untreated, AKI is a dangerous and often fatal illness that raises the risk of developing chronic renal disease and causes long-term kidney damage (CKD). Recent research has focused on determining the causes, risk factors, and treatment options for AKI. The three most prevalent causes of AKI, according to a 2019 systematic review published in the Journal of the American Society of Nephrology, are ischemia injury, nephrotoxic injury, and prerenal injury (Chawla et al. 2019). It demonstrates the need of identifying and treating these underlying causes early on in order to avert AKI and improve patient outcomes. According to a research published in the American Journal of Kidney Diseases in 2020, age, prior kidney disease, and particular medicines, such as nonsteroidal anti-inflammatory drugs (NSAIDs), are all important risk factors for AKI. This study also discovered that these risk variables should be continuously monitored and maintained in order to reduce the occurrence of AKI. Early identification and rapid treatment are critical for lowering the severity and duration of AKI and improving patient outcomes, according to a systematic review and meta-analysis published in the Clinical Journal of the American Society of Nephrology in 2018. (Furth et al. 2018). According to the authors, early identification and diagnosis of AKI can be aided by regular monitoring of renal function and the use of biomarkers such blood creatinine and urine output. According to a 2017 research in the Journal of Renal Nutrition, supportive care—which includes fluid management, electrolyte replacement, and nutrition intervention—helps patients with AKI feel better and experience less time in the hospital (Bhowmik et al. 2017).

The nutrition profile for individuals with acute renal illness is crucial for maintaining kidney function and avoiding additional injury. Recent study has highlighted the significance of customised dietary treatments in the treatment of acute renal damage (AKI). Fluid restriction was reported to be an effective therapy for minimising fluid buildup in patients with AKI in a comprehensive review and meta-analysis published in the Journal of Renal Care in 2018. (Hsu et al. 2018). It demonstrates that tailored fluid restriction based on patient-specific parameters including hydration status and urine output may be the most effective. A 2017 study published in the Journal of Renal Nutrition discovered that a low-potassium diet was beneficial in lowering potassium levels in the blood of AKI patients (Bhowmik et al. 2017). The study suggests closely monitoring potassium levels and tailored potassium limitation based on patient requirements. According to a research published in the American Journal of Kidney Diseases in 2020, phosphorus restriction was helpful in lowering blood phosphorus levels in AKI patients (van et al. 2020). The authors speculate that phosphorus restriction could benefit certain AKI patients, but they caution that more studies are required to completely comprehend how phosphorus restriction affects this population. Some medications and dietary supplements, including nonsteroidal anti-inflammatory drugs (NSAIDs) and vitamin D, may have a negative effect on kidney function and should be avoided or restricted in AKI patients, according to a comprehensive study that was published in the Journal of Nephrology in 2021 (Beng et al. 2021). The study suggests that AKI patients' medication and supplement use be closely monitored and controlled.

Patients with acute renal illness must eat a balanced and healthy diet to help them restore kidney function and general health. Individuals with acute renal illness may need to reduce their protein consumption to prevent additional kidney damage. However, it is still important for them to consume enough high-quality protein to support their overall health and recovery. Good protein sources for patients with the acute renal disease include lean meats, hens, fish, eggs, and dairy products. Fluid intake is critical for patients with acute renal disease, as their kidneys may not be able to effectively remove excess fluid from the body. It is important for these patients to monitor their fluid intake and limit their consumption of fluids if their doctor has advised them to do so. High sodium intake can increase fluid retention and put extra strain on the kidneys. AKI patients should reduce their sodium consumption and avoid high-sodium items including processed foods, canned soups, and fast food. Potassium is an essential element that aids in fluid equilibrium in the body. However, high potassium levels can be harmful to patients with acute renal disease and may lead to further damage to the kidneys. AKI sufferers should limit their potassium intake and avoid high-potassium foods such as bananas, oranges, and potatoes. Phosphorus is another important mineral that helps build and maintain strong bones. However, high phosphorus levels can also be harmful to AKI sufferers. Patients with AKI should reduce their phosphorus consumption and stay away from processed meals, meat, and dairy products, which are all high in phosphorus.

Overall patients with acute renal disease should follow a diet that is low in protein, sodium, potassium, and phosphorus. They should also consume enough high-quality protein and monitor their fluid intake. Patients with the acute renal disease need to work closely with a registered dietitian or a doctor to develop an individualized diet plan that meets their specific needs and supports their overall health and recovery.

Chronic Renal Disease

Patients with CRD require special dietary management to support their kidney function and overall health. Patients with CRD often need to limit their protein intake to reduce the workload on their kidneys (Kovesdy et al. 2013). However, it is important for these patients to still consume enough high-quality protein to support their overall health and recovery. Good protein sources for patients with CRD include lean meats, poultry, fish, eggs, and dairy products (Levey et al. 2003). Fluid management is important for patients with CRD, as their kidneys may not be able to effectively remove excess fluid from the body (Giamouzis et al. 2017). These patients may need to limit their fluid intake, particularly if their kidney function is severely compromised (Giamouzis et al. 2017). High sodium intake can increase fluid retention and put extra strain on the kidneys (Giamouzis et al. 2017). Patients with CRD should limit their sodium intake and avoid high-sodium foods such as processed foods, canned soups, and fast food (Giamouzis et al. 2017). Potassium is an important mineral that helps regulate fluid balance in the body (Levey et al. 2003). However, high potassium levels can be harmful to patients with CRD and may lead to further damage to the kidneys (Levey et al. 2003). Patients with CRD should limit their potassium intake and avoid high-potassium foods such as bananas, oranges, and potatoes (Levey et al. 2003). Phosphorus is another important mineral that helps build and maintain strong bones (Kovesdy et al. 2013). However, high phosphorus levels can also be harmful to patients with CRD (Kovesdy et al. 2013). Individuals who have CRD may need to watch their phosphorus intake and stay away from items high in phosphorus such dairy, meat, and processed foods (Kovesdy et al. 2013). When kidney function declines, patients with CRD may need to manage their consumption of additional vitamins and minerals, such as vitamin D and iron (Levey et al. 2003). They may need to take supplements or consume fortified foods to meet their nutritional needs (Levey et al. 2003).

Overall, patients with chronic renal disease should follow a diet that is low in protein, sodium, potassium, and phosphorus. They should also monitor their fluid intake and vitamin and mineral intake, as necessary. Patients with CRD need to work closely with a registered dietitian or a doctor to develop an individualized diet plan that meets their specific needs and supports their overall health and recovery.

Acute vs Chronic Renal Disease in Diet Profile

Acute renal disease is a sudden and severe impairment of renal function that may result from an injury, illness, or exposure to certain medications or toxins.

In this condition, the kidneys are still able to compensate and maintain fluid and electrolyte balance, but they may require temporary changes in the intake of certain nutrients. Individuals with acute renal disease may need to restrict their intake of fluids, sodium, and potassium. They may also need to increase their intake of protein, depending on the extent of the renal damage and their overall health status. On the other hand, chronic renal disease is a long-term, progressive reduction in renal function that over time causes the loss of kidney function. Individuals with chronic renal disease often require ongoing management of their dietary intake, as the disease progresses and the kidneys become less able to perform their functions. They may need to limit their intake of protein, phosphorus, potassium, and fluid, and may also need to take supplements of vitamins and minerals that their kidneys can no longer adequately retain. Furthermore, a low-sodium diet may also be necessary for patients who have chronic renal illness in order to assist control high blood pressure, which is a typical side effect of the condition.

9 Types of Diet Pattern for CRD Patients

The utilization of plant-based products has become progressively normal in the administration of CRD. Such eating regimens comprise basically plant-based food sources like organic products, vegetables, nuts, seeds, oils, entire grains, vegetables, and beans, and might possibly incorporate restricted measures of creature items like meat, fish, fish, eggs, or dairy. Various instances of plant-based counts calories incorporate veganism, the Mediterranean eating regimen, the Dietary Ways to deal with Stop Hypertension (Run) diet, and good dieting designs (Apetrii et al. 2021).

The DASH diet, also known as dietary approaches to stop hypertension, is an all-encompassing eating plan that emphasizes the consumption of vegetables, whole grains, fruits, low-fat dairy products, nuts, and moderate amounts of chicken, fish, and nuts, while limiting the intake of red and processed meats, salt, and sweetened beverages (Apetrii et al. 2021). On the other hand, the traditional Mediterranean diet places a strong emphasis on fresh foods such as fruits, vegetables, whole grain cereals, bread, beans, potatoes, nuts, and seeds, and recommends the use of extra virgin olive oil. It also suggests moderate consumption of dairy products, fish, and chicken, occasional use of wine during festivities, and rare consumption of red meat. Overall, both diets prioritize the consumption of healthy, nutrient-dense foods while limiting the intake of less nutritious options.

A healthy diet often includes a high intake of natural foods like fruits, vegetables, seafood, whole grains, cereals, and fibre while consuming moderate amounts of red meat, salt, and processed sweets. These dietary examples have been connected to decreased cardiovascular occasions and death rates in both sound grown-ups and those at high hazard for cardiovascular sickness (Apetrii et al. 2021) (Table 6.6).

Table 6.6 Different diet pattern studies with their interventions and results on CRD

Diet pattern	Reference	Study design	Participants	Intervention	Results
Plant-based	Banerjee et al. (2015)	Prospective cohort study	3972 patients with CRD	Dietary patterns (prudent, Western, and plant-based)	Decreased incidence of albuminuria and declining kidney function with a plant-based diet
High plant protein	Banerjee et al. (2015), Jhee et al. (2019)	Prospective cohort study	14,866 patients with CRD	High plant protein diet vs. low plant protein diet	A higher plant protein consumption is linked to a decreased risk of death from all causes
High protein	Jhee et al. (2019), Snelson et al. (2021), Li et al. (2019b), Jung et al. (2020)	Randomized controlled trial	164 healthy adults	High-protein diet vs. low-fat diet vs. carbohydrate-rich diet	No significant difference in kidney function between the three diet groups
Vegetarian very low-protein with Keto analogues	Bellizzi et al. (2022), Di Iorio et al. (2022)	Randomized controlled trial	120 patients with CRD	Keto analogue in a vegetarian extremely low-protein diet vs a traditional low-protein diet	Keto analogue group had a slower deterioration in renal function and a reduced likelihood of kidney failure
Whole grain	Miraghajani et al. (2019)	Systematic review and meta-analysis	13 observational studies and 12 clinical trials	Individuals with type 2 diabetes or those at risk for it	More consumption of whole grains is linked to a decreased incidence of type 2 diabetes
Legume	Marventano et al. (2017)	Systematic review and meta-analysis	26 observational studies	Participants with or at risk for cardiovascular disease	Increased consumption of legumes is linked to a decreased risk of cardiovascular disease

10 Correlation Between Ions Uptake and Progression of Chronic Renal Disease

The progression of CRD is associated with alterations in the renal handling of ions, including sodium, potassium, calcium, and phosphorus. As the disease advances, the kidney's ability to regulate these ions becomes diminished, leading to the development of several complications and co-morbidities. In people with CRD, sodium is a vital element that needs to be carefully regulated since high levels can cause fluid retention and higher blood pressure, both of which may speed up the disease's course (Cannata-Andia et al. 2018). Potassium is another ion that must be managed in individuals with CRD, as high levels can be harmful to the heart (Cannata-Andia et al. 2018). Calcium and phosphorus are also critical ions in the context of CRD. Calcium is crucial for maintaining bone health, while phosphorus is involved in various cellular processes (Cannata-Andia et al. 2018). In CRD, the kidney's ability to balance these ions becomes impaired, resulting in low calcium levels and increased phosphorus levels, which can result in the development of bone disease and other complications (Cannata-Andia et al. 2018). Alterations in the renal processing of ions are associated with the evolution of CRD. In order to create a specific strategy for regulating their ion levels and preserving their best health, people with CRD must collaborate closely with a healthcare provider. Since these changes can cause a wide range of problems and co-morbidities.

According to research, excessive sodium and low potassium levels are associated with the development and development of CRD. Sodium is a positively charged ion that is essential for maintaining fluid balance and blood pressure. Elevated degrees of sodium in the blood can be unsafe, as they can prompt liquid maintenance and expanded circulatory strain, the two of which are realized harmful factors for the turn of events and movement of CRD (Nordmann et al. 2019). Then again, potassium is an adversely charged particle that is engaged with different physiological cycles, including the guideline of circulatory strain and liquid equilibrium. Low degrees of potassium in the blood have been related to an expanded hazard of cardiovascular sickness, including CRD (Nordmann et al. 2019). According to research, people with CRD frequently have abnormal sodium and potassium levels, with excessive sodium and low potassium levels being frequent results (Nordmann et al. 2019). These alterations in ion levels can exacerbate the progression of CRD and contribute to the development of various complications and co-morbidities (Nordmann et al. 2019). Risk factors for the onset and progression of CRD include high salt and low potassium levels. As the disease progresses, alterations in ion levels can contribute to the development of various complications and co-morbidities, making it imperative for individuals with CRD to work closely with a healthcare professional to manage their ion levels and maintain optimal health (Nordmann et al. 2019).

11 Correlation Between Diet Pattern and Mortality Rate in Chronic Renal Disease

Diet is an indispensable component impacting the wellbeing results of people with CRD. Several research have been conducted to investigate the association between dietary patterns and the mortality rate in CRD. The findings of these research suggest that the quality of food intake might have a major influence on the health and survival of CRD patients. The experimental studies show that consuming a lot of animal protein (especially red and processed meats) increases the risk of mortality in CRD patients (Kovesdy et al. 2016). On the other side, a decreased death rate in CRD patients has been associated with a larger diet of plant-based protein, particularly legumes, nuts, and seeds (Levey et al. 2003). This suggests that the source of protein in the diet is a crucial factor in determining the health outcomes of CRD patients.

Fluid management is also an important aspect of diet in CRD patients, as their kidneys may not be able to effectively remove excess fluid from the body (Giamouzis et al. 2017). High sodium intake can increase fluid retention and put extra strain on the kidneys (Giamouzis et al. 2017). Patients with CRD should consume less salt and stay away from high-sodium items including processed meals, canned soups, and fast food to lower their risk of passing away (Levey et al. 2003). The balance of minerals in the diet can also affect the mortality rate of CRD patients. High potassium levels can be harmful to these patients and may lead to further damage to the kidneys (Levey et al. 2003). Patients with CRD should limit their potassium intake and avoid high-potassium foods, such as bananas, oranges, and potatoes, to reduce the risk of death (Levey et al. 2003). High phosphorus levels can also be harmful to CRD patients and may increase the risk of death (Kovesdy et al. 2016). To maintain their health, people with CRD may need to reduce their phosphorus consumption and stay away from high-phosphorus foods including dairy, meat, and processed foods. (Kovesdy et al. 2016).

The health and life of CRD patients depend on consuming a balanced and nourishing diet. A diet that is high in plant-based protein should be consumed by patients with CRD and low in animal protein, salt, potassium, and phosphorus. They should also monitor their fluid and mineral intake, as necessary, and work closely with a registered dietitian or doctor to develop an individualized diet plan that meets their specific needs and supports their overall health and recovery (Levey et al. 2003).

12 Conclusion

Nutraceuticals involving in the CRD therapy can complement traditional medical treatments, offering several benefits in managing the symptoms and slowing down the progression of the disease. Certain dietary supplements, such omega-3 fatty acids and antioxidants, enhance kidney function by reducing oxidative stress and inflammation. Others, like as calcium and vitamin D, are crucial for keeping bones strong and avoiding difficulties in CRD patients. However, it is important to note that while nutraceuticals can provide some therapeutic benefits, they should not be seen as a substitute for medical treatment. To choose the best course of action for treating their disease, patients with CRD should always consult carefully with their healthcare physician. This may involve a combination of conventional medical treatment, lifestyle changes, and the use of nutraceuticals, tailored to their individual needs. Overall, the use of nutraceuticals can provide a complementary approach to the management of CRD, helping patients to achieve better health results.

A well-balanced diet that emphasizes fruits and vegetables and limits animal protein can bring several benefits to individuals with CRD. It may alleviate the severity of symptoms in advanced stages and even decrease mortality. Reducing protein consumption may also lead to a reduction in sodium intake and improved blood pressure control. However, it can be challenging to determine the long-term health effects solely due to dietary changes, as other lifestyle factors like physical activity, avoiding tobacco, and moderate alcohol consumption may also play a role. Adherence to these dietary patterns may also be challenging for some patients, and a personalized approach should be taken to determine the best course of action. Comprehensive education, dietary advice, and routine monitoring by a renal dietitian are necessary to improve adherence. By effectively implementing a low- or very-low-protein diet, patients could save their kidneys in the long run.

The current chapter provided a systematic literature review on the nutraceuticals used in CRD prevention, such as antioxidants and anti-inflammatory compounds, which play a role in reducing oxidative stress and inflammation, both of which are important factors in the development and progression of CRD. Furthermore, certain nutraceuticals, including omega-3 fatty acids and phytochemicals, have demonstrated a protective impact on the kidneys by lowering oxidative stress and enhancing kidney performance. Research had shown that several metabolic pathways altering CRD, including those involved in glucose metabolism, lipid metabolism, and nitrogen metabolism provide a review for the development of effective treatments for CRD have been discussed. Genome and Proteome changes resulting in alterations in the genetic expression involved in kidney function and the production of proteins that have an impact on the disease. Micronutrients like vitamins, minerals and macronutrients like carbohydrates, proteins and omega-3 fatty acids etc. having efficiency in treatment are reviewed. Additionally, keto analogues with low-protein, high-fat diets, reported to improve kidney function in some patients with CRD, maintaining a healthy diet that is low in salt, fat, and sugar, and high in fiber and micronutrients have been proposed. Moreover, the correlation between ions uptake and progression of CRD indicated that sodium and potassium level i.e. high and low could increase the risk of developing CRD, while a balanced intake of these ions can help to reduce the risk describing the relation between diet pattern and mortality rate have been highlighted. Hence, this chapter not only enhances the understanding of reducing the role of pharmaceuticals while providing the efficiency of nutraceuticals for therapy in upcoming research and development.

References

Al Sunni A, Latif R. Effects of chocolate intake on perceived stress; a controlled clinical study. Int J Health Sci (Qassim). 2014;8(4):393–401. PMID 25780358, PMCID PMC4350893

Alge JL, Bekheirnia N, Willcockson AR, Qin X, Scherer SE, Braun MC, et al. Variants in genes coding for collagen type IV α-chains are frequent causes of persistent, isolated hematuria during childhood. Pediatr Nephrol. 2023;38(3):687–95. https://doi.org/10.1007/s00467-022-05627-w. PMID 35759000

Al-Sayed HA, Al-Sultan AI, Mahdy MA, et al. Diuretics use in chronic kidney disease: a systematic review and meta-analysis. Kidney Blood Press Res. 2020;45(5):460–9. https://doi.org/10.1159/000511422.

Anderson CE, Hamm LL, Batuman G, Kumbala DR, Chen CS, Kallu SG, et al. The association of angiogenic factors and chronic kidney disease. BMC Nephrol. 2018;19(1):117. https://doi.org/10.1186/s12882-018-0909-2. PMID 29783932

Anguiano L, Riera M, Pascual J, Valdivielso JM, Barrios C, Betriu A, et al. Circulating angiotensin-converting enzyme 2 activity in patients with chronic kidney disease without previous history of cardiovascular disease. Nephrol Dial Transplant. 2015;30(7):1176–85. https://doi.org/10.1093/ndt/gfv025. PMID 25813276

Apetrii M, Timofte D, Voroneanu L, Covic A. Nutrition in chronic kidney disease-the role of proteins and specific diets. Nutrients. 2021;13(3):956. https://doi.org/10.3390/nu13030956. PMID 33809492, PMCID PMC7999704

Arablou T, Aryaeian N, Khodaverdi S, Kolahdouz-Mohammadi R, Moradi Z, Rashidi N, et al. The effects of resveratrol on the expression of VEGF, TGF-β, and MMP-9 in endometrial stromal cells of women with endometriosis. Sci Rep. 2021;11(1):6054. https://doi.org/10.1038/s41598-021-85512-y. PMID 33723310

Awotidebe A, Monyeki MA, Moss SJ, Strydom GL, Amstrong M, Kemper HC. Relationship of adiposity and cardiorespiratory fitness with resting blood pressure of South African adolescents: the PAHL study. J Hum Hypertens. 2016;30(4):245–51. https://doi.org/10.1038/jhh.2015.81. PMID 26202691

Baek J, He C, Afshinnia F, Michailidis G, Pennathur S. Lipidomic approaches to dissect dysregulated lipid metabolism in kidney disease. Nat Rev Nephrol. 2022;18(1):38–55. https://doi.org/10.1038/s41581-021-00488-2. PMID 34616096

Bahrami A, Bayat H, Rajaei A, Mirmiran P. Effect of vitamin C supplementation on oxidative stress and lipid profile in chronic kidney disease: a systematic review and meta-analysis of clinical trials. J Ren Nutr. 2020;30(2):132–42. https://doi.org/10.1053/j.jrn.2019.07.008.

Banerjee T, Crews DC, Wesson DE, Tilea AM, Saran R, Ríos-Burrows N, et al. High dietary acid load predicts ESRD among adults with CKD. J Am Soc Nephrol. 2015;26(7):1693–700. https://doi.org/10.1681/ASN.2014040332. PMID 25677388

Batchelor EK, Kapitsinou P, Pergola PE, Kovesdy CP, Jalal DI. Iron deficiency in chronic kidney disease: updates on pathophysiology, diagnosis, and treatment. J Am Soc Nephrol. 2020;31(3):456–68. https://doi.org/10.1681/ASN.2019020213. PMID 32041774

Bawazeer MA, Al-Qahtani AH, Sulaiman A, et al. Vitamin D supplementation in chronic kidney disease: a systematic review and meta-analysis. Sci Rep. 2020;10(1):1. https://doi.org/10.1038/s41598-020-56906-5.

Bellizzi V. Low-protein diet or nutritional therapy in chronic kidney disease? Blood Purif. 2013;36(1):41–6. https://doi.org/10.1159/000350585. PMID 23735624

Bellizzi V, Cupisti A, Capitanini A, et al. Ketoacid analogues supplementation in chronic kidney disease and future perspectives. J Ren Nutr. 2022;32(1):15–22. https://doi.org/10.1053/j.jrn.2021.06.006.

Beng T, Wong J, Toh S. The effects of medication and supplement use on kidney function in patients with acute kidney injury: a systematic review. J Nephrol. 2021;34(1):61–8. https://doi.org/10.1007/s40620-020-00739-0.

Bhowmik B, Fenton J, McLaughlin J. The effect of a low-potassium diet on serum potassium levels in patients with acute kidney injury: a randomized controlled trial. J Ren Nutr. 2017;27(1):23–8. https://doi.org/10.1053/j.jrn.2016.07.005.

Bikbov B, Perico N, Remuzzi G. On behalf of the GBD genitourinary diseases expert group. disparities in chronic kidney disease prevalence among males and females in 195 countries: analysis of the global burden of disease 2016 study. Nephron. 2018;139(4):313–8. https://doi.org/10.1159/000489897. PMID 29791905

Bjørke-Monsen AL, Ulvik A, Nilsen RM, Midttun Ø, Roth C, Magnus P, et al. Impact of pre-pregnancy BMI on B vitamin and inflammatory status in early pregnancy: an observational cohort study. Nutrients. 2016;8(12):776. https://doi.org/10.3390/nu8120776. PMID 27916904, PMCID PMC5188431

Cannata-Andia JB, Egido J, Rodriguez-Manzaneque JC, Hernandez-Alejandro R. Chronic kidney disease and minerals. World J Nephrol. 2018;7(4):133–9.

Chauhan N, Shukla KK, Trivedi PC. Oxidative stress in chronic kidney disease and potential antioxidants. J Trace Elem Med Biol. 2019;54:51–60. https://doi.org/10.1016/j.jtemb.2018.12.007.

Chawla LS, Zerres K. Causes of acute kidney injury: a systematic review. J Am Soc Nephrol. 2019;30(1):3–4. https://doi.org/10.1681/ASN.2018080879.

Chen CH, Yang WC, Hsiao YH, Huang SC, Huang YC. High homocysteine, low vitamin B-6, and increased oxidative stress are independently associated with the risk of chronic kidney disease. Nutrition. 2016;32(2):236–41. https://doi.org/10.1016/j.nut.2015.08.016. PMID 26526964

Chen H, Yang T, Wang MC, Chen DQ, Yang Y, Zhao YY. Novel RAS inhibitor 25-O-methylalisol F attenuates epithelial-to-mesenchymal transition and tubulo-interstitial fibrosis by selectively inhibiting TGF-β-mediated Smad3 phosphorylation. Phytomedicine. 2018;42:207–18. https://doi.org/10.1016/j.phymed.2018.03.034. PMID 29655688

Chowdhury R, Stevens S, Gorman D. Vitamin D and chronic kidney disease: a review. J Ren Nutr. 2020;30(1):1–10. https://doi.org/10.1053/j.jrn.2019.06.008.

Chueakula N, Jaikumkao K, Arjinajarn P, Pongchaidecha A, Chatsudthipong V, Chattipakorn N, et al. Diacerein alleviates kidney injury through attenuating inflammation and oxidative stress in obese insulin-resistant rats. Free Radic Biol Med. 2018;115:146–55. https://doi.org/10.1016/j.freeradbiomed.2017.11.021. PMID 29195834

Cigarran-Guldris S, Gonzalez-Parra E, Martin-Malo A, Rodriguez-Benot A. Metabolic disturbances in chronic kidney disease. Lancet Diabetes Endocrinol. 2019;7(5):364–76.

Cupisti A, Kalantar-Zadeh K. Management of natural and added dietary phosphorus burden in kidney disease. Semin Nephrol. 2013;33(2):180–90. https://doi.org/10.1016/j.semnephrol.2012.12.018. PMID 23465504

Cushman WC, Williams DD, Dennison-Himmelfarb C, Glasser SP, Jaar BG, Fink NE, et al. Angiotensin receptor blockers in the management of hypertension and renal disease: a scientific statement from the American Heart Association. Hypertension. 2019;74(3):489–519. https://doi.org/10.1161/HYP.0000000000000080.

Daenen K, Andries A, Mekahli D, Van Schepdael A, Jouret F, Bammens B. Oxidative stress in chronic kidney disease. Pediatr Nephrol. 2019;34(6):975–91. https://doi.org/10.1007/s00467-018-4005-4. PMID 30105414

De Boer IH, Sachs M, Hoofnagle AN, Utzschneider KM, Kahn SE, Kestenbaum B, et al. Paricalcitol does not improve glucose metabolism in patients with stage 3-4 chronic kidney disease. Kidney Int. 2013;83(2):323–30. https://doi.org/10.1038/ki.2012.311. PMID 22913981

Di Iorio BR, De Santo NG. Clinical experience with very low-protein diets for chronic kidney disease patients. J Ren Nutr. 2022;32(2):161–9. https://doi.org/10.1053/j.jrn.2021.07.001.

Di Iorio B, Di Micco L, Torraca S, Sirico ML, Russo L, Pota A, et al. Acute effects of very-low-protein diet on FGF23 levels: a randomized study. Clin J Am Soc Nephrol. 2012;7(4):581–7. https://doi.org/10.2215/CJN.07640711. PMID 22362063

Farhat G, Drummond LR, Goulart RB, Uchida AH. Cocoa and cardiovascular health: effects on endothelial function and insulin resistance in humans. Nutrients. 2020;12(7):2192. https://doi.org/10.3390/nu12072192.

Fattinger K, Simon A, Heg D, Gutte H. Iron therapy in chronic kidney disease-associated anemia: a narrative review. Swiss Med Wkly. 2019;149:w20306. https://doi.org/10.4414/smw.2019.20306.

Feng S, Sun Y. Protective role of propofol in endometriosis and its mechanism. Exp Ther Med. 2018;16(4):3646–50. https://doi.org/10.3892/etm.2018.6648. PMID 30233720, PMCID PMC6143856

Ferri FF. Ferri's clinical advisor 2018 e-book: 5. Books in 1. Elsevier Health Sciences; 2017. p. 294–5. ISBN 9780323529570

Filiopoulos V, Hadjiyannakos D, Vlassopoulos D, Arvanitis M. Metabolic abnormalities in patients with chronic kidney disease. Hippokratia. 2016;20(1):7–14.

Filler G, Qiu Y, Kaskel F, McIntyre CW. Principles responsible for trace element concentrations in chronic kidney disease. Clin Nephrol. 2021;96(1):1–16. https://doi.org/10.5414/CN110322. PMID 33960297

Fouque D, Aparicio M. Eleven reasons to control the protein intake of patients with chronic kidney disease. Nat Clin Pract Nephrol. 2007;3(7):383–92. https://doi.org/10.1038/ncpneph0524. PMID 17592471

Fouque D, Horne R, Cozzolino M, Kalantar-Zadeh K. Balancing nutrition and serum phosphorus in maintenance dialysis. Am J Kidney Dis. 2014;64(1):143–50. https://doi.org/10.1053/j.ajkd.2014.01.429. PMID 24819675

Fouque D, Chen J, Chen W, Garneata L, Hwang SJ, Kalantar-Zadeh K, et al. Adherence to ketoacids/essential amino acids-supplemented low protein diets and new indications for patients with chronic kidney disease. BMC Nephrol. 2016;17(1):63. https://doi.org/10.1186/s12882-016-0278-7. PMID 27388899:63. This review showed the impact of low protein diet supplemented with ketoacid on metabolic, nutritional and renal parameters, and gave information how to improve adherence in diet prescription

Friedman DJ, Pollak MR. Apolipoprotein L1 and kidney disease in African Americans. Trends Endocrinol Metab. 2016;27(4):204–15. https://doi.org/10.1016/j.tem.2016.02.002. PMID 26947522

Fukasawa H, Furuya R, Kaneko M, Nakagami D, Ishino Y, Kitamoto S, et al. Clinical significance of trace element zinc in patients with chronic kidney disease. J Clin Med. 2023;12(4):1667. https://doi.org/10.3390/jcm12041667. PMID 36836202

Furlano M, Arlandis R, Venegas MDP, Novelli S, Crespi J, Bullich G, et al. MYH9 associated nephropathy. Nefrol (Engl Ed). 2019;39(2):133–40. https://doi.org/10.1016/j.nefro.2018.08.008. PMID 30471777

Furth S, Finkel K, Brookhart MA, Schmieder RE. Early detection and prompt intervention in acute kidney injury: a systematic review and meta-analysis. Clin J Am Soc Nephrol. 2018;13(8):1178–89. https://doi.org/10.2215/CJN.06320317.

Garibotto G, Sofia A, Saffioti S, Bonanni A, Mannucci I, Verzola D. Amino acid and protein metabolism in the human kidney and in patients with chronic kidney disease. Clin Nutr. 2010;29(4):424–33. https://doi.org/10.1016/j.clnu.2010.02.005. PMID 20207454

Garneata L, Stancu A, Dragomir D, Stefan G, Mircescu G. Ketoanalogue-supplemented vegetarian very low-protein diet and CKD progression. J Am Soc Nephrol. 2016;27(7):2164–76. https://doi.org/10.1681/ASN.2015040369. PMID 26823552

Gerrits T, Brouwer IJ, Dijkstra KL, Wolterbeek R, Bruijn JA, Scharpfenecker M, et al. Endoglin is an important mediator in the final common pathway of chronic kidney disease to end-stage renal disease. Int J Mol Sci. 2022;24(1):646. https://doi.org/10.3390/ijms24010646. PMID 36614087

Giamouzis G, Chappell MC. Fluid management in chronic kidney disease. Nat Rev Nephrol. 2017;13(3):157–68.

Go AS, Chertow GM, Fan D, McCulloch CE, Hsu CY. Chronic kidney disease and the risks of death, cardiovascular events, and hospitalization. N Engl J Med. 2004;351(13):1296–305. https://doi.org/10.1056/NEJMoa041031.

Goraya N, Wesson DE. Dietary management of chronic kidney disease: protein restriction and beyond. Curr Opin Nephrol Hypertens. 2012;21(6):635–40. https://doi.org/10.1097/MNH.0b013e328357a69b. PMID 23079747

Goraya N, Wesson DE. Dietary interventions to improve outcomes in chronic kidney disease. Curr Opin Nephrol Hypertens. 2015;24(6):505–10. https://doi.org/10.1097/MNH.0000000000000160. PMID 26335553

Griffin JW, Bradshaw PC. Amino acid catabolism in Alzheimer's disease brain: friend or foe? Oxidative Med Cell Longev. 2017;2017:5472792. https://doi.org/10.1155/2017/5472792. PMID 28261376

Guo Y, Ma J, Xiao L, Fang J, Li G, Zhang L, et al. Identification of key pathways and genes in different types of chronic kidney disease based on WGCNA. Mol Med Rep. 2019;20(3):2245–57. https://doi.org/10.3892/mmr.2019.10443. PMID 31257514

Gupta G, Dahiya R, Singh Y, Mishra A, Verma A, Gothwal SK, et al. Monotherapy of RAAS blockers and mobilization of aldosterone: a mechanistic perspective study in kidney disease. Chem Biol Interact. 2020;317:108975. https://doi.org/10.1016/j.cbi.2020.108975. PMID 32032593

Hager MR, Narla AD, Tannock LR. Dyslipidemia in patients with chronic kidney disease. Rev Endocr Metab Disord. 2017;18(1):29–40. https://doi.org/10.1007/s11154-016-9402-z. PMID 28000009

Hall JE, do Carmo JM, da Silva AA, Wang Z, Hall ME. Obesity-induced hypertension: interaction of neurohumoral and renal mechanisms. Circ Res. 2015;116(6):991–1006. https://doi.org/10.1161/CIRCRESAHA.116.305697. PMID 25767285

Heerspink HJL, Jager KJ, Hillege HL, de Zeeuw D, de Graeff PA. Angiotensin-converting enzyme inhibitors and angiotensin receptor blockers in diabetic kidney disease: a systematic review. Am J Kidney Dis. 2017;70(2):142–52. https://doi.org/10.1053/j.ajkd.2017.01.011.

Heyer CM, Sundsbak JL, Abebe KZ, Chapman AB, Torres VE, Grantham JJ, et al. Predicted mutation strength of nontruncating PKD1 mutations aids genotype-phenotype correlations in autosomal dominant polycystic kidney disease. J Am Soc Nephrol. 2016;27(9):2872–84. https://doi.org/10.1681/ASN.2015050583. PMID 26823553

Hirahashi J. Omega-3 polyunsaturated fatty acids for the treatment of IgA nephropathy. J Clin Med. 2017;6(7):70. https://doi.org/10.3390/jcm6070070. PMID 28753924

Hongsawong N, Chawprang N, Kittisakmontri K, Vittayananan P, Srisuwan K, Chartapisak W. Vitamin C deficiency and impact of vitamin C administration among pediatric patients with advanced chronic kidney disease. Pediatr Nephrol. 2021;36(2):397–408. https://doi.org/10.1007/s00467-020-04662-9. PMID 32683655

Hsu P, Li Y, Hung C. Fluid restriction for fluid management in patients with acute kidney injury: a systematic review and meta-analysis. J Ren Care. 2018;44(2):85–94. https://doi.org/10.1111/jorc.12266.

Hu X, Sun H, Gao Y, Li W, Hu J. Efficacy and safety of erythropoiesis-stimulating agents for anemia management in chronic kidney disease: a network meta-analysis. Ren Fail. 2020;42(6):633–44. https://doi.org/10.1080/0886022X.2020.1751936.

Hung RKY, Binns-Roemer E, Booth JW, Hilton R, Harber M, Santana-Suarez B, et al. Genetic variants of APOL1 are major determinants of kidney failure in people of African ancestry with HIV. Kidney Int Rep. 2022;7(4):786–96. https://doi.org/10.1016/j.ekir.2022.01.1054. PMID 35497797

Ibrahim MM, Al-Maskari F, Lawati JA, et al. The effect of angiotensin-converting enzyme inhibitors and angiotensin receptor blockers on chronic kidney disease progression and cardiovascular outcomes: a systematic review and meta-analysis. Am J Kidney Dis. 2019;74(3):311–21. https://doi.org/10.1053/j.ajkd.2019.03.013.

Isaka Y. Optimal protein intake in pre-dialysis chronic kidney disease patients with sarcopenia: an overview. Nutrients. 2021;13(4):1205. https://doi.org/10.3390/nu13041205. PMID 33917381

Jafari T, Fazeli B, Kia M, Ahmadzadeh A, Daneshi-Maskooni M. The effect of folic acid supplementation on the progression of diabetic nephropathy: a randomized controlled trial. J Ren Nutr. 2019;29(4):301–8. https://doi.org/10.1053/j.jrn.2018.09.005.

Jakše B, Jakše B, Pajek M, Pajek J. Uric acid and plant-based nutrition. Nutrients. 2019;11(8):1736. https://doi.org/10.3390/nu11081736. PMID 31357560

Jean G, Souberbielle JC, Chazot C. Vitamin D in chronic kidney disease and dialysis patients. Nutrients. 2017;9(4):328. https://doi.org/10.3390/nu9040328. PMID 28346348

Jhee JH, Kee YK, Park S, et al. High-protein diet with renal hyperfiltration is associated with rapid decline rate of renal function: a community-based prospective cohort study. Nephrol Dial Transplant. 2019;34(5):853–60. https://doi.org/10.1093/ndt/gfy206.

Jia J, Li X, Li Y, Li Y, Wang Y. Diuretics for the treatment of chronic kidney disease: a systematic review and meta-analysis. Clin Exp Nephrol. 2021;25(2):289–98. https://doi.org/10.1007/s10157-020-01967-9.

Jiang Z, Zhang X, Yang L, Li Z, Qin W. Effect of restricted protein diet supplemented with keto analogues in chronic kidney disease: a systematic review and meta-analysis. Int Urol Nephrol. 2016;48(3):409–18. https://doi.org/10.1007/s11255-015-1170-2. PMID 26620578

Jung CH, Choi KM. Impact of high-carbohydrate diet on metabolic parameters in patients with type 2 diabetes. Nutrients. 2020;12(12):3779. https://doi.org/10.3390/nu12123779.

Kalantar-Zadeh K, Fouque D. Nutritional management of chronic kidney disease. N Engl J Med. 2017;377(18):1765–76. https://doi.org/10.1056/NEJMra1700312. PMID 29091561

KDIGO. Kidney disease improving global outcomes. "KDIGO clinical practice guideline for the diagnosis, evaluation, prevention, and treatment of chronic kidney disease-mineral and bone disorder (CKD-MBD)" (PDF). Kidney Int. 2009;76(Suppl 113) Archived from the original (PDF) on 2016-12-13

Khanaghaei M, Ziamajidi N, Poorolajal J, Asadikaram G, Nikoyan P, Bajian M, et al. The influence of gastric bypass surgery on the concentration of high mobility group box 1, nuclear factor erythroid 2–related factor 2 and the genes expression of high mobility group box 1, nuclear factor erythroid2–related factor 2, interleukin 6, and tumor necrosis factor-alpha in the peripheral blood mononuclear cells of patients with morbid obesity. Mol Biol Rep. 2022;49(5):3745–55. https://doi.org/10.1007/s11033-022-07214-6. PMID 35107739

Khandelwal P, Murugan V, Hari S, Lakshmy R, Sinha A, Hari P, et al. Dyslipidemia, carotid intima-media thickness and endothelial dysfunction in children with chronic kidney disease. Pediatr Nephrol. 2016;31(8):1313–20. https://doi.org/10.1007/s00467-016-3350-4. PMID 26921213

Kim CS, Kim SW. Vitamin D and chronic kidney disease. Korean J Intern Med. 2014;29(4):416–27. https://doi.org/10.3904/kjim.2014.29.4.416. PMID 25045287

Kim TH, Shin S, Yoo SD, Shin BS. Effects of phytochemical P-glycoprotein modulators on the pharmacokinetics and tissue distribution of doxorubicin in mice. Molecules. 2018;23(2):349. https://doi.org/10.3390/molecules23020349. PMID 29414892, PMCID PMC6017107

Kistler BM, Moore LW, Benner D, Biruete A, Boaz M, Brunori G, et al. The International Society of Renal Nutrition and Metabolism commentary on the National Kidney Foundation and Academy of Nutrition and Dietetics KDOQI clinical practice guideline for nutrition in chronic kidney disease. J Ren Nutr. 2021;31(2):116–120.e1. https://doi.org/10.1053/j.jrn.2020.05.002. PMID 32737016

Klahr S, Levey AS, Beck GJ, Caggiula AW, Hunsicker L, Kusek JW, et al. The effects of dietary protein restriction and blood-pressure control on the progression of chronic renal disease. N Engl J Med. 1982;307(11):652–9.

Ko GJ, Obi Y, Tortorici AR, Kalantar-Zadeh K. Dietary protein intake and chronic kidney disease. Curr Opin Clin Nutr Metab Care. 2017;20(1):77–85. https://doi.org/10.1097/MCO.0000000000000342. PMID 27801685

Kovesdy CP, Kalantar-Zadeh K. Back to the future: restricted protein intake for conservative management of CKD, triple goals of renoprotection, uremia mitigation, and nutritional health. Back to the future. Int Urol Nephrol. 2016;48(5):725–9. https://doi.org/10.1007/s11255-016-1224-0. PMID 26886110

Kovesdy CP, Kopple JD, Kalantar-Zadeh K. Management of protein-energy wasting in non-dialysis-dependent chronic kidney disease: reconciling low protein intake with nutritional therapy. Am J Clin Nutr. 2013;97(6):1163–77. https://doi.org/10.3945/ajcn.112.036418. PMID 23636234

Krishnan ST, Devadhasan JP, Kim S. Recent analytical approaches to detect exhaled breath ammonia with special reference to renal patients. Anal Bioanal Chem. 2017;409(1):21–31. https://doi.org/10.1007/s00216-016-9903-3. PMID 27595582

Kumar V, Singhal PC. APOL1 and kidney cell function. Am J Physiol Ren Physiol. 2019;317(2):F463–77. https://doi.org/10.1152/ajprenal.00233.2019. PMID 31241995

Lee BT, Ahmed FA, Hamm LL, Teran FJ, Chen CS, Liu Y, et al. Association of C-reactive protein, tumor necrosis factor-alpha, and interleukin-6 with chronic kidney disease. BMC Nephrol. 2015;16(1):77. https://doi.org/10.1186/s12882-015-0068-7.

Levey AS, Coresh J, Balk E, Kausz AT, Levin A. National Kidney Foundation practice guidelines for chronic kidney disease: evaluation, classification, and stratification. Ann Intern Med. 2003;139(2):137–47. https://doi.org/10.7326/0003-4819-139-2-200307150-00009.

Li H, Sun H, Li L, Chen Y. Erythropoiesis-stimulating agents in the treatment of anemia in patients with chronic kidney disease: a network meta-analysis. Ren Fail. 2019a;41(5):539–48. https://doi.org/10.1080/0886022X.2019.1621352.

Li H, Li J, Li Y, Wang Y, Hu S. The effects of low-carbohydrate diets on cardiovascular risk factors: a meta-analysis. PLoS One. 2019b;14(12):e0226205. https://doi.org/10.1371/journal.pone.0226205.

Li G, Huang Y, Lu L, Xi Y, Chen C. Effects of green tea and black tea consumption on metabolic disorders and renal function: a systematic review and meta-analysis. Nutr J. 2021;20(1):1. https://doi.org/10.1186/s12937-021-00676-3.

Liao MT, Sung CC, Hung KC, Wu CC, Lo L, Lu KC. Insulin resistance in patients with chronic kidney disease. J Biomed Biotechnol. 2012;2012:691369. https://doi.org/10.1155/2012/691369.

Mafra D, Borges NA, Lindholm B, Shiels PG, Evenepoel P, Stenvinkel P. Food as medicine: targeting the uraemic phenotype in chronic kidney disease. Nat Rev Nephrol. 2021;17(3):153–71. https://doi.org/10.1038/s41581-020-00345-8. PMID 32963366

Mai M, Jiang Y, Wu X, Liu G, Zhu Y, Zhu W. Association of TGF-β1, IL-4, and IL-10 polymorphisms with chronic kidney disease susceptibility: a meta-analysis. Front Genet. 2020;11:79. https://doi.org/10.3389/fgene.2020.00079. PMID 32174964

Marventano S, Izquierdo Pulido M, Sánchez-González C, Godos J, Speciani A, Galvano F, et al. Legume consumption and CVD risk: a systematic review and meta-analysis. Public Health Nutr. 2017;20(2):245–54. https://doi.org/10.1017/S1368980016002299. PMID 28077199

Mirabelli M, Chiefari E, Arcidiacono B, Corigliano DM, Brunetti FS, Maggisano V, et al. Mediterranean diet nutrients to turn the tide against insulin resistance and related diseases. Nutrients. 2020;12(4):1066. https://doi.org/10.3390/nu12041066. PMID 32290535

Miraghajani M, Salehi-Abargouei A, Mirlohi M, et al. Whole-grain consumption and risk of type 2 diabetes: a systematic review and dose-response meta-analysis of cohort studies. Eur J Epidemiol. 2019;34(10):917–26. https://doi.org/10.1007/s10654-019-00559-7.

Mollazadeh H, Cicero AFG, Blesso CN, Pirro M, Majeed M. Potential role of bioavailable curcumin in weight loss and omental adipose tissue decrease: preliminary data of a randomized, controlled trial in overweight people with chronic kidney disease. Nutr J. 2019a;18(1):1. https://doi.org/10.1186/s12937-019-0497-6.

Mollazadeh H, Hosseinzadeh H, Sadeghnia HR. Paradoxical effects of nutraceuticals and dietary supplements on glucose regulation. Nutrients. 2019b;11(9):2022.

Mousa HA, Nahari OA, et al. Angiotensin receptor blockers in the management of hypertension and chronic kidney disease: a review. Curr Hypertens Rep. 2018;20(4):25. https://doi.org/10.1007/s11906-018-0826-4.

Naghavi M, Wang H, Lozano R, Davis A, Liang X, Zhou M. Global, regional, and national age-sex specific all-cause and cause-specific mortality for 240 causes of death, 1990–2013: a systematic analysis for the Global Burden of Disease Study 2013. Lancet. 2015;385(9963):117–71. https://doi.org/10.1016/S0140-6736(14)61682-2. PMCID 4340604. PMID 25530442 (GBD. Mortality and causes of death collaborators; 2013). Table 2, p. 137

National Institute of Diabetes and Digestive and Kidney Diseases. What is chronic kidney disease? June 2017.

Navarro-Gonzalez JF, Castillo-Garit JC, Alamo-Izquierdo LC, Hernandez-Garcia C. Omega-3 polyunsaturated fatty acids as a complementary therapy for chronic kidney disease: a review. Nutr Hosp. 2020;37(2):235–42. https://doi.org/10.20960/nh.01938.

Nazarko L, Iqbal J. Erythropoiesis-stimulating agents in the management of anemia in chronic kidney disease. Br J Nurs. 2018;27(20):S26–32. https://doi.org/10.12968/bjon.2018.27.20.S26.

Nerbass FB, Pecoits-Filho R, McIntyre NJ, McIntyre CW, Taal MW. High sodium intake is associated with important risk factors in a large cohort of chronic kidney disease patients. Eur J Clin Nutr. 2015;69(7):786–90. https://doi.org/10.1038/ejcn.2014.215. PMID 25293433

Nordmann AJ, et al. Salt and potassium intake and cardiovascular disease. Lancet. 2019;393(10170):661–75.

Pavkov ME, Nelson RG, Knowler WC, Cheng Y, Krolewski AS, Niewczas MA. Elevation of circulating TNF receptors 1 and 2 increases the risk of end-stage renal disease in American Indians with type 2 diabetes. Kidney Int. 2015;87(4):812–9. https://doi.org/10.1038/ki.2014.330. PMID 25272234

Raj SR, Shankar AU, Dubey A, Kumawat S. Potassium and magnesium supplementation in chronic kidney disease: a review. J Nephropathol. 2018;7(3):127–31. https://doi.org/10.15171/jnp.2018.30.

Ramanathan ASK, Vijayan M, Rajagopal S, Rajendiran P, Senguttuvan P. WT1 and NPHS2 gene mutation analysis and clinical management of steroid-resistant nephrotic syndrome. Mol Cell Biochem. 2017;426(1–2):177–81. https://doi.org/10.1007/s11010-016-2889-5. PMID 27885584

Remuzzi A, Remuzzi G. Chronic kidney disease: the global epidemic. Lancet. 2017;389(10064):347–58. https://doi.org/10.1016/S0140-6736(16)31484-9.

Ranganathan, N., & Anteyi, E. (2022). The role of dietary fiber and gut microbiome modulation in progression of chronic kidney disease. *Toxins, 14*(3), 183. https://doi.org/10.3390/toxins14030183

Romanova Y, Laikov A, Markelova M, Khadiullina R, Makseev A, Hasanova M, et al. Proteomic analysis of human serum from patients with chronic kidney disease. Biomol Ther. 2020;10(2):257. https://doi.org/10.3390/biom10020257. PMID 32046176

Rust P, Ekmekcioglu C. Impact of salt intake on the pathogenesis and treatment of hypertension. Hypertension: from basic research to clinical practice. 2017:61–84.

Sabatino A, Regolisti G, Gandolfini I, Delsante M, Fani F, Gregorini MC, et al. Diet and enteral nutrition in patients with chronic kidney disease not on dialysis: a review focusing on fat, fiber and protein intake. J Nephrol. 2017;30(6):743–54. https://doi.org/10.1007/s40620-017-0435-5. PMID 28884267

Saglimbene VM, Wong G, van Zwieten A, Palmer SC, Ruospo M, Natale P, et al. Effects of omega-3 polyunsaturated fatty acid intake in patients with chronic kidney disease: systematic review and meta-analysis of randomized controlled trials. Clin Nutr. 2020;39(2):358–68. https://doi.org/10.1016/j.clnu.2019.02.041. PMID 30905498

Sanchis-Gomar F, Lavie CJ, Perez-Quilis C, Henry BM, Lippi G. Angiotensin-converting enzyme 2 and antihypertensives (angiotensin receptor blockers and angiotensin-converting enzyme inhibitors) in coronavirus disease 2019. Mayo Clin Proc. 2020;95(6):1222–30. https://doi.org/10.1016/j.mayocp.2020.03.026. PMID 32376099

Saran R, Li Y, Robinson B, Abbott KC, Agodoa LY, Ayanian J, et al. US renal data system 2014 annual data report: epidemiology of kidney disease in the United States. Am J Kidney Dis. 2015;66(1):Svii, S1–305. https://doi.org/10.1053/j.ajkd.2015.05.001. PMID 26111994

Schrauben SJ, Jepson C, Hsu JY, Wilson FP, Zhang X, Lash JP, et al. Insulin resistance and chronic kidney disease progression, cardiovascular events, and death: findings from the chronic renal insufficiency cohort study. BMC Nephrol. 2019;20(1):60. https://doi.org/10.1186/s12882-019-1220-6. PMID 30786864

Scialla JJ, Anderson CA. Dietary acid load: a novel nutritional target in chronic kidney disease? Adv Chronic Kidney Dis. 2013;20(2):141–9. https://doi.org/10.1053/j.ackd.2012.11.001. PMID 23439373

Scialla JJ, Appel LJ, Wolf M, Yang W, Zhang X, Sozio SM, et al. Plant protein intake is associated with fibroblast growth factor 23 and serum bicarbonate levels in patients with chronic kidney disease: the chronic renal insufficiency cohort study. J Ren Nutr. 2012;22(4):379–388.e1. https://doi.org/10.1053/j.jrn.2012.01.026. PMID 22480598

Shah AP, Kalantar-Zadeh K, Kopple JD. Is there a role for ketoacid supplements in the management of CKD? Am J Kidney Dis. 2015a;65(5):659–73. https://doi.org/10.1053/j.ajkd.2014.09.029. PMID 25682182. This review thoroughly summarized the composition of ketoacid, the role on renal progression and metabolic control in CKD patients and the results of previous studies

Shah AP, Kalantar-Zadeh K, Kopple JD. Is there a role for ketoacid supplements in the management of CKD? Am J Kidney Dis. 2015b;65(5):659–73. https://doi.org/10.1053/j.ajkd.2014.09.029. PMID 25682182. This review thoroughly summarized the composition of ketoacid, the role on renal progression and metabolic control in CKD patients and the results of previous studies

Shao Y, Liu Y, Lu J, Shen X, He Q, Zhang Y, et al. The effects of low-protein diet supplemented with ketoacids on chronic kidney disease-mineral and bone disorder: a meta-analysis of randomized controlled trials. BMC Nephrol. 2019;20(1):1.

Siddarth M, Banerjee BD, Ghosh R, Singh N, Kalra OP. Gene–environment interaction and the risk of chronic kidney disease of uncertain etiology. In: Emerging trends in environmental biotechnology. CRC Press; 2022. p. 249–63.

Snelson M, Coughlan MT. Dietary protein, kidney function and hypertension. Nutrients. 2021;13(6):1816. https://doi.org/10.3390/nu13061816.

Soto C, Negreiros E, Tumlin JA, Lin JH. Loop diuretics in patients with chronic kidney disease: a systematic review. Am J Kidney Dis. 2018;72(6):751–8. https://doi.org/10.1053/j.ajkd.2018.06.017.

Stevens M, Oltean S. Modulation of VEGF-A alternative splicing as a novel treatment in chronic kidney disease. Genes. 2018;9(2):98. https://doi.org/10.3390/genes9020098. PMID 29462869

St-Jules DE, Woolf K, Pompeii ML, Kalantar-Zadeh K, Sevick MA, et al. Reexamining the phosphorus-protein dilemma: does phosphorus restriction compromise protein status? J Ren Nutr. 2016;26(3):136–40. https://doi.org/10.1053/j.jrn.2015.12.004. PMID 26873260

Streja E, Lau WL, Goldstein L, Sim JJ, Molnar MZ, Nissenson AR, et al. Hyperphosphatemia is a combined function of high serum PTH and high dietary protein intake in dialysis patients. Kidney Int Suppl. 2013;3(5):462–8. https://doi.org/10.1038/kisup.2013.96. PMID 25019031

Tallman DA, Khor BH, Karupaiah T, Khosla P, Chan M, Kopple JD. Nutritional adequacy of essential nutrients in low proten animal-and plant-based diets IN, The United States for chronic kidney disease patients. J Ren Nutr. 2022;33:249–60.

Taubert D, Roesen R, Lehmann C, Jung N, Schömig E. Effects of low habitual cocoa intake on blood pressure and bioactive nitric oxide: a randomized controlled trial. JAMA. 2007;298(1):49–60. https://doi.org/10.1001/jama.298.1.49.

Tay J, Thompson CH, Brinkworth GD, Noakes M. Effect of carbohydrate restriction on glucose metabolism in adults with chronic kidney disease: a randomized controlled trial. Am J Clin Nutr. 2015;102(4):765–72.

Tjempakasari A, Suroto H, Santoso D. Osteoblastogenesis of adipose-derived mesenchymal stem cells in chronic kidney disease patient with regular hemodialysis. Ann Med Surg (Lond). 2022;84:104796. https://doi.org/10.1016/j.amsu.2022.104796. PMCID 9758290. PMID 36536732

Valentijn FA, Knoppert SN, Pissas G, Rodrigues-Diez RR, Marquez-Exposito L, Broekhuizen R, et al. CCN2 aggravates the immediate oxidative stress–DNA damage response following renal ischemia–reperfusion injury. Antioxidants (Basel). 2021;10(12):12. https://doi.org/10.3390/antiox10122020. PMID 34943123

van der Meer P, Besselink N, Berden J, Homan van der Heide J, Krediet R. Phosphorus restriction in acute kidney injury: a randomized controlled trial. Am J Kidney Dis. 2020;76(1):97–104. https://doi.org/10.1053/j.ajkd.2019.08.018.

Venkatesh B, Delaney M. The role of erythropoiesis-stimulating agents in the management of anemia in chronic kidney disease. Kidney Int Suppl. 2017;7(2):43–8. https://doi.org/10.1016/j.kisu.2016.12.005.

Wang H, Naghavi M, Allen C, Barber RM, Bhutta ZA, Carter A, et al. Global, regional, and national life expectancy, all-cause mortality, and cause-specific mortality for 249 causes of death, 1980–2015: a systematic analysis for the Global Burden of Disease Study 2015. Lancet. 2016;388(10053):1459–544. https://doi.org/10.1016/S0140-6736(16)31012-1. PMCID 5388903. PMID 27733281 (GBD. Mortality causes of death collaborators; 2015)

Warady BA, Agarwal R, Bangalore S, Chapman A, Levin A, Stenvinkel P, et al. Alport syndrome classification and management. Kidney Med. 2020;2(5):639–49. https://doi.org/10.1016/j.xkme.2020.05.014. PMID 33094278

Wu MT, Wu CF, Liu CC, Tsai YC, Chen CC, Wang YH, et al. Melamine and oxalate coexposure induces early kidney tubular injury through mitochondrial aberrations and oxidative stress. Ecotoxicol Environ Saf. 2021;225:112756. https://doi.org/10.1016/j.ecoenv.2021.112756. PMID 34507040.

Xia L, Zhang J, Chen Q, Fan Y. Erythropoiesis-stimulating agents in the treatment of anemia in chronic kidney disease: a systematic review and meta-analysis. Medicine. 2021;100(1):e24300. https://doi.org/10.1097/MD.0000000000024300.

Yamabe N, Yokozawa T, Oya T, Kim M, Sakaue M. The green tea component, (−)-epigallocatechin gallate, attenuates reactive oxygen species-induced oxidative damage in cultured murine renal mesangial cells. J Med Food. 2013;16(3):277–83. https://doi.org/10.1089/jmf.2012.2554.

Zainordin NA, Eddy Warman NA, Mohamad AF, Abu Yazid FA, Ismail NH, Chen XW, et al. Safety and efficacy of very low carbohydrate diet in patients with diabetic kidney disease—a randomized controlled trial. PLoS One. 2021;16(10):e0258507. https://doi.org/10.1371/journal.pone.0258507. PMID 34644368

Zhang X, Qi Q, Bray GA, Hu FB, Sacks FM, Qi L. APOA5 genotype modulates 2-y changes in lipid profile in response to weight-loss diet intervention: the pounds lost trial. Am J Clin Nutr. 2012;96:917–22. https://doi.org/10.3945/ajcn.112.040907.

Zhang J, Wang C, Li L, Man Q, Meng L, Song P, et al. Omega-3 polyunsaturated fatty acid supplementation improves chronic kidney disease outcomes through regulation of mitochondrial function. EBioMedicine. 2018;32:257–71. https://doi.org/10.1016/j.ebiom.2018.05.030.

Zhang X, Wang Y, Li X, Li Y, Li Y. The renoprotective effects and mechanisms of angiotensin receptor blockers in chronic kidney disease. Front Pharmacol. 2021;12:606062. https://doi.org/10.3389/fphar.2021.606062.

Zhong X, Tu YJ, Li Y, Zhang P, Wang W, Chen SS, et al. Serum levels of WNT1-inducible signaling pathway protein-1 (WISP-1): a noninvasive biomarker of renal fibrosis in subjects with chronic kidney disease. Am J Transl Res. 2017;9(6):2920–32. PMID 28670380

Zhu W, Cai X, Wu Y, Chen X, Li Y, Hou J, et al. Effects of oral iron supplementation on exercise capacity in patients with heart failure with a reduced ejection fraction and iron deficiency: randomized clinical trial. JAMA Cardiol. 2018;3(3):200–6. https://doi.org/10.1001/jamacardio.2017.5205.

[illegible] The role of erythropoiesis-stimulating agents in the management of anemia in chronic kidney disease. [illegible]

[illegible] Global, regional, and national life expectancy, all-cause mortality, and cause-specific mortality for 249 causes of death [illegible]

[illegible]

Chapter 7
Treatment of Chronic Pancreatitis via Nutraceuticals

Divya Jindal, Pranav Pancham, and Manisha Singh

1 Introduction

Chronic pancreatitis is a complex and debilitating disease characterized by inflammation and progressive destruction of the pancreas, which is a vital organ located behind the stomach's upper left side of the abdomen (Pham and Forsmark 2018). This organ plays a crucial role in the digestion of food and the regulation of blood sugar levels. The pancreas produces enzymes necessary for the digestion of proteins, carbohydrates, and fats in the small intestine. Additionally, it releases hormones such as insulin and glaucoma aetiology, which regulate blood sugar levels (Karpińska and Czauderna 2022). In chronic pancreatitis, the pancreas becomes chronically inflamed, leading to progressive damage, and scarring leading to persistent excruciating abdominal pain which can be constant or intermittent with the possibility of radiating to the back, impaired digestion, and a host of other complications (Goulden 2013). This inflammation can be triggered by a variety of factors, including heavy alcohol consumption, gallstones, genetic predisposition, autoimmune disorders, and certain medications (Weiss et al. 2019). However, in some

D. Jindal
Indian Institute of Technology Bombay Monash Research Academy, Mumbai, India

P. Pancham
Western Sydney University, Campbelltown, NSW, Australia

M. Singh (✉)
Faculty of Health, Graduate School of Health, University of Technology Sydney, Sydney, NSW, Australia

ARCCIM, School of Public Health, Faculty of Health, University of Technology Sydney, Sydney, NSW, Australia

Department of Biotechnology, Jaypee Institute of Information Technology (JIIT), Noida, Uttar Pradesh, India

S. M. Jafari et al. (eds.), *Nutraceuticals: A New Millennium Approach for Preventive Medicine*, Food Bioactive Ingredients,
https://doi.org/10.1007/978-3-031-76481-3_7

cases, the cause of chronic pancreatitis remains unknown, which adds to the complexity of the disease. Significantly, it's been observed that pancreatitis pain often worsens after meals or during periods of excessive alcohol consumption. As the disease progresses, patients may experience weight loss, malnutrition, diarrhoea, and steatorrhea (fatty stools). These symptoms arise from impaired digestion and absorption of nutrients due to pancreatic insufficiency. Furthermore, chronic pancreatitis can lead to the development of complications such as pseudocysts, pancreatic duct strictures, diabetes mellitus, and even pancreatic cancer, further exacerbating the challenges faced by patients (Tandon et al. 2002). The effective management of chronic pancreatitis necessitates a collaborative approach involving gastroenterology consultation, endocrinological medications, balanced nutritional diet, pain management, and surgical interventions. The main objectives of treatment include relieving pain, enhancing digestion and nutritional well-being, and preventing or handling potential complications. Pain management often involves a combination of medications, including analgesics, pancreatic enzyme supplements, and sometimes opioids. However, due to the opioid crisis and concerns regarding addiction and tolerance, alternative approaches like nerve blocks, neuromodulation techniques, and psychological interventions are often being explored (Kosten and George 2002).

Moreover, nutritional support is one of the essential approaches in chronic pancreatitis, as malabsorption and malnutrition are common consequences of the disease (Cañamares-Orbís et al. 2022). Patients may require a specialised diet that is low in fat, high in protein, and supplemented with pancreatic enzymes (Rasmussen et al. 2013). In severe cases where oral intake is inadequate, enteral, or parenteral nutrition may be necessary. Controlling blood sugar levels is another critical management aspect, particularly in patients with concomitant diabetes mellitus (DM). Thus, close monitoring, dietary modifications, and appropriate medication regimens are essential to achieve optimal glycemic control. Although progress has been made in comprehending chronic pancreatitis, considerable challenges remain. Firstly, diagnosing the disease is intricate due to symptoms resembling other gastrointestinal disorders, and imaging studies may not always yield conclusive outcomes (Gilja et al. 2007). Consequently, there are often delays in diagnosis and the implementation of suitable treatment. Furthermore, managing chronic pancreatitis demands a personalized strategy that considers the patient's symptoms, complications, and nutritional requirements. The diverse presentation and progression of the disease further hinder the establishment of standardized treatment guidelines. Moreover, there is a paucity of effective therapies that can halt or reverse the progression of chronic pancreatitis. Current treatment strategies primarily focus on symptom control and complication management (Banks et al. 2010). Researchers are actively investigating potential targets for therapy, such as anti-inflammatory agents, antioxidants, and agents that promote pancreatic regeneration (Fig. 7.1).

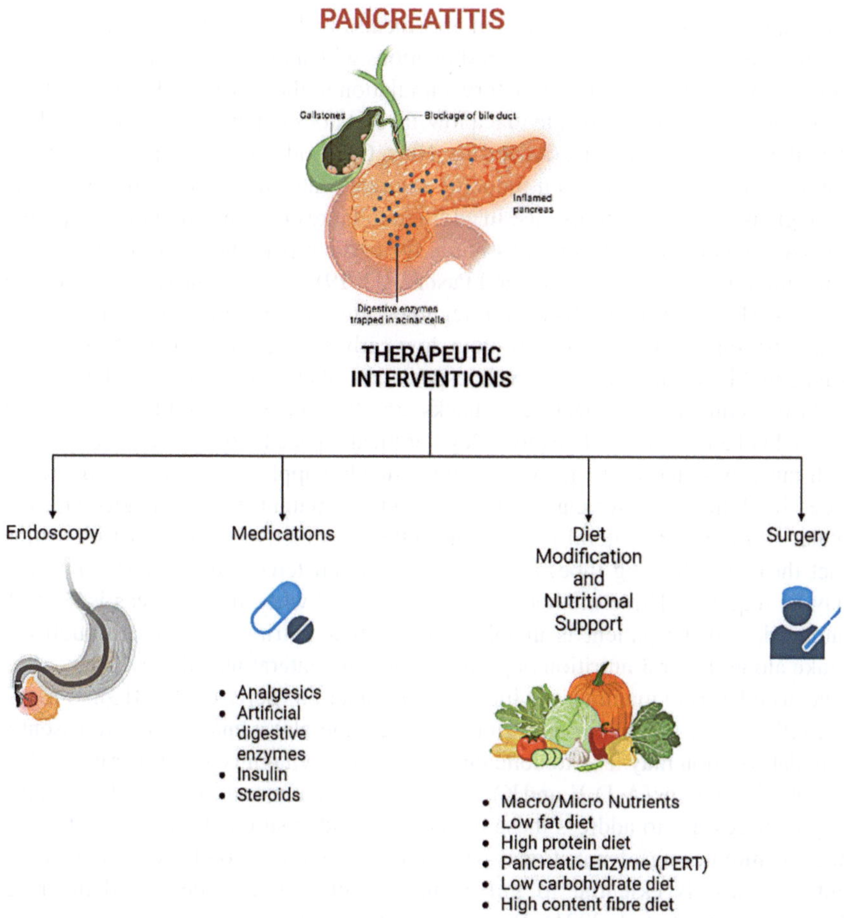

Fig. 7.1 Illustration showing the various pathological causes and therapeutic interventions for chronic pancreatitis

2 Nutritional Management of Chronic Pancreatitis

Nutritional management plays a crucial role in the overall treatment plan for patients with chronic pancreatitis. The primary goal of nutritional intervention is to ensure adequate nutrient intake, manage symptoms, and prevent complications associated with malnutrition. The pancreas produces digestive enzymes that help break down proteins, carbohydrates, and fats but in chronic pancreatitis, the production of these enzymes is reduced leading to malabsorption and malnutrition (Lakananurak and Gramlich 2020). Therefore, pancreatic enzyme replacement therapy (PERT) is commonly prescribed by clinicians, which involves taking pancreatic enzyme supplements with meals and snacks to facilitate the digestion and absorption of nutrients. The dosage of pancreatic enzymes is determined based on the severity of pancreatic insufficiency, the size of the meals, and individual patient response (Brennan and Saif 2019). Dietary modifications with a low-fat diet are often recommended to minimize the inflammatory stimulation in the pancreas (Pan et al. 2017). However, it is important to note that a low-fat diet does not mean eliminating all fats from the diet as healthy fats, (Avocados, Nuts, Olive oil, etc.), are still necessary for overall health. The focus is on reducing saturated fats and avoiding foods that are fried, greasy, or high in trans fats. In addition to fat restriction, a diet high in protein is usually recommended as it is essential for tissue repair and recovery, helping to maintain muscle mass (Carbone and Pasiakos 2019). Subsequently, lean sources of proteins (skinless poultry, fish, tofu, legumes, etc.) are preferred over fatty cuts of meat. For some patients, a high-protein, low-carbohydrate diet may be beneficial in managing blood sugar levels as well, especially if they have concurrent DM.

Also, frequent small meals and snacks are often recommended to ease the pancreatic load and improve digestion. Regular meals spaced throughout the day, along with nutritious snacks, help to maintain a steady supply of nutrients and prevent excessive hunger or overeating. In cases where malnutrition or inadequate oral intake is a concern, enteral (delivering nutrients directly into the gastrointestinal tract through a feeding tube) or parenteral (administered intravenously) nutrition may be required. These methods are typically used when the pancreas is severely damaged, and the patient is unable to meet their nutritional needs through oral intake alone. Enteral nutrition is preferred over parenteral nutrition whenever possible, as it helps to maintain gut integrity and function (Seres et al. 2013).

Furthermore, macronutrient (vitamins and minerals) management are essential as malabsorption may lead to deficiencies in certain vitamins and minerals, such as fat-soluble vitamins (A, D, E, and K), calcium, magnesium, and zinc. Supplementation may be necessary to address these deficiencies and ensure optimal nutrient status. Regular monitoring of nutritional parameters, including blood levels of vitamins and minerals, is essential to guide supplementation and identify deficiencies (Montoro-Huguet et al. 2021). Pancreatitis can lead to digestive problems, including diarrhoea. Soluble fibres, such as pectin and psyllium, can help absorb water and bulk up the stool, making it softer and easier to pass. Fibre can help regulate bowel movements by adding bulk and promoting regularity. This can be particularly helpful if pancreatitis has caused constipation or irregular bowel movements. Dietary fibres act as prebiotics, providing nourishment to beneficial gut bacteria. This helps maintain a healthy gut microbiota, which is crucial for overall digestive health and immune function (McRorie Jr. 2015). Certain types of soluble fibres, such as beta-glucan and oat bran, have been shown to reduce LDL cholesterol levels (Sima et al. 2018). This is important because people with pancreatitis may have an increased risk of developing heart disease. High-fibre foods have a lower glycemic index, meaning they cause a slower rise in blood sugar levels (Eleazu 2016). This can be beneficial for individuals with pancreatitis who may also have diabetes or impaired glucose tolerance. Fibre-rich foods tend to be more filling and can help control appetite, which is particularly useful for managing weight and preventing overeating, as obesity is a risk factor for pancreatitis (Ribichini et al. 2019). A well-balanced diet that is low in fat, high in protein, and supplemented with pancreatic enzymes is essential to alleviate symptoms, prevent malnutrition, and improve overall well-being (Cañamares-Orbís et al. 2022).

3 Digestive Enzymes: Restoring Pancreatic Function and Nutrient Absorption

Digestive enzymes play a crucial role in the process of breaking down food into smaller molecules that can be absorbed and utilized by the body, and they are released by the pancreas (Ianiro et al. 2016). However, in chronic pancreatitis or pancreatic insufficiency, the production and secretion of digestive enzymes are impaired, leading to difficulties in nutrient absorption and digestion (Hackert et al. 2014).

Digestive enzyme supplements have emerged as a valuable intervention for restoring pancreatic function and optimising nutrient absorption in such patients. Most recently, PERT which involves the administration of exogenous digestive enzymes to compensate for pancreatic insufficiency has been recognised as a potential therapeutic approach (Olivier et al. 2015). Pancreatic enzyme supplements typically contain a combination of lipases, proteases, and amylases (as listed in Table 7.1), which are encapsulated in enteric-coated tablets/capsules to be delivered in the acidic environment of the stomach (Ketwaroo and Graham 2019). The enteric coating allows the enzymes to reach the small intestine, where they can exert their digestive effects.

Table 7.1 This table highlights an array of digestive enzymes and their specific functions and sources in the body. They are essential in digesting various nutrients, aiding their absorption and promoting overall digestive well-being

S.No	Name of the associated enzymes	Functional properties	Food sources
1	Amylase	Converts carbohydrates into sugars	Saliva, pancreas, small intestines
2	Protease	Breaks down proteins into amino acids	Stomach, pancreas, small intestines
3	Lipase	Digest fats into fatty acids and glycerol	Pancreas, small intestines, liver
4	Lactase	Converts lactose into glucose and galactose	Small intestines
5	Peptidase	Breaks down proteins into amino acids	Small intestines
6	Sucrase	Converts lactose into glucose and fructose	Small intestines
7	Maltase	Converts maltose into glucose	Small intestines
8	Cellulase	Breaks down cellulose	Gut microbiome
9	Phytase	Liberates phosphate from phytate	Gut microbiome

The administration of pancreatic enzyme supplements with meals helps to restore the proper digestion and absorption of nutrients. These supplements should be taken at the beginning or during a meal, depending on the individual's digestive needs and the specific formulation of the enzyme supplement (Trang et al. 2014). The dosage of pancreatic enzymes is determined based on the severity of pancreatic insufficiency and the individual's response to treatment. One of the primary benefits of pancreatic enzyme replacement therapy is the improvement in nutrient absorption. By providing exogenous enzymes, PERT aids in the breakdown of fats, proteins, and carbohydrates into smaller molecules that can be readily absorbed by the small intestine (Boivin et al. 1990; Lowe 2002). This allows individuals with pancreatic insufficiency to regain the ability to absorb essential nutrients and prevent malnutrition. Moreover, proper nutrient absorption can contribute to weight gain and the restoration of healthy body composition in individuals who have experienced significant weight loss due to chronic pancreatitis. In addition to improving nutrient absorption, digestive enzyme supplements can help alleviate the symptoms associated with pancreatic insufficiency.

4 Anti-inflammatory Nutrients: Easing Pancreatic Inflammation

Certain nutrients possess anti-inflammatory properties that can help ease pancreatic inflammation in conditions like chronic pancreatitis. Including these nutrients in the diet can provide support for the pancreas and aid in managing inflammation-related symptoms. Omega-3 fatty acids, found in fatty fish like salmon and mackerel (Krupa and Parmar n.d.), as well as flaxseeds (Kajla et al. 2015) and chia seeds (Grancieri et al. 2019), have been shown to have anti-inflammatory effects. Incorporating these foods into the diet can help reduce inflammation in the pancreas. Turmeric, a spice commonly used in curry dishes, contains an active compound called curcumin that exhibits potent anti-inflammatory properties (Table 7.2). Adding turmeric to meals or consuming it as a supplement may help alleviate pancreatic inflammation (Sharifi-Rad et al. 2020). Ginger, another spice known for its anti-inflammatory effects, can be beneficial in reducing inflammation in the pancreas (Ballester et al. 2022). It can be consumed in various forms, including fresh ginger root, powdered ginger, or ginger tea. Green leafy vegetables, such as spinach, kale, and broccoli, are rich in antioxidants and other anti-inflammatory compounds. Including these vegetables in the diet can provide essential nutrients while helping to combat pancreatic inflammation (Pollock 2016). Berries, such as blueberries and strawberries, are packed with antioxidants that possess anti-inflammatory properties. These fruits can be incorporated into meals and smoothies or consumed as snacks (Kalt et al. 2020). However, it is important to consult with a healthcare professional or registered dietitian before making significant dietary changes, as individual needs and considerations may vary.

Table 7.2 Table representing the list of anti-inflammatory nutrients with their sources of extraction and therapeutic value

S. No	Anti-inflammatory nutrients	Subclass/type of anti-inflammatory nutrients	Source (plant/animal)	Therapeutic feature	References
1	Omega-3-fatty acids, eicosapentaenoic acid (EPA) and docosahexaenoic acid (DHA)	Poly Unsaturated fatty acids	Fatty fishes, flaxseeds, chia seeds, and walnuts	Alleviating pancreatic inflammation; reduces cytokine production and oxidative stress; improves blood flow	Oscarsson and Hurt-Camejo (2017)
2	Curcumin		*Curcuma longa*	Suppression of cell proliferation, invasion, and angiogenesis; induces apoptosis via deactivation of nuclear factor-kappa B (NF-κB); suppress several inflammatory cytokines, (TNF-α, interleukins (IL-1, IL-6, IL-8 and IL-1β) Increases the islet content of glutathione	Cho et al. (2007)
3	6-gingerol 6-shogaol	Exopeptidases Shogaols	Ginger	Inducing apoptosis;inhibiting STAT3 and NF-κB signalling;downregulating the expression of cyclin D1, survivin, c-Myc, and Bcl2 Inducing cell cycle arrest in the G0/G1-phase;decreasing the levels of cyclin A, cyclin D1, and cyclin E1;increasing the expression of caspase;inhibiting the mTOR signalling pathway	Xu et al. (2020)
4.	Olive oil	Polyphenols, oleocanthal	Plant-	Prevents production of pro-inflammatory COX-1 and COX-2 enzymes	Carpi et al. (2019)
	Nuts (walnut/almond)	Calcium, magnesium, zinc, vitamin E, fiber	Plant	Maintains cardiovascular health; contains alpha-linolenic acid (ALA), which boosts the immune system.	Ros (2010)

(continued)

Table 7.2 (continued)

S. No	Anti-inflammatory nutrients	Subclass/type of anti-inflammatory nutrients	Source (plant/animal)	Therapeutic feature	References
5	Berries/cherries (strawberries, blueberries, Raspberries, Blackberries)	Vitamins, Minerals & fiber	Plant	Contains anthocyanins antioxidants; reduces the risk of cancer development and progression; lowers the inflammatory markers	Basu et al. (2010)
6	Green tea	Epigallocatechin-3-gallate (EGCG)	Plant	Related to reducing the risk of heart disease, cancer, Alzheimer's, and obesity; lowers pro-inflammatory cytokine production and damage to the fatty acids	Chacko et al. (2010)

4.1 Omega-3 Fatty Acids

Omega-3 fatty acids have shown promise in easing pancreatic inflammation, particularly in conditions such as chronic pancreatitis. These essential fatty acids have anti-inflammatory properties that can help reduce the severity of inflammation and its associated symptoms. Omega-3 fatty acids, including eicosapentaenoic acid (EPA) and docosahexaenoic acid (DHA), are primarily found in fatty fish such as salmon, mackerel, sardines, and trout (Lei et al. 2015). They can also be obtained from plant-based sources like flaxseeds, chia seeds, and walnuts (Gebauer et al. 2006). These fatty acids play a crucial role in modulating the body's inflammatory response. Studies have suggested that omega-3 fatty acids can help alleviate pancreatic inflammation by suppressing the production of pro-inflammatory molecules called cytokines (Mori and Beilin 2004; Mayer et al. 2003). These cytokines play a significant role in promoting inflammation in the pancreas (Mayer et al. 2000). By reducing their production, omega-3 fatty acids help dampen the inflammatory response and promote a more balanced immune system. In addition to their anti-inflammatory effects, omega-3 fatty acids have also been shown to improve blood flow, reduce oxidative stress, and support overall pancreatic health (Wall et al. 2010). By enhancing blood flow to the pancreas, these fatty acids promote the delivery of oxygen and nutrients necessary for tissue repair and healing (Barchitta et al. 2019). To incorporate omega-3 fatty acids into the diet, it is recommended to consume fatty fish at least twice a week. If fish is not an option, plant-based sources like flaxseeds, chia seeds, and walnuts can be included. Omega-3 supplements derived from fish oil or algae are also available for those who have difficulty meeting their needs through diet alone.

It is important to note that omega-3 fatty acids should be consumed as part of a balanced diet and in consultation with a healthcare professional. While omega-3 fatty acids have shown anti-inflammatory effects, they should not be considered a standalone treatment for pancreatic inflammation (Weylandt et al. 2008) A comprehensive approach, including medication management and lifestyle modifications, is necessary for managing pancreatic conditions.

4.2 Curcumin

Curcumin, the active compound found in the spice turmeric, has gained attention for its potential to ease pancreatic inflammation. Turmeric has a long history of use in traditional medicine for its anti-inflammatory properties, and curcumin is considered its primary bioactive component (Hewlings and Kalman 2017). Studies have shown that curcumin possesses strong anti-inflammatory and antioxidant properties, which can help mitigate pancreatic inflammation (Jurenka 2009). It exerts its effects by modulating various signalling pathways involved in the inflammatory response, thereby reducing the production of pro-inflammatory molecules. In the context of pancreatic inflammation, curcumin has demonstrated promising results. It can inhibit the activation of nuclear factor-kappa B (NF-κB), a transcription factor that plays a pivotal role in promoting inflammation. By inhibiting NF-κB, curcumin helps downregulate the expression of pro-inflammatory cytokines and enzymes, thereby reducing pancreatic inflammation (Wang et al. 2019). Additionally, curcumin has been found to possess antioxidant properties, which can counteract oxidative stress. Oxidative stress is known to contribute to pancreatic damage and inflammation. By neutralizing harmful free radicals, curcumin helps protect pancreatic cells from oxidative damage and supports their overall health (Jakubczyk et al. 2020).

Turmeric can be added to various dishes, including curries, soups, and stir-fries. However, it is important to note that the bioavailability of curcumin is relatively low. Its absorption and utilization by the body can be enhanced by consuming it with black pepper or pairing it with fats, such as coconut oil (Han 2011). In addition to dietary sources, curcumin supplements are available. These supplements typically provide higher concentrations of curcumin than what can be obtained through food alone. However, it is important to consult with a healthcare professional before starting any new supplements, as they may interact with certain medications or have contraindications for specific health conditions. While curcumin shows promise in easing pancreatic inflammation, it is important to recognize that it should not be considered a standalone treatment for pancreatic conditions. A comprehensive approach, including medical management and lifestyle modifications, is essential for the management of pancreatic inflammation.

4.3 Ginger

Ginger, a commonly used spice with a long history of medicinal use, has been studied for its potential to ease pancreatic inflammation. It contains various bioactive compounds, including gingerols and shogaols, which contribute to its anti-inflammatory and antioxidant properties. Research suggests that ginger exhibits anti-inflammatory effects by inhibiting the production of pro-inflammatory cytokines and enzymes (Shahrajabian et al. 2019). By modulating inflammatory pathways, ginger can help reduce pancreatic inflammation and its associated symptoms. Moreover, ginger has been found to possess antioxidant activity, which helps protect pancreatic cells from oxidative stress. Oxidative stress plays a role in the development and progression of pancreatic inflammation. The antioxidants present in ginger can scavenge free radicals and minimize cellular damage, supporting the overall health of the pancreas. Studies specifically investigating the effects of ginger on pancreatic inflammation are limited, but preliminary research and anecdotal evidence suggest its potential benefits. Ginger has been traditionally used to alleviate digestive discomfort, including symptoms associated with pancreatic disorders (Akimoto et al. 2015).

To incorporate ginger into the diet, fresh ginger root can be peeled and grated, then added to various dishes, teas, or smoothies. Dried ginger powder can also be used as a convenient option. Ginger supplements are available, but it is important to consult with a healthcare professional before starting any new supplements, as they may interact with certain medications or have contraindications for specific health conditions. While ginger shows promise in easing pancreatic inflammation, it should be considered as part of a comprehensive treatment plan, including medical management and lifestyle modifications. It is not a substitute for professional medical advice or treatment.

5 Antioxidant Nutrients: Protecting Pancreatic Cells from Oxidative Stress

Antioxidant nutrients play a crucial role in protecting pancreatic cells from oxidative stress, a condition characterized by an imbalance between the production of harmful free radicals and the body's ability to neutralize them. Oxidative stress can contribute to pancreatic inflammation and damage, making antioxidant support important for pancreatic health (Robles et al. 2013). Vitamin C, also known as ascorbic acid, is a potent antioxidant that helps neutralize free radicals and reduce oxidative stress (Chambial et al. 2013). It supports the regeneration of other antioxidants, such as vitamin E, and plays a vital role in maintaining the health of pancreatic tissues (Traber and Stevens 2011). Good sources of vitamin C include citrus fruits, berries, kiwi, and leafy green vegetables (Doseděl et al. 2021). Vitamin E is another important antioxidant that protects pancreatic cells from oxidative damage. It helps maintain the integrity of cell membranes and prevents the oxidation of lipids. Nuts, seeds, vegetable oils, and leafy green vegetables are rich sources of vitamin E (Liao et al. 2022). Selenium is a trace mineral that acts as a cofactor for antioxidant enzymes, such as glutathione peroxidase. These enzymes help neutralize free radicals and reduce oxidative stress. Selenium can be found in foods like Brazil nuts, fish, shellfish, and whole grains (Zoidis et al. 2018). Beta-carotene is a precursor to vitamin A and possesses potent antioxidant properties. It helps protect pancreatic cells from oxidative damage and supports overall cellular health. Foods rich in beta-carotene include carrots, sweet potatoes, spinach, and other brightly coloured fruits and vegetables (Lavy et al. 2004). Flavonoids, a group of plant compounds with antioxidant activity, have been shown to have protective effects on pancreatic cells. Foods rich in flavonoids include berries, citrus fruits, apples, onions, and tea (Ullah et al. 2020) Incorporating a variety of antioxidant-rich foods into the diet can provide the necessary nutrients to support pancreatic health. To obtain a diverse array of antioxidants, consuming a wide range of fruits, vegetables, nuts, seeds, whole grains, and healthy fats is recommended (Table 7.3).

While dietary sources are the preferred way to obtain antioxidants, supplements can also be considered in consultation with a healthcare professional. However, it is important to note that high-dose antioxidant supplementation may not always be beneficial and can even have adverse effects in certain cases. Therefore, it is recommended to seek professional guidance before starting any antioxidant supplements.

Table 7.3 Table representing the list of antioxidant nutrients with their sources of extraction and therapeutic value

S. No	Antioxidant nutrients	Subclass/type of antioxidant nutrients	Source (plant/ animal based)	Therapeutic feature	References
1	Vitamin C	Carotenoid	Citrus fruits, tomatoes and tomato juice, and potatoes	Free radical scavenger Collagen synthesis Regeneration of other antioxidants Provides immunity	Sen and Chakraborty (2011)
2	Vitamin E	Tocopherols	Plant-based oils, nuts, seeds, fruits, and vegetables.	Lipid peroxidation prevention Enhances enzyme function	Howard et al. (2011)
3	Selenium		Seafood, organ meats, and Brazil nuts	Glutathione peroxidase activation maintaining oxidative stress DNA repair and cell growth regulation	Zoidis et al. (2018)

5.1 Vitamin C: Enhancing Pancreatic Defence Mechanisms

Vitamin C, also known as ascorbic acid, plays a significant role in enhancing pancreatic defence mechanisms against oxidative stress and inflammation. As a potent antioxidant, vitamin C helps neutralize harmful free radicals and protects pancreatic cells from oxidative damage. Here are some ways in which vitamin C enhances pancreatic defence mechanisms:

1. ***Free Radical Scavenging***: Vitamin C acts as a powerful antioxidant, donating electrons to neutralize free radicals and prevent them from causing cellular damage. By scavenging free radicals, vitamin C helps protect pancreatic cells from oxidative stress, which can contribute to inflammation and tissue damage (Niki 1991).
2. ***Regeneration of Other Antioxidants***: Vitamin C plays a key role in regenerating other antioxidants, such as vitamin E. After vitamin E neutralizes free radicals, it becomes oxidized. However, vitamin C can regenerate vitamin E, allowing it to continue its antioxidant function. This synergy between vitamins C and E helps maintain a robust antioxidant defence system in the pancreas (Kaźmierczak-Barańska et al. 2020).
3. ***Collagen Synthesis***: Vitamin C is essential for collagen synthesis, a crucial component of connective tissues, including those found in the pancreas. Collagen provides structural support to blood vessels, ensuring optimal blood flow and nutrient delivery to pancreatic cells. By promoting collagen synthesis, vitamin C supports pancreatic tissue integrity and overall health (Gref et al. 2020).
4. ***Immune Function:*** The pancreas contains immune cells that play a role in the defence against pathogens and inflammation. Vitamin C supports immune function by enhancing the activity of various immune cells, such as lymphocytes and phagocytes. This immune-boosting effect can aid in protecting the pancreas from infections and inflammatory processes (Carr and Maggini 2017).
5. ***Anti-inflammatory Effects:*** In addition to its antioxidant properties, vitamin C has anti-inflammatory effects. It can help modulate inflammatory pathways, reduce the production of pro-inflammatory molecules, and promote a more balanced immune response. Vitamin C contributes to the overall defence mechanisms against pancreatic damage by mitigating inflammation in the pancreas (Gęgotek and Skrzydlewska 2022).

To ensure adequate vitamin C intake, it is recommended to consume a varied diet rich in fruits and vegetables, which are excellent sources of this vitamin. Citrus fruits (such as oranges and grapefruits), berries, kiwi, peppers, and leafy green vegetables are particularly high in vitamin C. If needed, vitamin C supplements can be considered, but it is important to follow the recommended dosage and consult with a healthcare professional, as excessive intake can have adverse effects.

5.2 *Vitamin E: Preserving Pancreatic Health*

Vitamin E is a powerful antioxidant that plays a crucial role in preserving pancreatic health. It helps protect pancreatic cells from oxidative damage, supports the functioning of various enzymes, and contributes to overall tissue integrity. Here are some key ways in which vitamin E preserves pancreatic health:

1. ***Antioxidant Protection:*** Vitamin E acts as a potent antioxidant, neutralizing harmful free radicals and preventing oxidative damage to pancreatic cells. By donating electrons to free radicals, vitamin E helps stabilize them and prevents them from causing cellular harm. This antioxidant activity is particularly important in the pancreas, which is susceptible to oxidative stress due to its high metabolic activity (Rizvi et al. 2014).
2. ***Lipid Peroxidation Prevention:*** The pancreas contains lipids (fats) that are vulnerable to oxidation, leading to a process called lipid peroxidation. Vitamin E plays a critical role in inhibiting lipid peroxidation by scavenging free radicals that initiate this process. By preventing lipid peroxidation, vitamin E helps maintain the integrity of pancreatic cell membranes and protects against tissue damage (Niki 2021).
3. ***Enzyme Support:*** Vitamin E supports the activity of various enzymes involved in critical pancreatic functions. It helps preserve the integrity and activity of enzymes that are essential for proper digestion, nutrient absorption, and metabolism. By ensuring optimal enzyme function, vitamin E supports overall pancreatic health and efficient digestive processes (Coskun et al. 2020).
4. ***Anti-inflammatory Effects:*** Chronic inflammation can contribute to pancreatic damage and dysfunction. Vitamin E has been shown to possess anti-inflammatory properties, helping to reduce inflammation in the pancreas. By modulating inflammatory pathways and inhibiting the production of pro-inflammatory molecules, vitamin E can help mitigate the harmful effects of inflammation on pancreatic tissues (Singh et al. 2005).
5. ***Immune System Support:*** The immune system plays a vital role in pancreatic health, as it helps protect against infections and inflammation. Vitamin E supports immune function by enhancing the activity of immune cells, promoting a healthy immune response, and aiding in the defence against pathogens that can affect pancreatic tissues (Lewis et al. 2019).

Good dietary sources of vitamin E include nuts (such as almonds and sunflower seeds), seeds, vegetable oils (such as wheat germ oil and sunflower oil), spinach, and fortified cereals. Incorporating these foods into the diet can help ensure an adequate intake of vitamin E. Vitamin E supplements are also available, but it is important to consult with a healthcare professional before starting any new supplements to determine the appropriate dosage and consider individual needs.

5.3 Selenium: Shielding Pancreatic Tissues from Oxidative Damage

Selenium is a trace mineral that plays a critical role in shielding pancreatic tissues from oxidative damage. As an essential component of various antioxidant enzymes, selenium helps maintain the balance between oxidants and antioxidants in the body. Here are the key ways in which selenium preserves pancreatic health:

1. ***Glutathione Peroxidase Activation:*** Selenium is a crucial component of the enzyme glutathione peroxidase, which is responsible for neutralizing hydrogen peroxide and lipid hydroperoxides. By activating glutathione peroxidase, selenium helps break down and remove harmful oxidants from the pancreas. This process prevents oxidative damage to pancreatic cells and preserves their structural integrity (Baker et al. 1993).
2. ***Reducing Oxidative Stress:*** Selenium acts as a potent antioxidant by participating in redox reactions. It helps regenerate other antioxidants, such as vitamin C and vitamin E, which have a protective role against oxidative stress. By replenishing these antioxidants, selenium enhances the antioxidant defence system in the pancreas, reducing the burden of oxidative stress on pancreatic tissues (Tinggi 2008).
3. ***Anti-Inflammatory Effects:*** Chronic inflammation is closely linked to pancreatic damage and dysfunction. Selenium has been shown to possess anti-inflammatory properties, helping to reduce inflammation in the pancreas. By modulating the activity of inflammatory mediators and signalling pathways, selenium can help mitigate inflammation and protect pancreatic tissues from inflammatory damage (Mahmoodpoor et al. 2022).
4. ***Immune System Support:*** The immune system plays a vital role in pancreatic health, defending against infections and inflammation. Selenium supports immune function by influencing the activity of immune cells, including lymphocytes and macrophages. By optimizing immune responses, selenium aids in protecting pancreatic tissues from immune-mediated damage and contributes to overall pancreatic health (Huang et al. 2012).
5. ***DNA Repair and Cell Growth Regulation:*** Selenium is involved in DNA repair mechanisms and the regulation of cell growth and proliferation. These functions are crucial for maintaining the integrity of pancreatic cells and preventing the development of abnormal cell growth. By ensuring proper DNA repair and cell growth regulation, selenium helps safeguard the pancreas against cellular damage and dysfunction (Bera et al. 2013).

Selenium can be obtained through dietary sources such as Brazil nuts, seafood (particularly tuna, sardines, and shrimp), meat, poultry, eggs, and whole grains. However, the selenium content of foods can vary depending on the soil in which they are grown. In cases where dietary intake is insufficient, selenium supplements can be considered under the guidance of a healthcare professional. It is important to note that while selenium is essential for pancreatic health, excessive intake can be

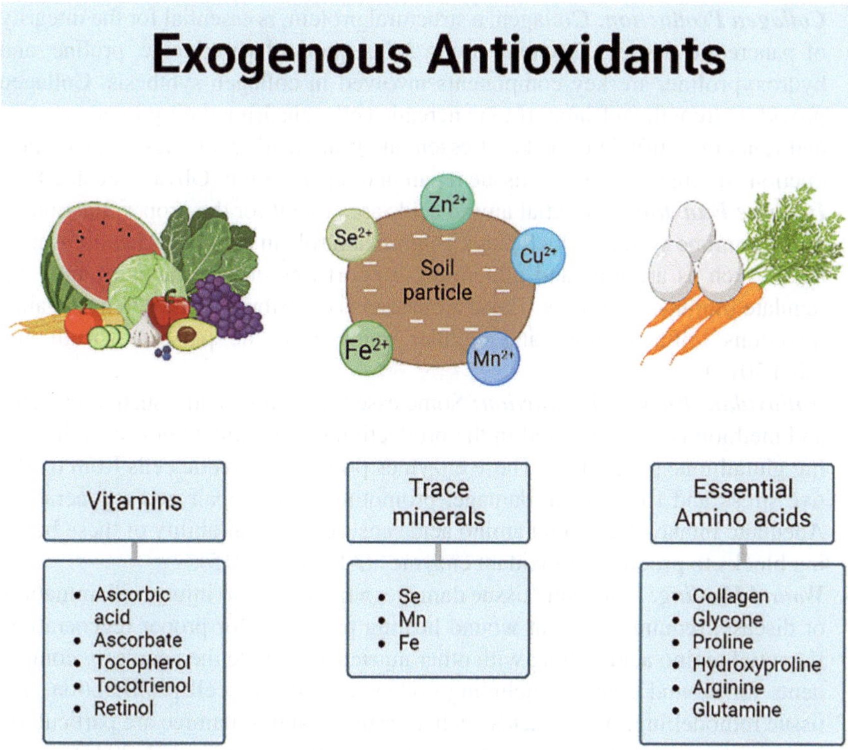

Fig. 7.2 Schematic illustration representing the exogenous antioxidants from the plant sources

harmful. High doses of selenium supplements should be avoided unless recommended by a healthcare professional, as they can lead to selenium toxicity (Fig. 7.2).

6 Essential Amino Acids: Supporting Pancreatic Tissue Repair and Regeneration

Essential amino acids play a crucial role in supporting pancreatic tissue repair and regeneration. These amino acids are the building blocks of proteins, which are essential for the growth, maintenance, and repair of tissues in the body. Here's how essential amino acids contribute to pancreatic health:

1. ***Protein Synthesis:*** Essential amino acids are required for the synthesis of proteins, including enzymes, hormones, and structural components of pancreatic cells. Proteins are vital for the repair and regeneration of pancreatic tissues, ensuring their optimal function. Adequate intake of essential amino acids supports the synthesis of new proteins needed for pancreatic tissue repair and maintenance (Sans et al. 2021).

2. ***Collagen Production:*** Collagen, a structural protein, is essential for the integrity of pancreatic tissues. Essential amino acids, particularly glycine, proline, and hydroxyproline, are key components involved in collagen synthesis. Collagen provides strength and support to pancreatic cells, ensuring their proper structure and function. Sufficient intake of essential amino acids promotes collagen production, aiding in pancreatic tissue repair and regeneration (Olivares et al. 2017).
3. ***Immune Function:*** Essential amino acids are crucial for the proper functioning of the immune system, which plays a significant role in pancreatic health. Amino acids, such as arginine and glutamine, support immune cell activity and help regulate immune responses. These amino acids contribute to the defence against infections, inflammation, and cellular damage in the pancreas (Ruth and Field 2013).
4. ***Antioxidant Enzyme Production:*** Some essential amino acids, such as cysteine and methionine, are involved in the production of antioxidant enzymes, including glutathione peroxidase. These enzymes protect pancreatic cells from oxidative stress and free radical damage, promoting tissue repair and regeneration. Adequate intake of essential amino acids ensures the availability of these building blocks to produce antioxidant enzymes (Al-Malki 2015).
5. ***Wound Healing:*** Pancreatic tissue damage, whether due to injury, inflammation, or disease, requires efficient wound healing processes for proper regeneration. Essential amino acids, along with other nutrients, provide the necessary components for wound healing, including collagen synthesis, cell proliferation, and tissue remodelling. Amino acids such as arginine and glutamine are particularly important for wound healing in the pancreas (Arribas-López et al. 2021).

Good dietary sources of essential amino acids include animal proteins such as meat, poultry, fish, eggs, and dairy products. Plant-based sources like legumes, soy products, quinoa, and nuts also provide essential amino acids, although some plant proteins may be lower in certain essential amino acids. A balanced diet that includes a variety of protein sources can ensure an adequate intake of essential amino acids. In certain cases, such as pancreatic diseases or conditions that affect protein digestion or absorption, healthcare professionals may consider amino acid supplements to support pancreatic tissue repair and regeneration. However, the use of supplements should be done under professional guidance to ensure appropriate dosing and monitoring.

6.1 Role of Branched-Chain Amino Acids (BCAAs) in Repair

Branched-chain amino acids (BCAAs) are a specific group of essential amino acids that play a significant role in supporting pancreatic tissue repair and regeneration. The three BCAAs—leucine, isoleucine, and valine—offer unique benefits to pancreatic health due to their specific metabolic properties. Here's how BCAAs contribute to pancreatic tissue repair and regeneration:

1. ***Protein Synthesis:*** BCAAs are essential for protein synthesis, which is crucial for pancreatic tissue repair and regeneration. Leucine plays a key role in activating the protein synthesis pathway through the mammalian target of rapamycin (mTOR) signalling. By stimulating protein synthesis, BCAAs support the growth and repair of pancreatic tissues (Blomstrand et al. 2006).
2. ***Insulin Secretion:*** BCAAs have been shown to affect insulin secretion, an essential process for maintaining glucose homeostasis. Isoleucine has been found to stimulate insulin secretion by acting as a signalling molecule in pancreatic beta cells. Proper insulin secretion is crucial for pancreatic health and function, and BCAAs play a role in supporting this process (Yoon 2016).
3. ***Energy Production:*** During times of pancreatic tissue damage or stress, BCAAs can serve as an alternative energy source. The oxidation of BCAAs provides energy for pancreatic cells, allowing them to maintain their function and support the repair process. This energy contribution is particularly important when pancreatic tissues are under metabolic stress or when normal glucose metabolism is impaired (Mikalayeva et al. 2021).
4. ***Antioxidant Protection:*** BCAAs, especially leucine, possess antioxidant properties and can help protect pancreatic cells from oxidative stress. Oxidative stress is a common feature of pancreatic diseases and can contribute to tissue damage and impaired regeneration. By reducing oxidative stress, BCAAs help create a favourable environment for pancreatic tissue repair and regeneration (Ichikawa et al. 2012).
5. ***Anti-inflammatory Effects:*** Inflammation plays a role in pancreatic tissue damage and repair processes. BCAAs have been shown to possess anti-inflammatory properties, helping to modulate inflammatory responses. By reducing excessive inflammation, BCAAs support a more balanced immune response and contribute to pancreatic tissue repair (Lee et al. 2017).

Good dietary sources of BCAAs include protein-rich foods such as meat, poultry, fish, dairy products, eggs, and legumes. Supplementing with BCAAs may be considered in certain cases, such as during periods of increased pancreatic tissue repair needs or in individuals with specific conditions that affect protein metabolism. However, it is important to consult with a healthcare professional before starting any supplements to determine appropriate dosing and individual needs.

6.2 Glutamine: Pancreatic Cell Growth and Repair

Glutamine is a conditionally essential amino acid that plays a crucial role in supporting pancreatic cell growth and repair. It is involved in various metabolic processes and functions as a key nutrient for the pancreas. Here's how glutamine contributes to pancreatic health:

1. ***Cell Growth and Division:*** Glutamine is essential for cell growth and division, including the regeneration of pancreatic cells. It provides the building blocks for

DNA and RNA synthesis, supporting the replication of pancreatic cells during tissue repair. Glutamine also contributes to the synthesis of proteins and other molecules necessary for cell growth and function (Yoo et al. 2020).

2. ***Energy Source:*** Glutamine serves as an important energy source for pancreatic cells. It can be converted to glucose through a process called gluconeogenesis, providing fuel for energy production. This is particularly significant when pancreatic tissues are under stress or when glucose metabolism is impaired. Glutamine helps meet the energy demands of the pancreas, allowing cells to function optimally during tissue repair and regeneration (Cruzat et al. 2018).
3. ***Antioxidant and Anti-inflammatory Effects:*** Glutamine exhibits antioxidant properties, helping to protect pancreatic cells from oxidative stress and damage. It also possesses anti-inflammatory effects, helping to reduce inflammation in the pancreas. By modulating oxidative stress and inflammation, glutamine supports a favourable environment for pancreatic cell growth, repair, and regeneration (Nemati et al. 2019).
4. ***Gut Barrier Function:*** The health of the gastrointestinal tract is closely linked to pancreatic health. Glutamine plays a role in maintaining the integrity of the gut barrier, which helps prevent the entry of harmful substances into the bloodstream. This is important for pancreatic health as a compromised gut barrier can contribute to inflammation and tissue damage. Glutamine supports the gut barrier function, aiding in overall pancreatic health and repair processes (Achamrah et al. 2017).
5. ***Immune System Support:*** Glutamine is essential for immune cell function and supports the immune response in the pancreas. It helps maintain the activity of immune cells, promoting their proper function in defending against infections and supporting tissue repair. Glutamine supports immune system balance and contributes to pancreatic health by ensuring optimal immune responses (Nemati et al. 2019).

Good dietary sources of glutamine include protein-rich foods such as meat, poultry, fish, dairy products, legumes, and nuts. In certain cases, such as during periods of increased pancreatic tissue repair needs or in individuals with specific conditions, glutamine supplementation may be considered under the guidance of a healthcare professional.

6.3 Arginine: Wound Healing in the Pancreas

Arginine is a semi-essential amino acid that plays a significant role in wound healing, including the healing of pancreatic tissues. It is involved in various physiological processes that support tissue repair and regeneration. Here's how arginine contributes to wound healing in the pancreas:

1. ***Nitric Oxide Production:*** Arginine serves as the precursor for nitric oxide (NO) synthesis in the body. Nitric oxide is a signalling molecule that plays a critical

role in the wound-healing process. It helps regulate blood flow, promote vasodilation, and stimulate the production of growth factors and cytokines that are essential for tissue repair. In the pancreas, nitric oxide contributes to improving blood supply and promoting healing in damaged pancreatic tissues (Wu et al. 2021).

2. ***Collagen Synthesis:*** Arginine plays a role in collagen synthesis, which is crucial for wound healing and tissue regeneration. Collagen is the main component of the extracellular matrix, providing structural support and promoting cell migration during the healing process. Arginine contributes to the production of collagen, aiding in the formation of new tissue in the pancreas and facilitating wound closure (Williams et al. 2002).
3. ***Immune Function:*** The immune system is closely involved in the wound healing process, including in pancreatic tissue repair. Arginine has immunomodulatory effects, influencing the activity of immune cells such as macrophages and lymphocytes. It helps regulate the inflammatory response, promoting a balanced immune reaction that is essential for efficient wound healing in the pancreas (Martí and Reith 2021).
4. ***Cell Proliferation and Migration:*** Arginine supports cell proliferation and migration, which are critical for tissue repair. It enhances the production of growth factors that promote cell division and stimulates the migration of cells to the site of injury. These effects contribute to the regeneration of pancreatic tissues, aiding in the healing process (Greene et al. 2013).
5. ***Antioxidant Activity:*** Arginine exhibits antioxidant properties, helping to protect pancreatic cells from oxidative stress during the wound healing process. It can scavenge free radicals and reduce oxidative damage, promoting a favourable environment for tissue repair and regeneration in the pancreas (Liang et al. 2018).

Good dietary sources of arginine include protein-rich foods such as meat, poultry, fish, dairy products, legumes, and nuts. In some cases, arginine supplementation may be considered to support wound healing, including pancreatic tissue repair. However, it is important to consult with a healthcare professional before starting any supplements to determine appropriate dosing and individual needs.

7 Micronutrient Supplementation: Replenishing Deficiencies in Chronic Pancreatitis

Micronutrient supplementation plays a crucial role in replenishing deficiencies that may arise in individuals with chronic pancreatitis. Chronic pancreatitis can lead to impaired digestion and absorption of nutrients, as well as increased nutrient losses due to malabsorption and pancreatic enzyme insufficiency (Nutrients listed in Table 7.4). Here's how micronutrient supplementation can help address deficiencies and support individuals with chronic pancreatitis:

Table 7.4 This table highlights the important micronutrients deficient in chronic pancreatitis, their functions, and the common food sources that address these deficiencies. Micronutrient supplementation is often crucial for individuals with chronic pancreatitis to ensure proper nutrient intake and overall health

S. No	Recommended micronutrients supplementation	Role in pancreatitis	Nutrient origins
1	Vitamin A	Promotes immune function and aids in tissue repair	Carrots, sweet potatoes, spinach
2	Vitamin D	Assists in the absorption of calcium and supports bone health	Fatty fish, fortified dairy products
3	Vitamin E	Supports as an antioxidant and aids in cellular upkeep	Nuts, seeds, vegetable oils
4	Vitamin K	Essential for blood clotting and bone integrity	Leafy greens, broccoli, eggs
5	B vitamins	Required for energy metabolism and maintenance of neural function	Whole grains, meat, dairy products
6	Zinc	Provides immunoprotection and aids in wound healing	Shellfish, beef, legumes
7	Magnesium	Essential for muscle and nerve function	Nuts, seeds, whole grains
8	Calcium	Required for bone integrity and nerve conduction	Dairy p1roducts, leafy greens

1. ***Fat-Soluble Vitamins:*** Chronic pancreatitis can affect the absorption of fat-soluble vitamins A, D, E, and K. Supplementation with these vitamins may be necessary to prevent deficiencies. Vitamin A is important for immune function and vision. Vitamin D supports bone health and immune function. Vitamin E acts as an antioxidant and supports cell integrity. Vitamin K is essential for blood clotting and bone health. Supplementation with these vitamins can help maintain adequate levels and support overall health (Martínez-Moneo et al. 2016).
2. ***B Vitamins:*** Chronic pancreatitis can also impair the absorption of B vitamins, including thiamine (B1), riboflavin (B2), niacin (B3), pyridoxine (B6), cobalamin (B12), and folate (B9). These vitamins play crucial roles in energy metabolism, nerve function, red blood cell production, and DNA synthesis. Supplementation can help address deficiencies and support the proper functioning of various body systems (Glasbrenner et al. 1991).
3. ***Minerals:*** Individuals with chronic pancreatitis may experience deficiencies in essential minerals such as calcium, magnesium, zinc, and selenium. Calcium and magnesium are important for bone health, muscle function, and nerve transmission. Zinc is involved in immune function, wound healing, and enzyme activity. Selenium acts as an antioxidant and supports thyroid function. Supplementation can help replenish these minerals and support their various functions in the body (O'Brien and Omer 2019).

4. ***Antioxidants:*** Chronic pancreatitis is associated with increased oxidative stress, which can lead to tissue damage and inflammation. Supplementation with antioxidants such as vitamin C, vitamin E, selenium, and other phytochemicals may help counteract oxidative damage and reduce inflammation in individuals with chronic pancreatitis (Ko et al. 2019).

It is important to note that the specific supplementation needs may vary among individuals with chronic pancreatitis, and recommendations should be individualized based on factors such as the severity of the condition, nutrient deficiencies, and overall health status. Consulting with a healthcare professional, such as a registered dietitian or gastroenterologist, is crucial to assess nutrient status, determine appropriate supplementation needs, and monitor the effectiveness of supplementation.

Supplementation should be accompanied by a well-balanced diet that focuses on meeting nutrient needs as much as possible through food sources. The goal is to optimize nutrient intake through dietary modifications and targeted supplementation to address deficiencies and support overall health and well-being in individuals with chronic pancreatitis.

7.1 Vitamin D: Strengthening Pancreatic Immunity

Vitamin D, often referred to as the "sunshine vitamin," plays a crucial role in supporting immune function and has been found to have a positive impact on pancreatic health and immunity. Here's how vitamin D strengthens pancreatic immunity:

1. ***Immune Cell Regulation:*** Vitamin D plays a vital role in modulating the immune system. It helps regulate the function of immune cells, including T cells, B cells, and macrophages, which are involved in immune responses and defence against pathogens. By promoting the proper functioning of immune cells, vitamin D supports a balanced immune response in the pancreas (Lewis et al. 2019; Prietl et al. 2013).
2. ***Anti-inflammatory Effects:*** Chronic inflammation is a characteristic feature of pancreatic diseases, including chronic pancreatitis. Vitamin D exhibits anti-inflammatory properties, helping to reduce inflammation in the pancreas. By modulating inflammatory responses, vitamin D supports a healthier immune environment in the pancreas and may help alleviate inflammation-related symptoms (Lewis et al. 2019).
3. ***Antimicrobial Activity:*** Vitamin D has antimicrobial properties and helps protect against infections. In the pancreas, vitamin D enhances the production of antimicrobial peptides, which are small proteins that help fight against various pathogens, including bacteria and viruses. By bolstering the immune defences in the pancreas, vitamin D contributes to a stronger immune response against potential infections (Hewison 2011).
4. ***Insulin Sensitivity:*** Vitamin D has been associated with improved insulin sensitivity and glucose regulation. Proper insulin sensitivity is essential for maintain-

ing blood sugar levels and preventing metabolic imbalances that can contribute to pancreatic dysfunction. By supporting insulin sensitivity, vitamin D helps maintain pancreatic health and function (Sung et al. 2012).

5. ***Pancreatic Cancer Prevention:*** Vitamin D has been linked to a reduced risk of pancreatic cancer, a malignancy that often has a poor prognosis. Studies have shown that vitamin D may inhibit the growth of pancreatic cancer cells, induce cell differentiation, and suppress tumour cell proliferation. While more research is needed to fully understand the mechanisms involved, vitamin D's potential protective effects against pancreatic cancer highlight its importance in maintaining pancreatic health (Bulathsinghala et al. 2010).

It is worth noting that vitamin D deficiency is common worldwide, especially in regions with limited sunlight exposure or individuals with reduced sun exposure. Dietary sources of vitamin D include fatty fish (such as salmon and mackerel), fortified dairy products, eggs, and certain mushrooms. However, it can be challenging to obtain adequate amounts of vitamin D through diet alone. In such cases, vitamin D supplementation may be recommended, particularly for individuals with chronic pancreatitis or other pancreatic conditions.

It is important to consult with a healthcare professional to assess vitamin D levels and determine the appropriate dosage for supplementation. They can help determine the optimal blood levels of vitamin D and provide guidance on safe and effective supplementation.

7.2 B Vitamins: Nourishing Pancreatic Cells

B vitamins play a vital role in nourishing pancreatic cells and supporting their proper functioning. These water-soluble vitamins are involved in numerous metabolic processes and cellular functions that are essential for pancreatic health. Here's how B vitamins nourish pancreatic cells:

1. ***Thiamine (B1):*** Thiamine is crucial for energy production and carbohydrate metabolism. It helps convert glucose into energy, which is essential for the normal functioning of pancreatic cells. Thiamine also plays a role in the synthesis of neurotransmitters and supports nerve function, which is important for proper pancreatic regulation (Subramanya et al. 2011).
2. ***Riboflavin (B2):*** Riboflavin is involved in energy metabolism and acts as a coenzyme in various cellular processes. It is necessary for the conversion of carbohydrates, fats, and proteins into energy that fuels pancreatic cells. Riboflavin also contributes to antioxidant defences, protecting pancreatic cells from oxidative stress (Ghosal and Said 2012).
3. ***Niacin (B3):*** Niacin is essential for energy metabolism and cellular respiration. It participates in the conversion of food into usable energy and supports the proper functioning of pancreatic cells. Niacin also plays a role in DNA repair

and synthesis, contributing to the overall health and integrity of pancreatic tissues (Chen and Wang 2018).

4. ***Pyridoxine (B6):*** Pyridoxine is involved in a wide range of enzymatic reactions, including protein metabolism, neurotransmitter synthesis, and immune function. It supports the proper functioning of pancreatic cells by participating in various metabolic pathways and assisting in synthesising important molecules (Srinivasan et al. 2019).
5. ***Cobalamin (B12):*** Cobalamin is necessary for the synthesis of DNA and red blood cell production. It plays a crucial role in nerve function and helps maintain the integrity of nerve cells, including those in the pancreas. Cobalamin supports the overall health and function of pancreatic cells (Yuan et al. 2021).
6. ***Folate (B9):*** Folate is essential for DNA synthesis and repair. It is involved in cell division and growth, supporting the regeneration and maintenance of pancreatic tissues. Adequate folate levels are important for the proper functioning of pancreatic cells and the overall health of the pancreas (Said et al. 2010).

Vitamin B is found in a variety of foods, including whole grains, legumes, leafy green vegetables, meat, fish, and dairy products. A balanced diet that includes these food sources can help ensure an adequate intake of B vitamins. However, certain factors, such as malabsorption or specific dietary restrictions, may increase the risk of B vitamin deficiencies in individuals with chronic pancreatitis. In such cases, supplementation may be recommended under the guidance of a healthcare professional.

7.3 Zinc: Supporting Pancreatic Enzyme Production

Zinc is an essential mineral that plays a vital role in supporting pancreatic enzyme production. Pancreatic enzymes are crucial for the digestion and absorption of nutrients. Here's how zinc supports pancreatic enzyme production:

1. ***Enzyme Synthesis:*** Zinc is involved in the synthesis and secretion of pancreatic enzymes, including amylase, lipase, and proteases. These enzymes are responsible for breaking down carbohydrates, fats, and proteins, respectively, into smaller molecules that can be absorbed by the body. Adequate zinc levels are necessary for the proper production and secretion of these enzymes by pancreatic cells (Wang et al. 2020).
2. ***Enzyme Activation:*** Zinc also plays a role in activating certain pancreatic enzymes. For example, zinc is required for the activation of pancreatic lipase, which is necessary for the breakdown and absorption of dietary fats. Without sufficient zinc, the activation of these enzymes may be impaired, leading to reduced digestive capacity (Brugger and Windisch 2016).
3. ***DNA and Protein Synthesis:*** Zinc is involved in DNA synthesis and protein synthesis, both of which are essential for the growth and maintenance of pancreatic tissues. Zinc supports the synthesis of proteins, including the enzymes pro-

duced by the pancreas. It also aids in DNA replication and repair, ensuring the proper functioning of pancreatic cells (Kelleher et al. 2011).

4. ***Immune Function:*** Zinc is crucial for maintaining a healthy immune system, and a well-functioning immune system is important for pancreatic health. The pancreas is susceptible to inflammation and damage, and a balanced immune response is necessary to protect and support its optimal functioning. Zinc supports immune cell function and helps regulate immune responses in the pancreas, contributing to its overall health (Wang et al. 2020).
5. ***Antioxidant Défense:*** The pancreas is susceptible to oxidative stress, which can lead to tissue damage and inflammation. Zinc acts as an antioxidant and helps protect pancreatic cells from oxidative damage. It participates in the antioxidant defence system, supporting the neutralization of free radicals and reducing oxidative stress in the pancreas (Prasad and Bao 2019).

Dietary sources of zinc include seafood, meat, poultry, dairy products, legumes, nuts, and seeds. However, individuals with chronic pancreatitis may have impaired absorption and increased zinc losses, which can lead to zinc deficiency. In such cases, zinc supplementation may be recommended under the guidance of a healthcare professional to ensure adequate zinc levels and support pancreatic enzyme production.

8 Targeted Nutrient Synergy: Enhancing Treatment Efficacy

Targeted nutrient synergy refers to the strategic combination of specific nutrients to enhance treatment efficacy for chronic pancreatitis. By synergistically targeting key mechanisms involved in the pathogenesis of the disease, this approach aims to optimize the management and outcomes of chronic pancreatitis. Here's how targeted nutrient synergy can enhance treatment efficacy for chronic pancreatitis:

1. ***Digestive Enzymes and Micronutrients:*** The supplementation of pancreatic enzymes, along with targeted micronutrient supplementation, can address nutrient deficiencies and optimize nutrient absorption. Pancreatic enzyme supplements aid in the digestion and absorption of macronutrients, while micronutrients such as fat-soluble vitamins, B vitamins, minerals, and antioxidants replenish deficiencies and support pancreatic health (Lakananurak and Gramlich 2020).
2. ***Anti-Inflammatory Nutrients and Antioxidants:*** Chronic pancreatitis involves inflammation and oxidative stress. By combining anti-inflammatory nutrients such as omega-3 fatty acids, curcumin, and ginger with antioxidant nutrients like vitamins C, E, and selenium, treatment can target both inflammatory processes and oxidative damage, reducing pancreatic inflammation and protecting pancreatic cells from further injury (Ko et al. 2019).
3. ***Amino Acids and Nutrient Support for Tissue Repair:*** Amino acids, including essential amino acids, branched-chain amino acids (BCAAs), glutamine, and arginine, support pancreatic tissue repair and regeneration. Combining these

amino acids with targeted nutrient support, such as vitamins and minerals, provides the building blocks necessary for tissue healing and supports optimal pancreatic function (Yoon 2016).

4. ***Micronutrients for Immune Support:*** Micronutrients such as vitamins D and C, zinc, and selenium are essential for immune function. By incorporating these immune-supportive nutrients into the treatment plan, the immune system can be strengthened, reducing the risk of infections and supporting overall pancreatic health (Jabłońska and Mrowiec 2021).
5. ***Antioxidant Nutrients and Micronutrients for Oxidative Stress Management:*** Chronic pancreatitis is associated with increased oxidative stress. By combining antioxidant nutrients such as vitamins C, E, and selenium with other micronutrients that support the antioxidant defence system, treatment can mitigate oxidative damage, reduce inflammation, and promote pancreatic cell health (Ko et al. 2019).
6. ***Nutritional Support and Dietary Modifications:*** In addition to targeted nutrient supplementation, dietary modifications tailored to individual needs are crucial for managing chronic pancreatitis. This may include adjusting macronutrient ratios, focusing on smaller, more frequent meals, avoiding alcohol and high-fat foods, and ensuring adequate hydration. Nutritional support provided by a registered dietitian can help optimize nutrient intake and support treatment outcomes (Yang 2021).

9 Lifestyle Considerations and Supportive Therapies: Holistic Approaches to Pancreatic Health

In addition to medical interventions and targeted nutrient strategies, holistic approaches to pancreatic health encompass lifestyle considerations and supportive therapies. These aspects focus on promoting overall well-being and improving the quality of life for individuals with chronic pancreatitis. Here are some key considerations:

1. ***Diet and Nutrition:*** A healthy and balanced diet is essential for managing chronic pancreatitis. Working with a registered dietitian can help develop a personalized meal plan that focuses on adequate nutrient intake, pancreatic enzyme supplementation, and dietary modifications tailored to individual needs. Emphasizing whole foods, limiting processed foods, avoiding triggers (such as alcohol and high-fat foods), and maintaining hydration are key factors in supporting pancreatic health (Lakananurak and Gramlich 2020).
2. ***Weight Management:*** Maintaining a healthy weight is crucial for individuals with chronic pancreatitis. Weight management strategies, including portion control, regular physical activity, and working with healthcare professionals, can help achieve and maintain a healthy weight. Excessive weight loss or obesity can further burden the pancreas and worsen symptoms (Kim 2021).
3. ***Stress Management:*** Chronic pancreatitis can be accompanied by physical and emotional stress. Engaging in stress-reducing practices such as meditation, deep

breathing exercises, and yoga, or engaging in hobbies and activities that promote relaxation can help manage stress levels. Seeking support from a therapist or joining support groups can also be beneficial in coping with the emotional challenges associated with chronic pancreatitis (Bremner et al. 2020).
4. ***Physical Activity:*** Regular physical activity is important for overall health and can have positive effects on pancreatic health. Engaging in moderate-intensity exercises, such as walking, swimming, or cycling, can help improve digestion, manage weight, enhance mood, and promote overall well-being. However, it's important to consult with a healthcare professional before starting or modifying an exercise routine, especially if there are specific limitations or concerns (Koehler and Drenowatz 2019).
5. ***Pain Management:*** Chronic pancreatitis can cause significant pain and discomfort. Alongside medical interventions, various supportive therapies can help manage pain. These may include acupuncture, massage therapy, heat therapy, or transcutaneous electrical nerve stimulation (TENS). Working closely with healthcare providers to develop a comprehensive pain management plan is essential for individualized care.
6. ***Alcohol and Smoking Cessation:*** Alcohol consumption and smoking can exacerbate symptoms and further damage the pancreas. Quitting smoking and minimizing or eliminating alcohol intake are crucial steps in managing chronic pancreatitis. Seeking professional support, such as counselling or rehabilitation programs, can aid in the cessation process (Palmer et al. 2018).
7. ***Regular Medical Monitoring:*** Regular check-ups and follow-ups with healthcare providers are essential for monitoring the progress of chronic pancreatitis, assessing nutrient status, adjusting treatment plans, and addressing any concerns or complications that may arise. Compliance with medical recommendations, including medication regimens and laboratory tests, is vital for optimal management (Banks et al. 2010; Mac Goey et al. 2019).

Holistic approaches to pancreatic health involve a multidimensional focus on lifestyle factors, emotional well-being, and supportive therapies. By addressing these aspects alongside medical interventions and targeted nutrient strategies, individuals with chronic pancreatitis can experience improved symptom management, enhanced quality of life, and better overall health outcomes. It is important to work closely with healthcare professionals to develop an individualized care plan that encompasses all aspects of pancreatic health.

10 Stress Management: Mitigating the Impact on Pancreatic Function

Stress management is crucial for mitigating the impact on pancreatic function in individuals with chronic pancreatitis. Stress can exacerbate symptoms, trigger inflammation, and affect overall well-being. Here are some strategies to help manage stress and support pancreatic health. Engaging in relaxation techniques such as deep breathing exercises, meditation, mindfulness, and progressive muscle relaxation can help reduce stress levels. These practices promote a state of calm and relaxation, which can positively impact pancreatic function and overall health. Regular physical activity is not only beneficial for physical health but also plays a significant role in stress management. Exercise helps release endorphins, which are natural mood-boosting chemicals. Cognitive-behavioural therapy (CBT) is another form of therapy that focuses on identifying and changing negative thought patterns and behaviours. It can help individuals with chronic pancreatitis manage stress, cope with challenges, and develop effective strategies to improve their emotional well-being. Creating a well-balanced schedule can help reduce stress and improve overall quality of life. Adopting a healthy lifestyle can contribute to stress reduction. Maintain a nutritious diet, stay hydrated, and limit caffeine and alcohol intake, as these can contribute to anxiety and stress. They can provide guidance, coping strategies, and support in managing stress and its impact on pancreatic health (Binker and Cosen-Binker 2014).

It is important to recognize that stress management is a personal journey, and what works for one individual may not work for another. Experiment with different techniques and strategies to find what suits you best. Remember, it is crucial to integrate stress management into your overall treatment plan for chronic pancreatitis and work closely with your healthcare team to ensure comprehensive care.

11 Conclusion

Chronic pancreatitis is a complex and debilitating disease wherein its management requires a multidisciplinary approach, with a focused treatment strategy to relieve pain, enhance digestion and nutritional well-being, and prevent potential complications. Pain management in chronic pancreatitis often involves a combination of medications, pancreatic enzyme supplements, and alternative approaches such as nerve blocks, neuromodulation techniques, and psychological interventions. However, nutritional support is crucial due to malabsorption and malnutrition associated with the disease. Thus, patients require specialized diets, depending upon their conditions (low in fat, high in protein) and supplemented with pancreatic enzymes. Moreover, diagnosing chronic pancreatitis is challenging which leads to delays in appropriate treatment. The personalized nature of the disease and its diverse presentation further complicate the establishment of standardized treatment guidelines. Additionally, there is a lack of effective therapies that can halt or reverse the progression of the disease, with current strategies primarily focused on symptom control and complication management. Nevertheless, researchers are actively exploring potential therapeutic targets, such as anti-inflammatory agents, antioxidants, and other agents that promote pancreatic regeneration. Continued research and advancements in understanding chronic pancreatitis are needed to improve diagnostic accuracy, develop personalized treatment approaches, and ultimately find interventions that can halt or even reverse the progression of this debilitating disease.

References

Pham A, Forsmark C. Chronic pancreatitis: review and update of aetiology, risk factors, and management. F1000Res. 2018;7

Karpińska M, Czauderna M. Pancreas-its functions, disorders, and physiological impact on the mammals' organism. Front Physiol. 2022;13:807632.

Goulden MR. The pain of chronic pancreatitis: a persistent clinical challenge. Br J Pain. 2013;7(1):8–22.

Weiss FU, Laemmerhirt F, Lerch MM. Etiology and risk factors of acute and chronic pancreatitis. Visc Med. 2019;35(2):73–81.

Tandon RK, Sato N, Garg PK, T. C. S. Group. Chronic pancreatitis: Asia–Pacific consensus report. J Gastroenterol Hepatol. 2002;17(4):508–18.

Kosten TR, George TP. The neurobiology of opioid dependence: implications for treatment. Sci Pract Perspect. 2002;1(1):13–20.

Cañamares-Orbís P, García-Rayado G, Alfaro-Almajano E. Nutritional support in pancreatic diseases. Nutrients. 2022;14(21)

Rasmussen HH, Irtun O, Olesen SS, Drewes AM, Holst M. Nutrition in chronic pancreatitis. World J Gastroenterol. 2013;19(42):7267–75.

Gilja OH, et al. Advanced imaging and visualization in gastrointestinal disorders. World J Gastroenterol. 2007;13(9):1408–21.

Banks PA, Conwell DL, Toskes PP. The management of acute and chronic pancreatitis. Gastroenterol Hepatol (N Y). 2010;6(2 Suppl 3):1–16.

Lakananurak N, Gramlich L. Nutrition management in acute pancreatitis: clinical practice consideration. World J Clin Cases. 2020;8(9):1561–73.

Brennan GT, Saif MW. Pancreatic enzyme replacement therapy: a concise review. JOP. 2019;20(5):121–5.

Pan LL, Li J, Shamoon M, Bhatia M, Sun J. Recent advances on nutrition in treatment of acute pancreatitis. Front Immunol. 2017;8:762.

Carbone JW, Pasiakos SM. Dietary protein and muscle mass: translating science to application and health benefit. Nutrients. 2019;11(5)

Seres DS, Valcarcel M, Guillaume A. Advantages of enteral nutrition over parenteral nutrition. Ther Adv Gastroenterol. 2013;6(2):157–67.

Montoro-Huguet MA, Belloc B, Domínguez-Cajal M. Small and large intestine (I): malabsorption of nutrients. Nutrients. 2021;13(4)

McRorie JW Jr. Evidence-based approach to fiber supplements and clinically meaningful health benefits, part 2: what to look for and how to recommend an effective fiber therapy. Nutr Today. 2015;50(2):90–7.

Sima P, Vannucci L, Vetvicka V. β-Glucans and cholesterol (review). Int J Mol Med. 2018;41(4):1799–808.

Eleazu CO. The concept of low glycemic index and glycemic load foods as a panacea for type 2 diabetes mellitus; prospects, challenges and solutions. Afr Health Sci. 2016;16(2):468–79.

Ribichini E, et al. Role of fibre in nutritional management of pancreatic diseases. Nutrients. 2019;11(9)

Ianiro G, Pecere S, Giorgio V, Gasbarrini A, Cammarota G. Digestive enzyme supplementation in gastrointestinal diseases. Curr Drug Metab. 2016;17(2):187–93.

Hackert T, Schütte K, Malfertheiner P. The pancreas: causes for malabsorption. Viszeralmedizin. 2014;30(3):190–7.

Olivier AK, Gibson-Corley KN, Meyerholz DK. Animal models of gastrointestinal and liver diseases. Animal models of cystic fibrosis: gastrointestinal, pancreatic, and hepatobiliary disease and pathophysiology. Am J Physiol Gastrointest Liver Physiol. 2015;308(6):G459–71.

Ketwaroo GA, Graham DY. Rational use of pancreatic enzymes for pancreatic insufficiency and pancreatic pain. Adv Exp Med Biol. 2019;1148:323–43.

Trang T, Chan J, Graham DY. Pancreatic enzyme replacement therapy for pancreatic exocrine insufficiency in the 21(st) century. World J Gastroenterol. 2014;20(33):11467–85.
Boivin M, Lanspa SJ, Zinsmeister AR, Go VL, DiMagno EP. Are diets associated with different rates of human interdigestive and postprandial pancreatic enzyme secretion? Gastroenterology. 1990;99(6):1763–71.
Lowe ME. The triglyceride lipases of the pancreas. J Lipid Res. 2002;43(12):2007–16.
Krupa FK, Parmar M. Omega-3 fatty acids. Treasure Island: StatPearls Publishing; n.d.
Kajla P, Sharma A, Sood DR. Flaxseed-a potential functional food source. J Food Sci Technol. 2015;52(4):1857–71.
Grancieri M, Martino HSD, Gonzalez de Mejia E. Chia (Salvia hispanica L.) seed total protein and protein fractions digests reduce biomarkers of inflammation and atherosclerosis in macrophages in vitro. Mol Nutr Food Res. 2019;63(19):e1900021.
Sharifi-Rad J, et al. Turmeric and its major compound curcumin on health: bioactive effects and safety profiles for food, pharmaceutical, biotechnological and medicinal applications. Front Pharmacol. 2020;11:01021.
Ballester P, Cerdá B, Arcusa R, Marhuenda J, Yamedjeu K, Zafrilla P. Effect of ginger on inflammatory diseases. Molecules. 2022;27(21)
Pollock RL. The effect of green leafy and cruciferous vegetable intake on the incidence of cardiovascular disease: a meta-analysis. JRSM Cardiovasc Dis. 2016;5:2048004016661435.
Kalt W, et al. Recent research on the health benefits of blueberries and their anthocyanins. Adv Nutr. 2020;11(2):224–36.
Oscarsson J, Hurt-Camejo E. Omega-3 fatty acids eicosapentaenoic acid and docosahexaenoic acid and their mechanisms of action on apolipoprotein B-containing lipoproteins in humans: a review. Lipids Health Dis. 2017;16(1):149.
Cho JW, Lee KS, Kim CW. Curcumin attenuates the expression of IL-1beta, IL-6, and TNF-alpha as well as cyclin E in TNF-alpha-treated HaCaT cells; NF-kappaB and MAPKs as potential upstream targets. Int J Mol Med. 2007;19(3):469–74.
Xu S, et al. 6-Gingerol induces cell-cycle G1-phase arrest through AKT-GSK 3β-cyclin D1 pathway in renal cell carcinoma. Cancer Chemother Pharmacol. 2020;85(2):379–90.
Carpi S, et al. The extra-virgin olive oil polyphenols oleocanthal and oleacein counteract inflammation-related gene and miRNA expression in adipocytes by attenuating NF-κB activation. Nutrients. 2019;11(12)
Ros E. Health benefits of nut consumption. Nutrients. 2010;2(7):652–82.
Basu A, Rhone M, Lyons TJ. Berries: emerging impact on cardiovascular health. Nutr Rev. 2010;68(3):168–77.
Chacko SM, Thambi PT, Kuttan R, Nishigaki I. Beneficial effects of green tea: a literature review. Chin Med. 2010;5:13.
Lei QC, et al. The role of omega-3 fatty acids in acute pancreatitis: a meta-analysis of randomized controlled trials. Nutrients. 2015;7(4):2261–73.
Gebauer SK, Psota TL, Harris WS, Kris-Etherton PM. N-3 fatty acid dietary recommendations and food sources to achieve essentiality and cardiovascular benefits. Am J Clin Nutr. 2006;83(6 Suppl):1526s–35s.
Mori TA, Beilin LJ. Omega-3 fatty acids and inflammation. Curr Atheroscler Rep. 2004;6(6):461–7.
Mayer K, et al. Short-time infusion of fish oil-based lipid emulsions, approved for parenteral nutrition, reduces monocyte proinflammatory cytokine generation and adhesive interaction with endothelium in humans. J Immunol. 2003;171(9):4837–43.
Mayer J, Rau B, Gansauge F, Beger HG. Inflammatory mediators in human acute pancreatitis: clinical and pathophysiological implications. Gut. 2000;47(4):546–52.
Wall R, Ross RP, Fitzgerald GF, Stanton C. Fatty acids from fish: the anti-inflammatory potential of long-chain omega-3 fatty acids. Nutr Rev. 2010;68(5):280–9.
Barchitta M, et al. Nutrition and wound healing: An overview focusing on the beneficial effects of curcumin. Int J Mol Sci. 2019;20(5)

Weylandt KH, et al. Reduction of inflammation and chronic tissue damage by omega-3 fatty acids in fat-1 transgenic mice with pancreatitis. Biochim Biophys Acta. 2008;1782(11):634–41.
Hewlings SJ, Kalman DS. Curcumin: a review of its effects on human health. Food Secur. 2017;6(10)
Jurenka JS. Anti-inflammatory properties of curcumin, a major constituent of Curcuma longa: a review of preclinical and clinical research. Altern Med Rev. 2009;14(2):141–53.
Wang Y, Bu C, Wu K, Wang R, Wang J. Curcumin protects the pancreas from acute pancreatitis via the mitogen-activated protein kinase signalling pathway. Mol Med Rep. 2019;20(4):3027–34.
Jakubczyk K, Drużga A, Katarzyna J, Skonieczna-Żydecka K. Antioxidant potential of curcumin-a meta-analysis of randomized clinical trials. Antioxidants (Basel). 2020;9(11)
Han HK. The effects of black pepper on the intestinal absorption and hepatic metabolism of drugs. Expert Opin Drug Metab Toxicol. 2011;7(6):721–9.
Shahrajabian MH, Sun W, Cheng Q. Clinical aspects and health benefits of ginger (Zingiber officinale) in both traditional Chinese medicine and modern industry. Acta Agric Scand Sect B—Soil Plant Sci. 2019;69(6):546–56.
Akimoto M, Iizuka M, Kanematsu R, Yoshida M, Takenaga K. Anticancer effect of ginger extract against pancreatic cancer cells mainly through reactive oxygen species-mediated autotic cell death. PLoS One. 2015;10(5):e0126605.
Robles L, Vaziri ND, Ichii H. Role of oxidative stress in the pathogenesis of pancreatitis: effect of antioxidant therapy. Pancreat Disord Ther. 2013;3(1):112.
Chambial S, Dwivedi S, Shukla KK, John PJ, Sharma P. Vitamin C in disease prevention and cure: an overview. Indian J Clin Biochem. 2013;28(4):314–28.
Traber MG, Stevens JF. Vitamins C and E: beneficial effects from a mechanistic perspective. Free Radic Biol Med. 2011;51(5):1000–13.
Doseděl M, et al. Vitamin C-sources, physiological role, kinetics, deficiency, use, toxicity, and determination. Nutrients. 2021;13(2)
Liao S, et al. Vitamin E and metabolic health: relevance of interactions with other micronutrients. Antioxidants (Basel). 2022;11(9)
Zoidis E, Seremelis I, Kontopoulos N, Danezis GP. Selenium-dependent antioxidant enzymes: actions and properties of selenoproteins. Antioxidants (Basel). 2018;7(5)
Lavy A, Karban A, Suissa A, Yassin K, Hermesh I, Ben-Amotz A. Natural beta-carotene for the prevention of post-ERCP pancreatitis. Pancreas. 2004;29(2):e45–50.
Ullah A, et al. Important flavonoids and their role as a therapeutic agent. Molecules. 2020;25(22)
Sen S, Chakraborty R. The role of antioxidants in human health. In: Oxidative stress: diagnostics, prevention, and therapy, vol. 1083. (ACS Symposium Series, no. 1083): American Chemical Society; 2011. p. 1–37.
Howard AC, McNeil AK, McNeil PL. Promotion of plasma membrane repair by vitamin E. Nat Commun. 2011;2:597.
Niki E. Action of ascorbic acid as a scavenger of active and stable oxygen radicals. Am J Clin Nutr. 1991;54(6 Suppl):1119s–24s.
Kaźmierczak-Barańska J, Boguszewska K, Adamus-Grabicka A, Karwowski BT. Two faces of vitamin C-antioxidative and pro-oxidative agent. Nutrients. 2020;12(5)
Gref R, et al. Vitamin C–squalene bioconjugate promotes epidermal thickening and collagen production in human skin. Sci Rep. 2020;10(1):16883.
Carr AC, Maggini S. Vitamin C and immune function. Nutrients. 2017;9(11)
Gęgotek A, Skrzydlewska E. Antioxidative and anti-inflammatory activity of ascorbic acid. Antioxidants (Basel). 2022;11(10)
Rizvi S, Raza ST, Ahmed F, Ahmad A, Abbas S, Mahdi F. The role of vitamin E in human health and some diseases. Sultan Qaboos Univ Med J. 2014;14(2):e157–65.
Niki E. Lipid oxidation that is, and is not, inhibited by vitamin E: consideration about physiological functions of vitamin E. Free Radic Biol Med. 2021;176:1–15.
Coskun M, Kayis T, Gulsu E, Alp E. Effects of selenium and vitamin E on enzymatic, biochemical, and immunological biomarkers in galleria mellonella L. Sci Rep. 2020;10(1):9953.

Singh U, Devaraj S, Jialal I. Vitamin E, oxidative stress, and inflammation. Annu Rev Nutr. 2005;25:151–74.

Lewis ED, Meydani SN, Wu D. Regulatory role of vitamin E in the immune system and inflammation. IUBMB Life. 2019;71(4):487–94.

Baker RD, Baker SS, LaRosa K, Whitney C, Newburger PE. Selenium regulation of glutathione peroxidase in human hepatoma cell line Hep3B. Arch Biochem Biophys. 1993;304(1):53–7.

Tinggi U. Selenium: its role as an antioxidant in human health. Environ Health Prev Med. 2008;13(2):102–8.

Mahmoodpoor A, et al. The effects of selenium supplementation on inflammatory markers in critically ill patients. SN Applied Sciences. 2022;4(12):326.

Huang Z, Rose AH, Hoffmann PR. The role of selenium in inflammation and immunity: from molecular mechanisms to therapeutic opportunities. Antioxid Redox Signal. 2012;16(7):705–43.

Bera S, De Rosa V, Rachidi W, Diamond AM. Does a role for selenium in DNA damage repair explain apparent controversies in its use in chemoprevention? Mutagenesis. 2013;28(2):127–34.

Sans MD, Crozier SJ, Vogel NL, D'Alecy LG, Williams JA. Dietary protein and amino acid deficiency inhibit pancreatic digestive enzyme mRNA translation by multiple mechanisms. Cell Mol Gastroenterol Hepatol. 2021;11(1):99–115.

Olivares O, et al. Collagen-derived proline promotes pancreatic ductal adenocarcinoma cell survival under nutrient-limited conditions. Nat Commun. 2017;8:16031.

Ruth MR, Field CJ. The immune modifying effects of amino acids on gut-associated lymphoid tissue. J Anim Sci Biotechnol. 2013;4(1):27.

Al-Malki AL. Suppression of acute pancreatitis by L-lysine in mice. BMC Complement Altern Med. 2015;15:193.

Arribas-López E, Zand N, Ojo O, Snowden MJ, Kochhar T. The effect of amino acids on wound healing: a systematic review and meta-analysis on arginine and glutamine. Nutrients. 2021;13(8)

Blomstrand E, Eliasson J, Karlsson HK, Köhnke R. Branched-chain amino acids activate key enzymes in protein synthesis after physical exercise. J Nutr. 2006;136(1 Suppl):269s–73s.

Yoon MS. The emerging role of branched-chain amino acids in insulin resistance and metabolism. Nutrients. 2016;8(7)

Mikalayeva V, Pankevičiūtė M, Žvikas V, Skeberdis VA, Bordel S. Contribution of branched-chain amino acids to energy production and mevalonate synthesis in cancer cells. Biochem Biophys Res Commun. 2021;585:61–7.

Ichikawa K, et al. Branched-chain amino acid-enriched nutrients stimulate antioxidant DNA repair in a rat model of liver injury induced by carbon tetrachloride. Mol Biol Rep. 2012;39(12):10803–10.

Lee JH, et al. Anti-inflammatory and anti-genotoxic activity of branched-chain amino acids (BCAA) in lipopolysaccharide (LPS) stimulated RAW 264.7 macrophages. Food Sci Biotechnol. 2017;26(5):1371–7.

Yoo HC, Yu YC, Sung Y, Han JM. Glutamine reliance in cell metabolism. Exp Mol Med. 2020;52(9):1496–516.

Cruzat V, Macedo Rogero M, Noel Keane K, Curi R, Newsholme P. Glutamine: metabolism and immune function, supplementation and clinical translation. Nutrients. 2018;10(11)

Nemati A, Alipanah-Moghadam R, Molazadeh L, Naghizadeh Baghi A. The effect of glutamine supplementation on oxidative stress and matrix metalloproteinase 2 and 9 after exhaustive exercise. Drug Des Devel Ther. 2019;13:4215–23.

Achamrah N, Déchelotte P, Coëffier M. Glutamine and the regulation of intestinal permeability: from bench to bedside. Curr Opin Clin Nutr Metab Care. 2017;20(1):86–91.

Wu G, Meininger CJ, McNeal CJ, Bazer FW, Rhoads JM. Role of L-arginine in nitric oxide synthesis and health in humans. Adv Exp Med Biol. 2021;1332:167–87.

Williams JZ, Abumrad N, Barbul A. Effect of a specialized amino acid mixture on human collagen deposition. Ann Surg. 2002;236(3):369–74.; discussion 374–5.

Martí ILAA, Reith W. Arginine-dependent immune responses. Cell Mol Life Sci. 2021;78(13):5303–24.
Greene JM, Feugang JM, Pfeiffer KE, Stokes JV, Bowers SD, Ryan PL. L-arginine enhances cell proliferation and reduces apoptosis in human endometrial RL95-2 cells. Reprod Biol Endocrinol. 2013;11(1):15. Accessed 26 Feb 2013
Liang M, et al. L-arginine induces antioxidant response to prevent oxidative stress via stimulation of glutathione synthesis and activation of Nrf2 pathway. Food Chem Toxicol. 2018;115:315–28.
Martínez-Moneo E, et al. Deficiency of fat-soluble vitamins in chronic pancreatitis: a systematic review and meta-analysis. Pancreatology. 2016;16(6):988–94.
Glasbrenner B, Malfertheiner P, Büchler M, Kuhn K, Ditschuneit H. Vitamin B12 and folic acid deficiency in chronic pancreatitis: a relevant disorder? Klin Wochenschr. 1991;69(4):168–72.
O'Brien SJ, Omer E. Chronic pancreatitis and nutrition therapy. Nutr Clin Pract. 2019;34(S1):S13–26.
Ko KH, An JM, Son MS, Chung JB, Hahm KB. Antioxidant therapy in chronic pancreatitis-promises and pitfalls. Ann Transl Med. 2019;7(Suppl 3):S115.
Prietl B, Treiber G, Pieber TR, Amrein K. Vitamin D and immune function. Nutrients. 2013;5(7):2502–21.
Hewison M. Antibacterial effects of vitamin D. Nat Rev Endocrinol. 2011;7(6):337–45.
Sung CC, Liao MT, Lu KC, Wu CC. Role of vitamin D in insulin resistance. J Biomed Biotechnol. 2012;2012:634195.
Bulathsinghala P, Syrigos KN, Saif MW. Role of vitamin D in the prevention of pancreatic cancer. J Nutr Metab. 2010;2010:721365.
Subramanya SB, Subramanian VS, Sekar VT, Said HM. Thiamin uptake by pancreatic acinar cells: effect of chronic alcohol feeding/exposure. Am J Physiol Gastrointest Liver Physiol. 2011;301(5):G896–904.
Ghosal A, Said HM. Mechanism and regulation of vitamin B2 (riboflavin) uptake by mouse and human pancreatic β-cells/islets: physiological and molecular aspects. Am J Physiol Gastrointes Liver Physiol. 2012;303(9):G1052–8.
Chen TH, Wang JJ. Niacin pretreatment attenuates ischemia and reperfusion of pancreas-induced acute pancreatitis and remote lung injury through suppressing oxidative stress and inflammation and activation of SIRT1. Transplant Proc. 2018;50(9):2860–3.
Srinivasan P, Ramesh V, Wu J, Heskett C, Chu BD, Said HM. Pyridoxine and pancreatic acinar cells: transport physiology and effect on gene expression profile. Am J Physiol Cell Physiol. 2019;317(6):C1107–c1114.
Yuan J, et al. Vitamin B(12) attenuates acute pancreatitis by suppressing oxidative stress and improving mitochondria dysfunction via CBS/SIRT1 pathway. Oxidative Med Cell Longev. 2021;2021:7936316.
Said HM, Mee L, Sekar VT, Ashokkumar B, Pandol SJ. Mechanism and regulation of folate uptake by pancreatic acinar cells: effect of chronic alcohol consumption. Am J Physiol Gastrointest Liver Physiol. 2010;298(6):G985–93.
Wang M, Phadke M, Packard D, Yadav D, Gorelick F. Zinc: roles in pancreatic physiology and disease. Pancreatology. 2020;20(7):1413–20.
Brugger D, Windisch WM. Subclinical zinc deficiency impairs pancreatic digestive enzyme activity and digestive capacity of weaned piglets. Br J Nutr. 2016;116(3):425–33.
Kelleher SL, McCormick NH, Velasquez V, Lopez V. Zinc in specialized secretory tissues: roles in the pancreas, prostate, and mammary gland. Adv Nutr. 2011;2(2):101–11.
Prasad AS, Bao B. Molecular mechanisms of zinc as a pro-antioxidant mediator: clinical therapeutic implications. Antioxidants (Basel). 2019;8(6)
Jabłońska B, Mrowiec S. Nutritional support in patients with severe acute pancreatitis-current standards. Nutrients. 2021;13(5)
Yang AL. Nutrition and acute pancreatitis. J Clin Med. 2021;10(4)
Kim JY. Optimal diet strategies for weight loss and weight loss maintenance. J Obes Metab Syndr. 2021;30(1):20–31.

Bremner JD, et al. Diet, stress and mental health. Nutrients. 2020;12(8)
Koehler K, Drenowatz C. Integrated role of nutrition and physical activity for lifelong health. Nutrients. 2019;11(7)
Palmer M, et al. The effectiveness of smoking cessation, physical activity/diet and alcohol reduction interventions delivered by mobile phones for the prevention of non-communicable diseases: a systematic review of randomised controlled trials. PLoS One. 2018;13(1):e0189801.
Mac Goey P, Dickson EJ, Puxty K. Management of the patient with acute pancreatitis. BJA Educ. 2019;19(8):240–5.
Binker MG, Cosen-Binker LI. Acute pancreatitis: the stress factor. World J Gastroenterol. 2014;20(19):5801–7.

Chapter 8
Prevention of Chronic Obstructive Pulmonary Disease via Nutraceuticals

Renuka Soni, Shanjua Beri, Kushi Anand, Vineet Kumar Gowasmi, Seid Mahdi Jafari, and Saurabh Kumar Jha

1 Introduction

Chronic obstructive pulmonary disease (COPD) is a leading cause of mortality and mobility worldwide after cardiovascular diseases. A dramatic surge in the COPD burden and comorbidities associated has been ascertained due to COVID-19 pandemic. COPD is defined as progressive obstruction in the lung airways on exposure to toxic inhalants. COPD is caused by a mixture of pulmonary disorders such as emphysema, obstructive bronchitis, and chronic inflammation in airways and parenchyma of lungs. The COPD symptoms includes persistent cough with sputum production and dyspnea. Tobacco and cigarette smoking, air pollution, inhalation of toxic gases and chemicals are the major risk factors in the onset and development of COPD (Adeloye et al. 2015; Singh et al. 2022; Collins et al. 2019). Even though there are models to examine the variations in the inflammatory influx brought on by

R. Soni
Department of Biotechnology, Sharda University, Greater Noida, Uttar Pradesh, India

S. Beri · S. K. Jha (✉)
Department of Zoology, Kalindi College, University of Delhi, New Delhi, India

K. Anand
Kristu Jayanti College (Autonomous), Bangalore, Karnataka, India

V. K. Gowasmi
Department of Biotechnology, Delhi Technological University, New Delhi, India

S. M. Jafari
Department of Food Materials and Process Design Engineering, Gorgan University of Agricultural Sciences and Natural Resources, Gorgan, Iran

Halal Research Center of IRI, Iran Food and Drug Administration, Ministry of Health and Medical Education, Tehran, Iran

S. M. Jafari et al. (eds.), *Nutraceuticals: A New Millennium Approach for Preventive Medicine*, Food Bioactive Ingredients,
https://doi.org/10.1007/978-3-031-76481-3_8

this exposure (to smoke, fumes, and biomass), the pathogenetic pathways involved in the onset and development of COPD have not yet been fully elucidated. In recent years, the focus of researchers is to recognize and uncovers the risk factors and genetic predisposition involved in the onset of COPD (Ingebrigtsen et al. 2011). Genetically predisposed individuals who experience a non-reversible bronchial blockage due to prolonged exposure to harmful gases and gaseous particles are at high risk of developing COPD (Vogelmeier et al. 2017).

Progression of this disease negatively affects the quality of life and physic functioning of individual with a frequent hospitalization and high costs of healthcare facilities. Numerous studies in recent decades have shown that the pathogenesis of COPD is also influenced by the nutritional components of various diets. Poor diet and nutrient deficiency are addition factors that worsen the recovery in COPD patients. Additionally, recent research has started to highlight the potential role of diet in altering gene expression through epigenetic changes that may predispose the subject to developing a non-reversible bronchial obstruction (Adcock 2017). The modification of DNA, lipids, or proteins as a result of lung damage caused by oxidative stress can start cellular responses that promote lung inflammation and lead to tissue degradation.

The associations between diet and COPD addresses a complex interfacing network that might impact the presence of a persistent airway stress in COPD patients. As a result, these interactions might also affect COPD treatment. Healthy eating habits and adequate nutritional intake help reduce COPD-related oxidative stress, pulmonary and systemic inflammation. Therefore, dietary intake of balanced and essential nutrition's is necessary to lower the risk of pulmonary failure and death among COPD patients. Healthy dietary patterns and specific nutrient intake attenuates the oxidative stress, pulmonary and systemic inflammation associated with COPD. The dietary guidelines for patients who are at risk of developing COPD are consistent with the dietary guidelines for the general population. However, it is suggested to consume a diet that is low in fat, high in dietary fibre, complex carbohydrates, fruits, and vegetables. Dietary proposals in patients with cutting edge COPD ought to be resolved independently for every patient, considering the level of hunger, as well as the financial states of individual patients (Collins et al. 2019; Mekal et al. 2021). The chapter discusses in detail about the pathophysiology, progression and treatment of COPD focussing on the association of dietary intake and decline in COPD.

2 Pathophysiology of COPD

Pathophysiology of COPD deals with the physiological evolution and progression of the disease. It is a heterogenous reversible disease caused by the long-term obstruction and inflammation in the airways of lungs (Leap et al. 2021). Inhalation of the toxic gases or noxious inhalants present in cigarettes causes inflammation in the airways due to the immune response against the toxin. As a result of immune response, an epithelial barrier prevents the entering of foreign substance through the lungs and removes the pathogen from the upper airways. The humoral memory response against the foreign toxic inhalant develops slowly. The process of invasion, immune response, inflammation and remodelling of lung parenchyma on interaction with toxic inhalant leads to bronchiolitis, emphysema, chronic bronchitis and subsequent development of Chronic obstructive pulmonary disease (Hogg et al. 2004). Apart from cigarettes smoking, the other major risk factors of COPD involve poor-lung growth due to aging, inflammatory response due to inhalation of toxic agents, exposure to fumes, dust and chemical agents, air pollution burning of wood and biomass fuels, and asthma (Lange et al. 2015; Eisner et al. 2010). SARS-Cov-2 has also appeared to be a major risk factor of morbidity and mortality in COPD patients in the past two years. The clinical representation of COPD is highly variable. About 60% of COPD patients have a history of cigarette, smoking or inhalation of toxic gases for prolonged period. The commonly reported signs and symptoms of COPD includes mucus production, chronic cough, obstruction in airway, dyspnea, low forced expiratory volume in one second (FEV_1), chest tightness and whizzing. In severe cases of COPD emphysema in associated with chronic bronchitis is also observed (Lange et al. 2015; Kessler et al. 2011). Chronic obstruction of airways often occurs as two overlapping chronic lung inflammatory process (Fig. 8.1):

- Emphysema, a condition refers to the destruction of parenchymal, and loss of alveolar units (air passages of lungs)

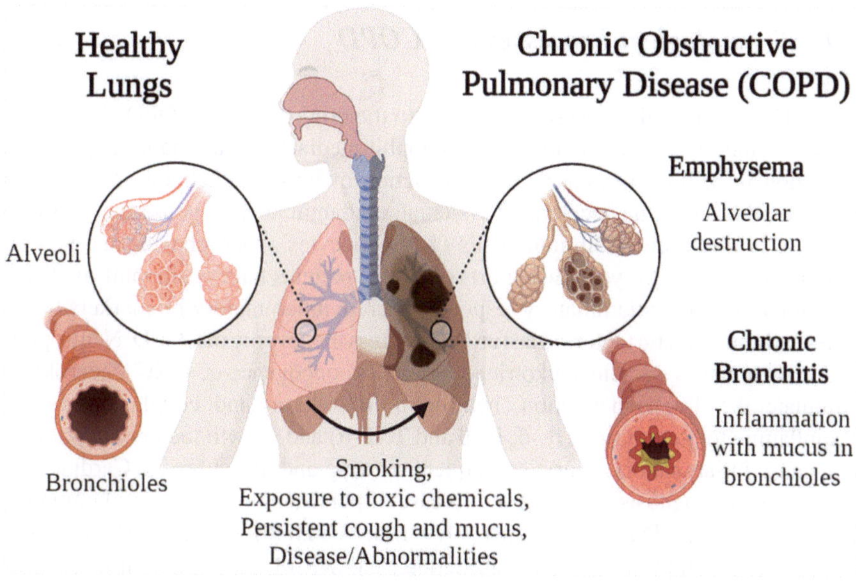

Fig. 8.1 Pathophysiology of COPD

- Chronic bronchitis, refers to the inflammation, mucus production and remodelling of bronchial tubes

Prolonged inflammation, oxidative stress, protease imbalance, signal transduction pathways, and chronic inflammation leads to bronchial epithelial goblet cell metaplasia including submucosal bronchial gland hyperplasia and hypertrophy. The obstructed airflow hastens the deterioration of lung function, that further increases acute lung exacerbations. The mucus attachments block the lung airways routes and cause steady colonization of the respiratory airway with bacterial pathogens. Patients with ongoing bronchitis have lower physical activity and quality of life (Tian and Wen 2015). Emphysema is the persistent, abnormal enlargement of the respiratory airspaces, or alveoli, accompanied by destruction of the alveolar walls without apparent macroscopic fibrosis. On microscopy, there are more CD8 T lymphocytes and macrophages. The loss of epithelial and endothelial cells damages alveolar walls in addition to bulla, or emphysematous airspaces larger than 1 cm are observed. The destruction of the alveolar walls results in the loss of the lungs' elastic recoil. This prompts airflow hindrance, decreased regions for gas trade, and expanded dead space (Berg and Wright 2016; Higham et al. 2019). Pulmonary hypertension is also one of the significant reasons behind COPD comorbidity as it develops in later stages of COPD. Intimal thickening of the arteries adjacent to the bronchioles is caused by smooth muscle proliferation and the deposition of elastin and collagen, which results in pulmonary arterial constriction in the presence of chronic hypoxia. These arteries are unable to fully dilate in response to exercise, acetylcholine, or an increase in airflow (Berg and Wright 2016).

2.1 Physiological Processes and COPD

Several physiological processes contribute to the progression of COPD. These processes include local, systematic and eosinophil mediated inflammation, apoptosis, alteration in cellular responses and destruction of lung parenchyma. Majority of COPD patients have an accelerated systematic inflammatory response that cause an increase in the levels of immune cells (T-lymphocytes, macrophages, and neutrophils) in the lung airways. Few cohorts of COPD patients have eosinophil-mediated inflammation. The inflammatory responses secrete inflammatory biomarkers (cytokines and chemotactic factors) that further progress the severity of COPD. Neutrophils and macrophages secrete leukotriene B4, IL-8, oncogenes-α, CXC chemokines stimulate the T-cells activation, neutrophil infiltration, and cellular migration. Proinflammatory cytokines (IL-6, IL-β and TNF-α) and growth factors are released that cause inflammatory injury, subsequent scarring and lung fibrosis. Cardiac failure and poor-prognosis are observed in case of persistent systematic inflammation in COPD patients (Duffy and Criner 2019; MacNee 2006). Oxidative stress modulates the DNA, lipids and proteins, initiating a cascade of cellular and inflammatory responses releasing proteases, elastase, cathepsin A, E, L, G and S, metalloproteinases. The inflammatory response triggers the activation of redox-sensitive transcription factors like unfolded protein response, induction of autophagy, and nuclear factor (NF)-B. These proteases cause mucus production, alveolar and lung tissue destruction thereof. Elevated serum levels of IL-6 and C-reactive protein (CRP) are associated with the high risk of COPD and loss of lung function in individuals with a history of smoking (Szmidt et al. 2020).

2.2 Environmental and Genetic Causes of COPD

In addition to smoking, poor ventilation and cooking fumes from indoor wood stoves, environmental pollution, dust and gases from underground tunnel workers from coal and gold mining, gas from cadmium mining have all been linked to an increase in COPD prevalence. Patients with lower socioeconomic status are more likely to develop the disease (Higham et al. 2019). About 80% of COPD cases prevail dur to cigarette smoking. However, only 15–20% smokers develop COPD, suggesting the involvement of genetic factors in COPD development. Free radicals in cigarettes damage the lungs through oxidative stress. COPD may be caused by defects in the mechanisms that protect against this damage. Epigenetics changes in humans are associated with both the environment and genetic changes. It is found that epigenetics is involved in the pathogenesis of COPD. Several studies have demonstrated that in lung tissue, sputum and blood, abnormal changes in miRNAs, chromatin remodelling, histone modifications, and DNA methylation are present in COPD patients. Subsequently, persistent respiratory circumstances have epigenetic markers that upholds the advancement of novel respiratory medication diagnostics and therapeutics (Yang and Schwartz 2011). The epigenetic mechanisms and their associated signalling pathways in COPD can be regulated through the beneficial impacts of nutrient. Numerous studies demonstrated that nutrients can regulate transcription and translational processes and alter the structure of chromatin, thus affecting metabolic traits. This field of study known as nutriepigenomics or nutritional epigenomics studies dealing with preventive measures of nutrients in COPD via epigenetic regulation (Cherneva and Kostadinov 2019). Moreover, Low BMI is associated with the poor prognosis in COPD patients (Higham et al. 2019).

2.3 COPD and Gut-Lung Axis

Microbiota of our whole body and the gut have developed a strong symbiotic relationship that is essential for maintaining health, vitamin synthesis, nutrient metabolism, prevention of pathogen colonisation, and immune response control. Despite having similar embryonic origin, gut and lung differ anatomically and microbiologically. The transient microorganisms that make up the lung microbiota mostly come from the upper respiratory tract (URT), which includes the nose, oropharynx, nasopharynx, paranasal sinuses, and nasal passages. However, a balanced microbiota in gut and lungs is necessary for the regulation of immune system and overall growth of the body. Gut-lung axis is defined as the cross-talk between the lungs and gut microbes. The bloodstream links the microbiota of the gut and lung, making it easier for microbial metabolites, hormones, cytokines, and endotoxins to interact with one another (Jandhyala et al. 2015; Mjösberg and Rao 2018; Qu et al. 2022a). The pathophysiology of COPD is significantly influenced by imbalances in the microbiota of the lungs and the gut. Asthma, malnutrition, smoking, chronic lung infections and medications all have the potential to affect the microbiota in both the gut and lungs. The gut-lung axis is disrupted in COPD, which results in a dysfunctional immune response, disturbances in metabolites, increased production of pro-inflammatory cytokines, altered microbiota composition, microbiota diversity, and poor lung infection clearance or gastrointestinal symptoms or diseases. Use of antibiotics in COPD patients disturbs the long-term micro flora of both gut and lungs, thereby increasing the antibiotic resistance. COPD is profoundly connected with the dysbiosis in the stomach and lungs, which can be actuated by epigenetic changes or managed through NLRP3 inflammasome (Sun et al. 2020; Qu et al. 2022b). Zang L study reports that respiratory infection and COPD, asthma, cystic fibrosis, emphysema diseases such as can be prevented by intervening the gut microbiota with natural products, balanced diet, nutrient supplementation, and probiotics (Zhang et al. 2020).

3 Malnutrition and COPD

Malnutrition refers to an imbalance (deficiency or excess) of nutrients and energy levels in the human body. Decreased nutritional intake and an increased bodily need for vitamins, minerals, energy, and proteins refers to malnutrition (Stratton et al. 2003). The risk and prevalence of malnutrition are higher in COPD patients. An increase in the chances of weight and nutrient loss is directly proportional to the COPD severity among COPD patients (Ingadottir et al. 2018). Cachexia and Sarcopenia are two conditions that develop in patients with prolonged stable COPD with nutritional deficiency (Collins et al. 2010; Jones et al. 2015). Cachexia is defined as a yearly weight loss of 5% accompanied by anorexia, fatigue, decrease in the fat-free mass index (FFMI), and muscle strength increased. It is quite common in COPD patients and is referred to as pulmonary cachexia. BMI and FFMI in COPD patients are quite low, thus contributing to the poor prognosis and increases hospital stay. Serious consequences are observed in decreased levels of FFM. Whereas Sarcopenia is a regular sickness of COPD patients and is defined by low skeletal muscle strength and mass. About 16% of COPD patients are sarcopenic, while 20–40% exhibit decline in muscle mass only. Notably, mortality is higher in COPD patients in case of MSS (malnutrition-sarcopenia syndrome) when both the conditions are combined (Cederholm et al. 2019). Sarcopenia, weight loss, muscle loss and malnutrition contribute to an increase in fatality among COPD patients and a high protein diet with cereals and fibres have proven to elevate this COPD severity. Thus, a nutrition-based treatment for COPD may be a viable non-pharmaceutical strategy for managing the disease progression.

Patients with COPD may also develop osteoporosis as a result of anorexia, which causes them to consume fewer dairy products and engage in less physical activity daily. Osteoporosis is also highly prevalent among COPD patients. Low bone mineral density and low serum vitamin D levels are strongly correlated in COPD patients. The regulation of calcium metabolism depends on vitamin D. Patients with pulmonary fibrosis and other pulmonary conditions have also been found to have decreased lung function due to low serum levels of vitamin D. Serum levels of the GC-globulin (GC gene-encoded vitamin D binding protein (VDBP)), appear to have an impact on vitamin D levels. VDBP stabilizes the vitamin D molecule ensuring its transportation to the target tissues due to very short lifespan if Vitamin D in serum. The role of this carrier protein is unclear in declining the lung function; however, it regulates immunological responses like the chemotaxis of neutrophils and macrophage activation (Franco et al. 2009; Tramontano and Palange 2023).

Figure 8.2 briefly illustrates the mechanism of malnutrition in COPD patients. These factors contribute to malnutrition and subsequently increase rate of mortality and morbidity in COPD individuals. Medications, low-grade systemic inflammation, tissue hypoxia, muscle loss and atrophy, aging and changes in metabolism are the primary factors that contribute to nutritional impairment and poor muscle

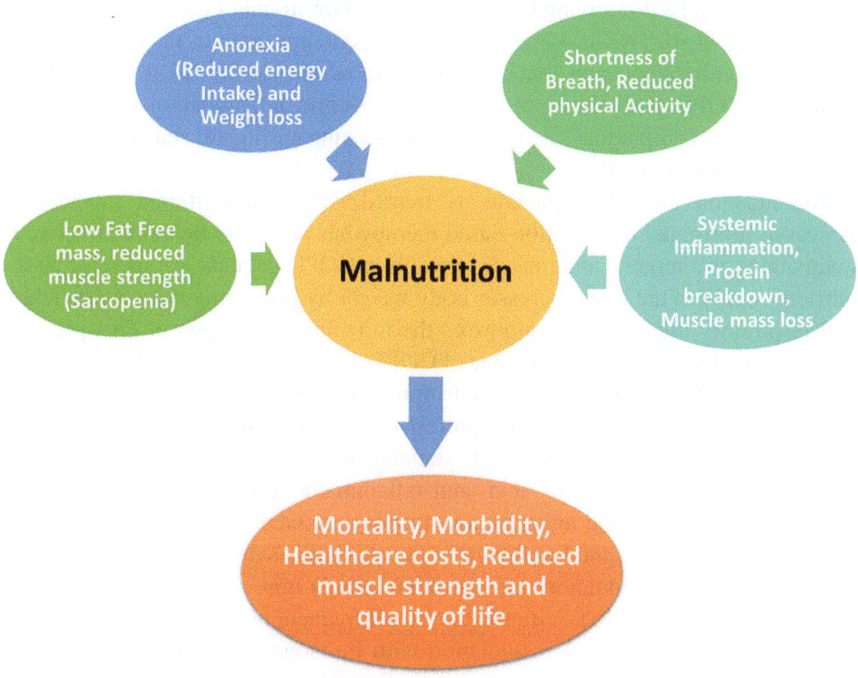

Fig. 8.2 Mechanism of malnutrition in COPD patients

energetics in COPD patients. The chronic obstruction in the airways causes difficulty in breathing as a result the COPD patient is in a hypermetabolism state with an increased resting energy metabolism consuming more calories. As a result of an obstruction in airflow and increased load on breathing, blood flow to the critical organs (heart, ventilatory muscles, central nervous system) is maintained by the body. This results in the development of hypoxia and nutrition-deficient condition in the skeleton tissues, thus weakening muscle strength and bones. Hypoxia adversely influence muscle capability in association with the ongoing hypoxemia and the frequently concurrent paleness that promotes further muscle hypoxia because of a decrease in oxygen conveyance that led to elevated protein turnover and lacking muscle regeneration. Muscle endurance is affected by impaired muscle mass and oxidative capacity as a result of all these mechanisms (Tramontano and Palange 2023).

Aging is directly linked to the free-fat mass loss, particularly muscle tissue. Diminished muscle strength and basal metabolic rate due to depletion in muscle mass result in a decreased physical movement and exercise. A reduction in nutritious diet altogether results in loss of appetite, systemic inflammation and physical activity (Tramontano and Palange 2023; Langen et al. 2013; Schols et al. 1991; Barreiro and Gea 2016). Systematic inflammation occurs in COPD patients due to the increased serum levels of inflammatory markers such as chemokines, TNF-α, IL-1, IL-6, and IL-8. Several studies have reported that an increased level of IL-6 is directly linked to loss of appetite, and weight loss is associated with elevated levels serum levels of TNF-α among COPD patients (Barreiro and Gea 2016; Rawal and Yadav 2015). IL-6, TNF-α, Leptin, and adiponectin are some of the bioactive proteins secreted by adipocytes that are referred to as adipokines. It has been demonstrated in COPD individuals, a low-grade systemic inflammation is caused by dysregulation of adipokines, also they are influencing nutritional status and regulate appetite (Breyer et al. 2012).

Malnutrition in COPD patients is treated with glucocorticoid medication. However, nutritional intervention-based therapy has proven to be more effective in maintaining the appetite and muscle strength in COPD patients. An increase in the nutrient supply to the body increases body weight by improving muscle efficiency, and respiratory muscles and suppresses the inflammatory response. This prevents further progression and exacerbations of COPD (Rawal and Yadav 2015; Mete et al. 2018). In COPD patients with malnutrition, traditional high-carbohydrate diet is avoided and a high protein and fat is preferred. High protein in the diet significantly improves the growth and strength of skeletal muscles (Rawal and Yadav 2015). Omega-3 polyunsaturated fatty acids anti-inflammatory properties and antioxidants in fruits highly reduce the inflammatory response in COPD patients. Vitamin supplementation (Vit A, C, and E) efficiently elevate COPD symptoms and improves its prognosis. Immunomodulatory effects of Vitamin D improve the muscle's overall strength (Mete et al. 2018). A diet with a small frequent nutritional meal, and multivitamin supplementation in association with Pulmonary rehabilitation (physical exercise, yoga) is a recommended treatment for malnutrition in COPD patients.

4 Dietary Nutrients as Metabolic Modulator in COPD

4.1 Fruits and Vegetables

The dietary quality, nutritional status, and oxidative-inflammatory pathogenic underpinnings of COPD provided the justification for confirming the benefits of antioxidant and anti-inflammatory dietary components for the respiratory system. Consistent epidemiologic evidence from longitudinal and cross-sectional studies reported potential beneficial effects of a high intake of antioxidant nutrients, and of foods rich in antioxidants, mostly fresh, hard fruits and, to a lesser extent, vegetables, on lung function and decline in incidence of COPD associated mortality (Rodríguez-Rodríguez et al. 2016; Walda et al. 2002; Kaluza et al. 2018; Scoditti et al. 2019) (Table 8.1).

After one week of diet supplementation of Beetroot juice, a dietary source of nitrates improved the mitochondrial respiration and energy production via nitric oxide formation, increased plasma nitrate levels and decreased diastolic blood pressure in COPD patients, though no effects on submaximal exercise's oxygen consumption, physical activity level, or walking capacity was observed (Friis et al. 2017). Nonetheless, one more randomized preliminary detailed that COPD patients following an eating routine wealthy in leafy foods showed a yearly expansion in FEV1 contrasted and the benchmark group following a free eating routine north of 3 years ($p = 0.03$), after change for actual work, liquor consumption, co-morbidities,

Table 8.1 Recommended consumption of nutrients in COPD patients

Nutrient	Recommended consumption in COPD (per day)
Carbohydrates (gm)	<200
Lipids (gm)	30
Proteins (g/kg of body weight)	1.5
Energy (Kcal/mg/body weight)	30
Vitamin D (μg)	15–20
Vitamin A (μg)	60.7
Vitamin E (mg)	72.4
Vitamin B12 (μg)	37.9
Vitamin C (mg)	100
Folate (μg)	96.6
Sodium (mg)	58.6
Iron (mg)	37.9
Magnesium (mg)	83.9
Fibres (gm)	30–40

Rondanelli et al. (2020)

and fuel recurrence (Keranis et al. 2010). These findings suggest that fruit and vegetable consumption is a significant factor in pulmonary function and COPD risk. It is important to note that eating fresh fruit may be one part of a healthier lifestyle that also includes less smoking, more exercise, eating less Western food (like meat) and more vegetables.

4.2 *Vitamin C*

Oxidative overload in COPD is quite common and is typically brought on by infection and hypoxia. Oxidative overload plays a crucial role in the impairment and remodelling of lung tissue. Emphysema in COPD is primarily caused by dysregulated antiproteases in lung tissue caused by elevated oxidative stress. Vitamin C is a potent antioxidant that is commonly consumed as a natural supplement and dietary intake of Vitamin C, is associated with the higher levels of FEV1. Many in vivo and clinical based research studies over the past years have showed that Vitamin C has high potential in elevating COPD exacerbations and improving the overall function of lungs. According to in vitro data from endothelial cell lines, vitamin C could block the production of IL-8 and inhibit NF-B activation by IL-1 and TNF-α by means that were not dependent on the antioxidant activity of vitamin C (Bowie and O'Neill 2000). Observational examinations in youngsters showed utilization of fruits rich in L-ascorbic acid (Vitamin-C), directly related to diminished wheezing (Forastiere et al. 2000). Experimental and observational research suggests that Vitamin C plays an important role in the COPD pathogenesis and progression. A study by Koike et al. found that vitamin C supplementation prevented smoke-induced emphysema, reduced oxidative stress, and restored damaged lung tissue in knock-out mice that were unable to synthesize vitamin C (Koike et al. 2014). Also, in one of the epidemiological studies conducted in the United Kingdom of over 7000 grown-ups aged 45–74 times reported that increase in plasma concentration of Vitamin-C has preventive effects in COPD patients by lowering the risk of obstructive airways (Sargeant et al. 2000). Vitamin C is important for lung function and intake of vitamin C-rich whole foods like fruit and vegetables is effective in reducing the COPD exacerbations. Vitamin C in association with Vitamin E protect lungs against oxidative damage reducing COPD related moralities. Vitamin C supplementation in COPD patients improves their nutritional status by promoting lung function and serum antioxidant levels by decreasing oxidative damage to the lung.

4.3 *Vitamin D*

Epidemiological investigations show promising relationship between vitamin D and lung wellbeing; However, little is known about the mechanisms that underlie these effects. Vitamin D can be obtained through food or through supplements; sunlight being the fundamental source of vitamin D levels (van Iersel et al. 2022). Low levels of vitamin D in the blood are linked to decreased lung function, increased use of corticosteroids, and more frequent exacerbations in children. Studies have shown a correlation between higher intake of vitamin D and improved lung function, as evidenced by higher FEV1 and lower prevalence of COPD in the general population. Conversely, individuals with lower serum vitamin D levels, especially smokers, were found to have increased odds of developing COPD. Vitamin D deficiency was linked to decreased FEV1 and FVC in certain populations, with variations noted between genders and medical condition (Afzal et al. 2014). While several studies have reported positive associations between serum vitamin D levels and FEV1 in various populations, some cross-sectional studies did not confirm these findings. Prospective analyses, however, revealed that lower vitamin D concentrations were associated with a faster decline in lung function over time and a higher risk of developing COPD. In patients with COPD, studies consistently showed lower dietary intake and serum levels of vitamin D compared to healthy individuals. Vitamin D deficiency was inversely associated with lung function parameters in COPD patients, and supplementation appeared to have beneficial effects, particularly in those who were deficient (Sluyter et al. 2017).

Overall, while evidence is mixed regarding the association between vitamin D intake and serum levels and the risk of COPD in the general population, there is strong evidence supporting the beneficial effects of vitamin D supplementation on lung function decline in COPD patients, especially in those with existing deficiency.

4.4 Whole Grains

Intake of Whole grains in a diet are beneficial for maintaining good lung health and preventing COPD associated mortality. Whole grains are plentiful in selenium, vitamin E, flavonoids, phytic corrosive, phenolic acids, and fundamental unsaturated fats, which may additively or synergistically add to wholegrains' beneficiary impact on both the respiratory and non-respiratory illnesses (Adeloye et al. 2015). Fibre present in Whole grains, fruits, and vegetables contribute to their antioxidant and anti-inflammatory. Optimal level of fibre intake in diet is directly associated with the high serum levels of Adiponectin, an insulin-sensitizing adipocytokine having anti-inflammatory properties, and lower serum levels of cytokines (IL-6, TNF-α) and C-Reactive protein (CRP) (Singh et al. 2022; Collins et al. 2019). Intake of higher fibre diet decreases the prevalence of COPD by 40% in a large cohort of the population. Considered in terms of the types of fibre (fruit, vegetables, and cereals), the beneficial association specifically among current smokers and ex-smokers was found to be strongest in diet having cereal fibre (Ingebrigtsen et al. 2011).

4.5 Coffee

Coffee's widespread use has sparked a growing interest in its potential contribution to respiratory health. Coffee's constituents, caffeine and polyphenols, have been linked to improvements in lung function and decreased mortality from respiratory diseases, but less in case of COPD patients. Caffeine has anti-inflammatory properties and acts as a bronchodilator improving respiratory health. Phenolic compounds in coffee have antioxidant, anti-inflammatory activity that further add to its contribution to a healthy lung function (Vogelmeier et al. 2017).

4.6 PUFAs

Polyunsaturated fatty acids (PUFAs) are a part of n-3 (omega-3) family that includes linolenic acid, its long-chain derivatives docosahexaenoic acid and eicosapentaenoic acid. The endogenous production of omega-3 fatty acids is quite low and its nutritional intake relies upon exogenous sources. The main source of PUFAs are seafoods and fish. Omega-3 fatty acids have potent anti-inflammatory properties that have beneficial effects in several diseases including COPD, cardiovascular disease, cancer, rheumatoid arthritis, and diabetes (Adcock 2017). Metabolism of PUFAs produces leukotriene, prostaglandin and eicosanoids which are strong mediators of bronco-constriction, thrombosis, vaso-constriction, and inflammation. Cytochrome P450 enzymes, which are highly expressed in the lungs, produce some fatty acid compounds such as EPA and DHA metabolites that are potent anti-inflammators, bronchodilators, and vasodilators. The eicosanoids resolvins and protectins, metabolites of long-chain n-3 PUFAs, eliminates inflammatory mediators and promote healing of inflamed lung parenchyma tissue (Mekal et al. 2021).

The synthesis of generally less biologically active eicosanoids is favoured after the dietary intake of n-3 PUFA that is further accompanied by a partial substitution of n-6 PUFA in cell membranes, thereby competing for the metabolizing enzymes. Apart from the synthesis of beneficial lipid metabolites, n-3 PUFA mediates anti-inflammatory by the direct regulation of inflammatory gene expression (cyclooxygenase-2, matrix degrading enzymes, cytokines, and adhesion molecules) through nuclear transcription factors regulation (pro-inflammatory NF-B), which is involved in the pathogenesis of lung inflammation.

5 COPD and Dietary Patterns

All the essential nutrients have the ability to regulate the oxidative stress and inflammatory processes that contributes to the onset of COPD, thus making them potential future therapeutic targets. Studies have demonstrated a link between certain dietary habits and the risk of COPD. However, specific dietary modifications as a clinically relevant tool to promote lung health and congruently indicated that the pattern of dietary intake is an important factor in the pathogenesis and prevention of COPD. Cross sectional and cohort studies typically use food questionnaires to collect diet-related data, which may not be as accurate as circulating concentrations. On the other hand, because it incorporates the variety, quantities, proportions, and a combination of the various foods and beverages that are included in the diet, the analysis of dietary patterns may be more effective than that of individual nutrients (Table 8.2). Also, the evaluation of the effects of individual food and nutrient intake on COPD is also enhanced by examining dietary patterns. As diet and its nutrients vary from region to region, dietary patterns may vary, preventing comparisons between nations and continents (Scoditti et al. 2019; Friis et al. 2017). A diet high in fruits, vegetables, fish, and whole-grain products is a hallmark of a healthy diet. This dietary pattern has been linked to a lower risk of COPD. Diets high in fruits and vegetables have a positive correlation with FEV1 and FVC in the general population and daily fish consumption reduced FEV1 decline. A lower prevalence of COPD has been linked to higher scores in a healthy dietary pattern. In another cohort study by Kaluza et al. (2018) on mammography founds that woman who consumed more fruit per day had a lower risk of COPD than women who consumed less fruit. Moreover, it was observed that consuming vegetables for an extended period of time did not maintain this risk difference (Bowie and O'Neill 2000).

5.1 *Western Diet*

A higher incidence of COPD has been linked to western diet pattern that emphasizes food intake of refined grains, red and processed meats, chips, eggs, and flavoured drinks. Cured and red meats, desserts, and refined grains are all examples of

Table 8.2 Dietary pattern and outcome in COPD patients

Dietary pattern	Nutrients involved	Outcome
Western diet	Processed meat (cured, red meat), coffee, fat, boiled vegetables, eggs, cereals, salt, beer, potatoes, refined grains and sugar, chips and fizzy drinks	Respiratory symptoms (cough, phlegm, wheeze), FEV1, FEV1/FVC, COPD prevalence
DASH dietary	Sweetened beverages, salt, whole grains, processed red meat, nuts, legumes, fruits and vegetables	COPD and cough
"Meat dim sum" dietary pattern	Eggs, rice, desserts preserved food, processed red meat and noodle-based dishes	Cough and phlegm
Mediterranean diet	Red wine, nuts, legumes, fish, dairy products, fruits and vegetables, extra virgin olive oil, seafood, poultry.	Oxidative stress, inflammatory response and impaired lung function

Scoditti et al. (2019), Friis et al. (2017), Keranis et al. (2010)

western diet foods that contributes to the development of COPD. Consuming processed meat has been linked to a high risk of COPD exacerbation readmission. The Western diet has been linked to frequent complaints of wheezing, coughing, and phlegm, as well as a decline in COPD patients' lung function and lower predicted concentrations of FEV1 and FVC and the ratio of FEV1/FVC. Likewise, an eating regimen wealthy in cereals, refined sugars, salt, and lager moreover is related with lower FEV1, a more serious risk for COPD, and a critical decrease in lung capability (Scoditti et al. 2019; Forastiere et al. 2000).

This positive association between COPD risk and Western-style dietary patterns may have been generated by multiple mechanisms, even though the role of processed meat in the pathogenesis of COPD is largely unknown and there are no definitive mechanistic studies. In the first place, handled or restored meat has a high level of nitrosamine and nitrate or nitrite compounds. These nitrites might cause DNA harm, hindrance of mitochondrial breath, and nitrosative pressure by expanding provocative cycles in the lung parenchyma and aviation routes related with this nitrosative pressure. Eminently, this cycle can have advantageous impacts at physiological focuses by shielding the vascular tissues from incendiary cells. A beneficial effect of dietary nitrite on the destruction of the alveolar structure and subsequent inflammation of the airways was found in a mouse study. As a result, pulmonary emphysema, which is characterized by proteolytic destruction of the alveolar structure, was prevented from developing by these nutrients.

Second, Western diets are low in n-3 PUFAs and high in n-6 PUFAs and saturated fatty acids (SFAs). Increased systemic inflammation has been linked to a decline in pulmonary function when SFAs and trans fats are consumed. The utilization of pentadecanoic corrosive (C15:0), an odd chain SFA, usefully affected patients with COPD, working on the FEV1/FVC proportion. This fatty acid is considered a biomarker of fat consumption in dairy products, particularly milk, which contains approximately 1.2% pentadecanoic acid, and it is found in lower concentrations in fish, lamb, and beef. In addition, the Western diet, which is high in high-energy macronutrients and high in SFA and n-6 PUFA, encourages obesity, which is bad for health. SFA and n-6 PUFA have negative effects on inflammation and are linked to asthma and other lung diseases (Koike et al. 2014).

5.2 *Dietary Pattern "Meat—Dim-Sum"*

The "meat—dim-sum" diet emphasizes the consumption of rice, noodles, preserved foods, pork, chicken, and fish. Subjects who followed a "meat—dim-sum" dietary pattern had a higher risk of coughing up phlegm in a male and female population. Consuming a lot of different kinds of meat raises the risk of COPD (Sargeant et al. 2000).

5.3 Mediterranean Diet

Having a high intake of extra virgin olive oil, fruit, vegetables, fresh produce, nuts, and legumes; a low intake of red meat and prepared foods; and a moderate intake of fish and seafood, poultry, fermented dairy products, and red wine (with meals) are all characteristics of the Mediterranean diet. This eating pattern favours poly- and monounsaturated fats, especially those derived from the oleic acid found in olive oil, over "trans" and saturated fats, which are higher in cholesterol. It is also rich in phytoestrogens, folates, and selenium (Rondanelli et al. 2020) Over the past ten years, a correlation between lung function and adherence to the Mediterranean diet has been noted. Low FVC and FEV1 values in smokers and older persons, as well as a higher prevalence of decreased lung function, have been linked to poor adherence to the Mediterranean diet. It has been demonstrated that several elements of this dietary pattern, including fruit and vegetables, vitamins C and E, flavonoids, ß-carotene, fatty acids, and various minerals, have a protective influence on oxidative and inflammatory processes and may also have a protective effect on lung function (Rondanelli et al. 2020). It is well-known that N-3 PUFAs are important nutrients with a variety of physiological roles as well as positive impacts on health, such as anti-inflammatory and antioxidative qualities. Consuming n-3 PUFAs (eicosapentanoic acid and docosahexanoic acid), which are found in fatty fish (such as tuna, mackerel, and salmon) and in the oil derived from these species, may help to slow the decline in respiratory function.

Finally, compared to a Western diet high in saturated fat, the Mediterranean diet intake over a 4-week period reduced serum levels of advanced glycation end products. Lower spirometric values and abnormal ventilatory patterns were linked to the deposition of these products. The effect of the Mediterranean diet on lung function revealed sexual dimorphism, with healthy women's central airways being smaller than men's, and female smokers' FEV1 declines being more pronounced than those of male smokers (Scoditti et al. 2019).

5.4 DASH Diet

The DASH dietary pattern is the example of a one of the major healthy eating patterns, calls for a higher intake of fruit, legumes, vegetables, nuts, low fat dairy products, and whole grains and a lower intake of red meat, processed foods, sweetened beverages, and salt. A case-control study showed that COPD patients adhered to a DASH diet pattern less frequently than the control group without COPD. Respiratory function tests did not reveal a significant difference across tertials of a DASH diet, even though patients with COPD had FEV1, FVC, and FEV1/FVC concentrations that were significantly lower compared with the control group prior to the intervention research. Adherence to a DASH dietary pattern only considerably reduced cough as a symptom (Walda et al. 2002).

6 Conclusion

Identification of modifiable risk factors for COPD prevention and therapy is highly demanded given the alarmingly rising incidence of COPD worldwide. Based on the information now available, increased knowledge of nutrition and dietary determinants affecting respiratory health may be of importance for public health due to their implications on disease modification. There are many chances for changes and interventions given how poorly the existing dietary practises in the general population and in people with respiratory diseases are reported to be. A two-hit lifestyle burden (smoking and an unhealthy diet) is presently on the rise due to the rising smoking habit in emerging nations and the inexorable global phenomena of Westernisation of lifestyle variables, which includes a more processed and convenience-oriented diet. Numerous scientific organisations endorse the cautious/Mediterranean diets as good eating patterns because of the significant evidence linking them to enhanced cardiometabolic health, including a decreased risk of CV disease, diabetes, and obesity. Published studies also consistently demonstrate the detrimental effects of the Western diet, which is high in refined foods, saturated fat, meat, and sugar, on lung function and the risk of developing COPD. In contrast, specific dietary factors and diets, primarily the prudent/Mediterranean diets, which are high in plant-based foods and healthy fats, have been shown to preserve lung function and reduce the risk of developing COPD or its long-term progression. It's interesting to note that the size of diet's influence on lung function is reportedly equivalent to that of chronic smoking (van Iersel et al. 2022), highlighting the possibility that good dietary practises may have a significant impact on the development of COPD as well as the related metabolic and CV risk. Asthma risk and lung function were positively impacted by particular dietary patterns and/or nutrients in several trials, but not COPD risk or asthma, which clearly suggests a genuine underlying effect rather thana generalised and most likely confounded effect. While asthma has an immunological aetiology, COPD and CV disorders both have a systemic inflammatory pathophysiology. Therefore, nutritional targeting of oxidative balance and extreme inflammation may provide a special opportunity to preventor cure COPD and its associated cardiovascular co-morbidities. To establish the efficacy and mechanisms of food in preventing and treating COPD, further animal research and human intervention trials are required. The manipulation of the microbiota by food is a recent field of study that has promise for improving health and reducing illness risk. To address the inter-individual variability in response to food and enhance a personalised nutrition intervention to prevent/treat COPD, the field of nutrigenetics (i.e., the interaction between genetic variations and diet) also merits consideration.

In conclusion, the multifaceted interplay between nutrition, lifestyle factors, and respiratory health, particularly in the context of COPD prevention and management, underscores the significance of dietary interventions as a critical component of public health strategies. The escalating global incidence of COPD, coupled with the pervasive adoption of Westernized dietary patterns, necessitates urgent attention to the role of nutrition in mitigating disease burden. While smoking cessation and pharmacologic therapies remain cornerstone interventions in COPD management, the integration of dietary modifications represents a pivotal frontier in disease prevention and management. Embracing holistic approaches that encompass lifestyle modifications, including dietary interventions, holds promise for alleviating the burgeoning global burden of COPD and improving respiratory health outcomes. By harnessing the potential of nutrition as a modifiable risk factor, concerted efforts can be directed towards mitigating the socioeconomic and healthcare burdens associated with COPD worldwide.

References

Adcock IM. Role of epigenetic modifications in pathology of COPD. Tanaffos. 2017;16(Suppl 1):S2.

Adeloye D, Chua S, Lee C, Basquill C, Papana A, Theodoratou E, et al. Global Health Epidemiology Reference Group (GHERG). Global and regional estimates of COPD prevalence: Systematic review and meta-analysis. J Glob Health. 2015;5(2):020415.

Afzal S, Lange P, Bojesen SE, et al. Plasma 25-hydroxyvitamin D, lung function and risk of chronic obstructive pulmonary disease. Thorax. 2014;69:24–31.

Barreiro E, Gea J. Amino acid and protein metabolism in pulmonary diseases and nutritional abnormalities: a special focus on chronic obstructive pulmonary disease. Chapter 12. In: Dardevet D, editor. The molecular nutrition of amino acids and proteins. Elsevier; 2016. p. 145–59.

Berg K, Wright JL. The pathology of chronic obstructive pulmonary disease: progress in the 20th and 21st centuries. Arch Pathol Lab Med. 2016;140(12):1423–8.

Bowie AG, O'Neill LA. Vitamin C inhibits NF-kappa B activation by TNF via the activation of p38 mitogen-activated protein kinase. J Immunol. 2000;165(12):7180–8.

Breyer MK, Rutten EP, Locantore NW, Watkins ML, Miller BE, Wouters EF, et al. Dysregulated adipokine metabolism in chronic obstructive pulmonary disease. European J Clinc Invest. 2012;42(9):983–91.

Cederholm T, Jensen GL, Correia MI, Gonzalez MC, Fukushima R, Higashiguchi T, et al. GLIM criteria for the diagnosis of malnutrition–a consensus report from the global clinical nutrition community. Journal of cachexia, sarcopenia and muscle. 2019;10(1):207–17.

Cherneva RV, Kostadinov D. Epigenetic targets for therapeutic approaches in COPD and asthma. Nutrigenomics—possible or illusive. Folia Med (Plovdiv). 2019;61(3):358–69.

Collins PF, Stratton RJ, Kurukulaaratchy R, Warwick H, Cawood AL, Elia M. Prevalence of malnutrition in outpatients with chronic obstructive pulmonary disease. Proceedings of the Nutrition Society. 2010;69(OCE2):E147.

Collins PF, Yang IA, Chang YC, Vaughan A. Nutritional support in chronic obstructive pulmonary disease (COPD): an evidence update. J Thorac Dis. 2019;11(Suppl 17):S2230–7.

Duffy SP, Criner GJ. Chronic Obstructive Pulmonary Disease: Evaluation and Management. Med Clin North Am. 2019;103(3):453–61.

Eisner MD, Anthonisen N, Coultas D, et al. An Official American Thoracic Society Public Policy Statement: novel risk factors and the global burden of chronic obstructive pulmonary disease. Am J Respir Crit Care Med. 2010;182:693–718.

Forastiere F, Pistelli R, Sestini P, Fortes C, Renzoni E, Rusconi F, et al. Consumption of fresh fruit rich in vitamin C and wheezing symptoms in children. SIDRIA Collaborative Group, Italy (Italian Studies on Respiratory Disorders in Children and the Environment). Thorax. 2000;55(4):283–8.

Franco CB, Paz-Filho G, Gomes PE, Nascimento VB, Kulak CA, Boguszewski CL, et al. Chronic obstructive pulmonary disease is associated with osteoporosis and low levels of vitamin D. Osteoporosis international. 2009;20:1881–7.

Friis AL, Steenholt CB, Løkke A, Hansen M. Dietary beetroot juice—effects on physical performance in COPD patients: a randomized controlled crossover trial. Int J Chron Obstruct Pulmon Dis. 2017;12:1765–73.

Higham A, Quinn AM, Cancado JED, Singh D. The pathology of small airways disease in COPD: historical and future directions. Respir Res. 2019;20(1):49.

Hogg JC, Chu F, Utokaparch S, et al. The nature of small-airway obstruction in chronic obstructive pulmonary disease. N Engl J Med. 2004;350:2645–53.

Ingadottir AR, Beck AM, Baldwin C, Weekes CE, Geirsdottir OG, Ramel A, et al. Two components of the new ESPEN diagnostic criteria for malnutrition are independent predictors of lung function in hospitalized patients with chronic obstructive pulmonary disease (COPD). Clinical Nutrition. 2018;37(4):1323.

Ingebrigtsen TS, Thomsen SF, van der Sluis S, Miller M, Christensen K, Sigsgaard T, et al. Genetic influences on pulmonary function: a large sample twin study. Lung. 2011;189(4):323–30.

Jandhyala SM, Talukdar R, Subramanyam C, Vuyyuru H, Sasikala M, Nageshwar RD. Role of the normal gut microbiota. World J Gastroenterol. 2015;21(29):8787–803.

Jones SE, Maddocks M, Kon SS, Canavan JL, Nolan CM, Clark AL, et al. Sarcopenia in COPD: prevalence, clinical correlates and response to pulmonary rehabilitation. Thorax. 2015;70(3):213–8.

Kaluza J, Harris HR, Linden A, Wolk A. Long-term consumption of fruits and vegetables and risk of chronic obstructive pulmonary disease: a prospective cohort study of women. Int J Epidemiol. 2018;47(6):1897–909.

Keranis E, Makris D, Rodopoulou P, Martinou H, Papamakarios G, Daniil Z, et al. Impact of dietary shift to higher-antioxidant foods in COPD: a randomised trial. Eur Respir J. 2010;36(4):774–80.

Kessler R, Partridge MR, Miravitlles M, Cazzola M, Vogelmeier C, Leynaud D, et al. Symptom variability in patients with severe COPD: a pan-European cross-sectional study. Eur Respir J. 2011;37(2):264–72.

Koike K, Ishigami A, Sato Y, Hirai T, Yuan Y, Kobayashi E, et al. Vitamin C prevents cigarette smoke-induced pulmonary emphysema in mice and provides pulmonary restoration. Am J Respir Cell Mol Biol. 2014;50(2):347–57.

Lange P, Celli B, Agusti A, et al. Lung-function trajectories leading to chronic obstructive pulmonary disease. N Engl J Med. 2015;373:111–22.

Langen RC, Gosker HR, Remels AH, Schols AM. Triggers and mechanisms of skeletal muscle wasting in chronic obstructive pulmonary disease. Int J Biochem Cell Biol. 2013;45:2245–56.

Leap J, Arshad O, Cheema T, Balaan M. Pathophysiology of COPD. Crit Care Nurs Q. 2021;44(1):2–8.

MacNee W. ABC of chronic obstructive pulmonary disease: pathology, pathogenesis, and pathophysiology. BMJ. 2006;332(7551):1202–4.

Mekal D, Czerw A, Deptala A. Dietary Behaviour and Nutrition in Patients with COPD Treated with Long-Term Oxygen Therapy. Int J Environ Res Public Health. 2021;18(23):12793.

Mete B, Pehlivan E, Gülbaş G, Günen H. Prevalence of malnutrition in COPD and its relationship with the parameters related to disease severity. Int J Chron Obstruct Pulmon Dis. 2018;13:3307–12.

Mjösberg J, Rao A. Lung inflammation originating in the gut. Science. 2018;359(6371):36–7.

Qu L, Cheng Q, Wang Y, Mu H, Zhang Y. COPD and Gut-Lung Axis: How Microbiota and Host Inflammasome Influence COPD andRelated Therapeutics. Front Microbiol. 2022a;13:868086.

Qu L, Cheng Q, Wang Y, Mu H, Zhang Y. COPD and Gut-Lung Axis: How Microbiota and Host Inflammasome Influence COPD and Related Therapeutics. Front Microbiol. 2022b;13:868086.

Rawal G, Yadav S. Nutrition in chronic obstructive pulmonary disease: a review. J Transl Int Med. 2015;3(4):151–4.

Rodríguez-Rodríguez E, Ortega RM, Andrés P, Aparicio A, González-Rodríguez LG, López-Sobaler AM, et al. Antioxidant status in a group of institutionalised elderly people with chronic obstructive pulmonary disease. Br J Nutr. 2016;115(10):1740–7.
Rondanelli M, Faliva MA, Peroni G, Infantino V, Gasparri C, Iannello G, et al. Food Pyramid for Subjects with Chronic Obstructive Pulmonary Diseases. Int J Chron Obstruct Pulmon Dis. 2020;15:1435–48.
Sargeant LA, Jaeckel A, Wareham NJ. Interaction of vitamin C with the relation between smoking and obstructive airways disease in EPIC Norfolk. European Prospective Investigation into Cancer and Nutrition. Eur Respir J. 2000;16(3):397–403.
Schols AM, Fredrix EW, Soeters PB, Westerterp KR, Wouters EF. Resting energy expenditure in patients with chronic obstructive pulmonary disease. Am J Clin Nutr. 1991;54:983–7.
Scoditti E, Massaro M, Garbarino S, Toraldo DM. Role of diet in chronic obstructive pulmonary disease prevention and treatment. nutrients. 2019;11(6):1357.
Singh D, Mathioudakis AG, Higham A. Chronic obstructive pulmonary disease and COVID-19: interrelationships. Curr Opin Pulm Med. 2022;28(2):76–83.
Sluyter JD, Camargo CA, Waayer D, et al. Effect of monthly, high-dose, long-term vitamin D on lung function: a randomized controlled trial. Nutrients. 2017;9:1353.
Stratton RJ, Green CJ, Elia M. Disease-related malnutrition: an evidence-based approach to treatment. Cabi. 2003;
Sun Z, Zhu QL, Shen Y, Yan T, Zhou X. Dynamic changes of gut and lung microorganisms during chronic obstructive pulmonary disease exacerbations. Kaohsiung J Med Sci. 2020;36(2):107–13.
Szmidt MK, Kaluza J, Harris HR, Linden A, Wolk A. Long-term dietary fibre intake and risk of chronic obstructive pulmonary disease: a prospective cohort study of women. Eur J Nutr. 2020;59(5):1869–79.
Tian P, Wen F. Clinical significance of airway mucus hypersecretion in chronic obstructive pulmonary disease. J Transl Int Med. 2015;3(3):89–92.
Tramontano A, Palange P. Nutritional State and COPD: Effects on Dyspnoea and Exercise Tolerance. Nutrients. 2023;15(7):1786.
van Iersel LEJ, Beijers RJHCG, Gosker HR, Schols AMWJ. Nutrition as a modifiable factor in the onset and progression of pulmonary function impairment in COPD: a systematic review. Nutr Rev. 2022;80(6):1434–44.
Vogelmeier CF, Criner GJ, Martinez FJ, Anzueto A, Barnes PJ, Bourbeau J, et al. Global strategy for the diagnosis, management, and prevention of chronic obstructive lung disease 2017 report: GOLD executive summary. Arch Bronconeumol. 2017;53(3):128–49.
Walda IC, Tabak C, Smit HA, Räsänen L, Fidanza F, Menotti A, et al. Diet and 20-year chronic obstructive pulmonary disease mortality in middle-aged men from three European countries. Eur J Clin Nutr. 2002;56(7):638–43.
Yang IV, Schwartz DA. Epigenetic control of gene expression in the lung. Am J Respir Crit Care Med. 2011;183(10):1295–301.
Zhang L, Valizadeh H, Alipourfard I, Bidares R, Aghebati-Maleki L, Ahmadi M. Epigenetic Modifications and Therapy in Chronic Obstructive Pulmonary Disease (COPD): An Update Review. COPD. 2020;17(3):333–42.

Chapter 9
Nutraceuticals in Inflammatory Autoimmune Diseases

Lovely M. Annamma, Sabrin Ali Azim, and Naglaa S. Ashmawy

1 Introduction

A broad category of illnesses known as autoimmune disorders (AD) are caused by the body's immune system attacking its own tissues. The body's natural defensive mechanism views its own cells as alien body cells in those disorders. This mistaken attack on the normal cells causes various body organs to be affected. There are over 80 types of autoimmune disorders. Rheumatoid arthritis (RA), multiple sclerosis (MS), systemic lupus erythematosus (SLE), Hashimoto's thyroiditis, inflammatory bowel disease (IBD), and psoriasis are the most prevalent inflammatory autoimmune illnesses. Both the pathogenesis and the etiology of AD are currently poorly understood. Environmental and genetic variables work together to set off the development of inflammatory autoimmune disorders (Genetic risk and a primary role for cell-mediated immune mechanisms in multiple sclerosis 2011; Cotsapas and Hafler 2013).

While medical interventions play a crucial role in managing these conditions, emerging research highlights the impact of nutrition and diet on disease progression, severity of the symptoms, and the overall quality of life (Manzel et al. 2014).

L. M. Annamma
College of Dentistry, Ajman University, Ajman, United Arab Emirates

S. A. Azim
College of Dentistry, Gulf Medical University, Ajman, United Arab Emirates

N. S. Ashmawy (✉)
Pharmaceutical Sciences Department, College of Pharmacy, Gulf Medical University, Ajman, United Arab Emirates

Department of Pharmacognosy, Faculty of Pharmacy, Ain Shams University, Abbassia, Cairo, Egypt
e-mail: naglaa.saad@pharma.asu.edu.eg

S. M. Jafari et al. (eds.), *Nutraceuticals: A New Millennium Approach for Preventive Medicine*, Food Bioactive Ingredients,
https://doi.org/10.1007/978-3-031-76481-3_9

The relationship between inflammatory autoimmune diseases and nutrition has shown a growing interest among researchers as well as clinicians. It is now widely recognized that dietary factors can modulate the immune system, affect inflammation levels, and significantly contribute to disease development or attenuation (Vieira et al. 2014). Understanding the interplay between nutrition and autoimmune diseases holds promise for optimizing patient care through targeted dietary interventions.

This book chapter aims to investigate the complex connections between inflammatory autoimmune diseases and nutrition, exploring the scientific evidence linking dietary components, such as specific nutrients, bioactive compounds, and dietary patterns, to disease outcomes. By examining the latest research findings, we seek to shed light on how nutrition and different diet regimens can influence the inflammatory processes underlying autoimmune diseases and impact disease progression, symptom management, and even the risk of developing additional comorbidities. Additionally, the potential mechanisms through which specific dietary factors exert their effects on the immune system and inflammation pathways will be explored.

Additionally, the book chapter will evaluate critically the state of nutritional therapies now used in the management and control of autoimmune disorders that are inflammatory, covering the data supporting dietary changes, supplements, and the function of specialty diets.. We will also address challenges and gaps in research, emphasizing the need for further investigations and clinical trials to establish robust dietary recommendations for patients.

As the scientific understanding of the relationship between nutrition and inflammatory autoimmune diseases continues to expand, it becomes increasingly important to bridge the gap between research findings and clinical practice. By enhancing our knowledge in this field, we can empower healthcare professionals, patients, and individuals at risk of developing these conditions to make informed choices about their diet and lifestyle, potentially improving disease outcomes and overall well-being.

As various reports have studied the effects of nutrients and the diet type on the development and progression of AD. This chapter could be to highlight the latest evidence-based nutrient management options for the most commonly occurring inflammatory autoimmune disorders.

2 Effect of Diet Type on Autoimmune Diseases

It was reported that autoimmune diseases have a higher prevalence in Western societies in comparison to developing countries and the Eastern world. It was observed that socioeconomic improvement is associated directly with the consumption of high amounts of Western diet which led to a higher prevalence of autoimmune diseases (Venter et al. 2020; Rose 2016).

There are many reasons explaining the relationship between the Western diet and AD development. High standards of hygiene are always found in Western countries which reduces the occurrence of infection diseases, but this also can limit the exposure to beneficial microbes which play significant part for the suitable functioning of the defense mechanism (Bach 2002; Okada et al. 2010; Rook 2012). Furthermore, lifestyle factors such as psychosocial stress as well as smoking may act as additional risk factors for autoimmune diseases. It was reported that the lack of physical activity, together with high-calorie consumption through fast food, contributes to the high prevalence of obesity and AD in developed societies (Manzel et al. 2014).

Western diet (WD) is a term referring to certain dietary habits where foods of high fats, cholesterol, simple sugars, proteins, salts as well as processed food are consumed (Gioia et al. 2020). While the Mediterranean diet (MD) as defined by UNESCO in 2010 it is not limited to a dietary regimen, but it involves a collection of cultural traditions to establish sound nutrition as the foundation of a healthy life pattern from all perspectives. MD is described as a diet with high amounts of olive oil, vegetables, fruits as well as whole grains together with a controlled intake of fish, meat, as well as dairy products. MD incorporates also various spices, condiments, and water (Mazzucca et al. 2021).

The mechanisms by which specific nutrients alter immune function and have an anti-inflammatory effect activity have been extensively reported. Certain nutrients like polyunsaturated fatty acids (PUFA) which are abundant in sea food as well as plant oil seeds have been reported to decrease inflammation by reducing cytokine production and leukocyte migration (Kostoglou-Athanassiou et al. 2020).

Moreover, a diet which is abundant in fruits and vegetables can provide the body with polyphenols, which are commonly found in tea, coffee, as well as some legumes. Epigallocatechin gallate, capsaicin, and curcumin are common examples of polyphenols having potent anti-inflammatory characteristics. Those compounds can modulate pathways for signaling that control the expression of genes known as the pro-inflammatory, like cyclooxygenase (COX), phospholipase A2 (PLA2) and lipoxygenase (LOX). Numerous studies have been carried out to evaluate the beneficial effects of these polyphenols on the autoimmune diseases (ADs) (Nelson et al. 2020).

Isoflavones, another type of polyphenol, found in soybeans, black beans, and olive oil have been also reported for their anti-inflammatory and antioxidant characteristics and help alleviate inflammatory symptoms (Klack et al. 2012; Terahara 2015). Additionally, minerals, found naturally in milk, soybean, and mollusks as zinc and selenium have been shown to improve symptoms of AD by decreasing antibodies levels such as anti-dsDNA (Klack et al. 2012; Constantin et al. 2019).

The abundant ingestion of vegetables, fruits, as well as whole grains in the diet of the Mediterranean (MD) is very advantageous because of its richness in fiber content, which is essential to supporting the gut microbiome. While fiber itself is not considered as a direct source of energy for humans, it undergoes bacterial fermentation in the gut, producing fatty acids which serve as an energy substrate for colonocytes, thus protecting the intestinal barrier integrity (García-Montero et al. 2021). The microbiota derived from fatty acids have a significant impact on the growth and differentiation of regulatory T cells, which are vital for maintaining the immunological balance.. Dietary composition has a major impact on the communication that occurs between the microbiota and the immune system of the body (Arpaia et al. 2013).

It was reported that in the Mediterranean region, the consumption of processed food as well as food additives has been expanding especially in young generations in comparison to elderly people who prefer to stick to the Mediterranean diet (Castelló et al. 2022; Russo et al. 2021).

3 Mechanisms by Which the Western Diet Affects AD

3.1 Effect of Western Diet on Gut Microbiome and Regulation of T Cells

There is an increase in the incidence of AD in the last three decades which may be caused by the increased intestinal permeability resulted from the consumption of fast food as well as industrial food additives (Mazzucca et al. 2021). This increase in permeability led to the entrance of food toxins, bacteria and immunogenic antigens (Aaron et al. 2019). Moreover, the absence of fibers in the Western diet leads to deprivation of the intestinal gut microbiota which will stimulate the induction of the enzymes that cause the intestinal mucin layer to break down, resulting in a condition known as "leaky gut" (Desai et al. 2016). This condition induces an autoimmune response, through different mechanisms, among them the molecular similarity either through the interaction of self-molecules with certain chemical products, resulting in the creation of new antigenic molecules, or between food and antigens (Lerner and Matthias 2015). Moreover, people who have a genetic predisposition may experience the onset and progression of autoimmune diseases (ADs) due to leaky gut conditions.

Besides increasing inflammation, a leaky gut has an additional effect related to the risk of inadequate absorption of crucial macronutrients and micronutrients. This is significant because deficiencies in vitamins, such as vitamin D, which have been linked to an increased risk of developing some autoimmune diseases, such as multiple sclerosis (MS) and rheumatoid arthritis (RA) (Gioia et al. 2020).

3.2 Effects of Western Diet on Obesity

Increased fat content in the Western diet is associated with the development of obesity, leading to white adipose tissue accumulation and systemic inflammation (Luo et al. 2016). In addition to being a tissue used to store energy, white adipose tissue also releases adipokines, which are distinct pro-inflammatory mediators like TNF-α, IL-6, and C-reactive protein. It has been reported that these adipokines contribute to obesity-related chronic inflammation (Conde et al. 2011) which disrupts The immunity system's capacity to function properly (Haneishi et al. 2023). The high consumption of saturated fats found in WD is reported to influence the innate immune system by means of toll-like receptor activation, which subsequently trigger pro-inflammatory signaling pathways (Li et al. 2020).

Western diet not just affects T lymphocyte differentiation, but it also affects how T cells operate.. High Saturated fat intake by T cells and added sugars, common in Western diets, has been shown to impair T cell responsiveness and cytokine production. Studies have indicated that obesity can lead to exhaustion of adipose tissue T cells (Porsche et al. 2021). Additionally, excess fructose consumption, prevalent in Western diets due to the widespread has been linked to T cell dysfunction, impairing their ability to respond to pathogens and regulate inflammation (Ma et al. 2022).

In mouse models of Multiple Sclerosis, high infiltration of T cells as well as macrophages was observed as a result of the effect of high-fat diets on the body's adaptive immune system, which was correlated with the deterioration in the clinical parameters (Mazzucca et al. 2021). Diet-induced obesity has been observed to exacerbate various autoimmune disease models, including inflammatory bowel disease (Manzel et al. 2014). Furthermore, Leptin, an adipose-derived hormone, was reported to stimulate T-cell proliferation and promote TH1 responses (Kiernan et al. 2023). Fasting has been observed to decrease circulating leptin levels and improve the severity of autoimmune conditions (Javanbakht et al. 2023).

3.3 *Effects of Western Diet on Sodium Intake and TH17 Cells*

The Western diet involves the consumption of high amounts of processed foods as well as fast food which contain high salt content. Salt intake through this type of diet can contain over 100 times more sodium in comparison to homemade meals. Les effets d'une consommation excessive de sel on autoimmune diseases was reported (Jobin et al. 2021). Studies have shown that a high-sodium diet, commonly found in the Western diet, can influence the immune system and promote the differentiation of TH17 cells. TH17 cells are a subgroup of the T helper cells that play a significant role in the immune response, particularly in the defense against certain infections and autoimmune disorders (Jobin et al. 2021; Kleinewietfeld et al. 2013). However, an increased activation or dysregulation of the TH17 cells can play a role in chronic inflammation and pathologies autoimmunes (Milovanovic et al. 2020; Guo 2016). Excessive sodium intake has been found to produce more of interleukin-17, which is a pro-inflammatory cytokine formed by the TH17 cells. IL-17 is known to play a role in the pathogenesis as well as the progression of arthrose (RA) (Salgado et al. 2015), several sclerosis (MS) (Cortese et al. 2017), and inflammatory bowel disease. The increased production of IL-17 driven by a high-sodium Western diet can contribute to the emergence of these conditions or their worsening (Manzel et al. 2014).

Furthermore, the Western diet's impact on the gut microbiota may also influence TH17 cell differentiation. The gut microbiota regulates significantly the immune system which affects the overall health. It has been observed that a high-sodium diet can disrupts the gut microbiome, resulting in changes to the microbial communities and this potentially affects the immune cell function, including TH17 cells.

Reducing sodium intake and adopting a more balanced and wholesome diet can help mitigate the adverse effects of the Western diet on TH17 cell activation and chronic inflammation. Emphasizing whole foods, such as lean proteins, fruits, whole grains as well as vegetables can provide sufficient nutrients without increased sodium consumption. Additionally, incorporating probiotic-rich foods and prebiotic fibers can support healthy gut microbiota and promote immune system balance (Richards et al. 2016).

In conclusion, the high sodium content of the Western diet may activate and dysregulate TH17 cells, which may result in autoimmune disorders and chronic inflammation. Reducing sodium intake and emphasizing whole foods in your diet can help mitigate these effects and support the immune system's overall balance and wellbeing.

4 Inflammatory Autoimmune Diseases and Nutrition

4.1 Rheumatoid Arthritis and Nutrition

Rheumatoid arthritis (RA) is a disease, where hands, wrists, and knees joints are attacked by the individual's immune system, and this causes inflammation of the joint lining causing joint tissue damage. On account of the resulting damage, chronic pain, imbalance, and joint deformity ensue. Sometimes, in addition to joints, other tissues may also be impacted, including the eyes and lungs. X-rays, MRIs, ultrasounds, and blood tests can all be used to diagnose RA. Treatments for RA-related pain and other symptoms are currently available (Radu and Bungau 2021).

Management of RA can be performed by either using modern medicine like medication and surgical intervention or by utilizing an alternative medicine, and nutrient management. The commonest medication given for RA to reduce symptoms is called DMARD (disease-modifying antirheumatic drugs). Biological response modifiers are the second-in-line treatment option.

In cases where the DMARD alone cannot alleviate the symptoms of RA, biological drugs such as abatacept, adalimumab, etanercept, tofacitinib, upadacitinib, and tocilizumab are prescribed along with DMARD. Glucocorticoids and nonsteroidal anti-inflammatory drugs (NSAIDs) are also used to treat RA symptoms. Self-management strategies are highly recommended including physical activity, weight loss, and protecting joints by limiting repetitive joint actions (Sepriano et al. 2020; Shams et al. 2021).

It is reported that the combination of DMARD with biological treatments has shown side effects including infections like Herpes Zoster, increased probability of malignant effects along with a higher chance of developing venous thromboembolism. Some other side effects are also reported including gastrointestinal perforations, nasopharyngitis, and headaches (Sepriano et al. 2023).

The research findings indicate that a significant portion of individuals with RA have experimented with herbal remedies, including thyme, chamomile, borage, lavender, ginger, and cinnamon, to alleviate symptoms. These natural herbs have been noted for their potential anti-inflammatory properties (Fazeli et al. 2021). Moreover, Curcumin supplementation when used in doses ranging between 250–1500 mg for 8–12 weeks has been reported to induce a decrease in both Erythrocyte sedimentation rate (ESR) and C Reactive Protein (CRP) in rheumatoid arthritis patients aged 18 and up arthritis and ulcerative colitis (Ebrahimzadeh et al. 2021). The underlying mechanism of action includes suppression of the pro-inflammatory signaling pathways and induction of the activation of the intracellular antioxidant defense system (Chandran and Goel 2012; Jacob et al. 2019). Furthermore, Curcumin has an antioxidant effect and acts on fibroblast-like synoviocytes by inhibiting Tumor Necrosis Factor- α (TNF- α) and proinflammatory pathways. These effects have improved the inflammation-associated symptoms in RA patients (Kloesch et al. 2013). It has been demonstrated that several other spices, like cinnamon and ginger, reduce the Disease Activity Score. (DAS28) in RA patients thus reducing the inflammatory symptoms (Aryaeian et al. 2019; Letarouilly et al. 2020; Shishehbor et al. 2018). In contrast, extensive research has been conducted on the impact of vitamin D supplements on individuals with RA. Vitamin D plays an important function in calcium and phosphate metabolism, contributing to the maintenance of bone mass through the regulation of calcium absorption from the gastrointestinal system (Holick and M. 2011). Its synthesis occurs in the skin through the action of ultraviolet rays, and it exhibits immunomodulatory effects (Bikle 2011). It has been reported that the deficiency of vitamin D can be an underlying factor in various autoimmune disorders (Merlino et al. 2004; Song et al. 2012). The mechanism by which vitamin D improves RA symptoms is reported to be by inhibiting the proinflammatory T helper type 1 profile and suppressing proliferation and immunoglobulin production (Correale et al. 2009). It has been also observed that vitamin D can suppress the differentiation of precursor B cells to plasma cells (Chen et al. 2007). Higgins et al. also stated that vitamin D affected the innate as well as the adaptive immune systems through differentiation of T-cells (Higgins et al. 2013). It has been also reported that vitamin D receptor agonist can suppress collagen-induced arthritis. Hence, Vitamin D supplementation induces immune tolerance, and since RA patients are subjected to osteoporosis, vitamin D helps in maintaining bone mass (Higgins et al. 2013). Vitamin D deficiency has been reported to be common between RA patients and this deficiency may be part of corticosteroids side effects. Corticosteroids are given to RA patients to reduce pain and inflammation. Corticosteroids and immunosuppressants have been reported to reduce calcium absorption as well as altering vitamin D metabolism (Adorini 2005). Hence supplementing Vitamin D in RA indirectly reduces inflammation, minimizes pain, strengthens bone disabilities, and improves the quality of life.

4.2 Systemic Lupus Erythematosus and Nutrition

Systemic Lupus Erythematosus (SLE) is an auto-inflammatory disease affecting both the innate and the adaptive human immune systems. It can affect various body parts including joints, skin, brain, kidneys, lungs, as well as blood vessels. It is more common in females and the disease can be very debilitating, causing significant psychological distress, and scarring (Aparicio-Soto et al. 2017). SLE can cause damage to multiple organs in the body due to circulating antibodies which leads to chronic inflammation together with cardiovascular diseases (Schoenfeld et al. 2013), metabolic syndrome, thyroid diseases (Gergianaki et al. 2021) and cancer (Noble et al. 2016). Diagnosis of SLE is performed through the performance of confirmatory tests for antibodies against nuclear proteins (DNA & RNA). SLE treatment includes antimalarials, biologic therapies, synthetic DMARDs, corticosteroids, intravenous immunoglobulins, thalidomide, laser and other topical applications such as calcineurin inhibitors, sunscreens, and salbutamol creams. Hydroxychloroquine is reported to be used as the first line of the treatment (Mok et al. 2021).

Unwanted side effects are commonly observed with the current anti-inflammatory treatments used to manage SLE. The SLE medications when used over long periods can cause adverse effects such as osteoporosis and organ damage in the case of glucocorticoids (Ruiz-Irastorza et al. 2012; Abdelrahman et al. 2023) hence alternative medications are being a resort for the patients to avoid the side effects of modern medications. Dietary management is safe and effective along with conventional treatment, evidence of effects on SLE as such is not clear. The literature has reported many supplements such as vitamins, omega-3, fish oil, and curcumin to be beneficial in improving the symptoms associated with SLE (de Medeiros et al. 2019; Islam et al. 2020). Omega-3 has been proven to reduce the parameters including Erythrocyte Sedimentation Rate (ESR), C-reactive protein, and various inflammatory markers, and oxidative stress as well as improve lipid levels and glycemic index (Jiao et al. 2022). The mechanism of action of omega-3 in alleviating the symptoms of SLE has been reported to be through the reduction of inflammation by regulating the inflammatory mediators. SLE patients are reported to have altered lipid profiles which can cause cardiovascular complications. Studies have demonstrated the beneficial impact of omega-3 fatty acids on vascular tone, attributed to their ability to decrease oxidative stress (Duarte-Garcia et al. 2020). Another nutrient that has been reported to decrease inflammation is curcumin. It is the main active constituent found in turmeric and it has been recognized for its anti-inflammatory and antioxidant properties. In patients of SLE with a renal involvement, a study reported that supplementation with curcumin led to a decrease in proteinuria, hematuria, and systolic blood pressure. This suggests that curcumin can have beneficial effects in managing renal manifestations of SLE (Jiao et al. 2022). Moreover, a study reported by Momtazi et al. demonstrated that curcumin possesses anti-inflammatory properties by modulating the inflammatory pathways in the body. These findings highlighted the potential of curcumin as a therapeutic agent in reducing inflammation associated with SLE (Momtazi-Borojeni et al. 2018).

Additionally, vitamins D, C, A, E, and B have been observed to be of a great value to SLE patients within the recommended dosage as an adjunct to pharmacotherapy. Vitamin D supplementation was reported to reduce inflammatory characteristics of SLE by action on inflammatory markers (Jiao et al. 2022). Vitamin D has been reported to reduce apoptosis and modify the cell cycle progression in SLE (Zou et al. 2023). Vitamin D deficiency has been reported in SLE patients due to avoidance of sunlight, renal failure, and also because of the effects of glucocorticoids (Correa-Rodríguez et al. 2021). Thus, vitamin D supplements are given to SLE patients to avoid as well as treat its deficiency Studies have indicated a noteworthy impact of vitamin D supplements on the management of patients with systemic lupus erythematosus (SLE) by suppressing inflammatory damage and preventing adverse bone events during the treatment of SLE with medications (Ao et al. 2021). In regards to vitamin E, it has been reported for its beneficial effect on SLE patients, where vitamin E supplements have antioxidant properties that can regulate antibody production and improve the symptoms of SLE (Constantin et al. 2019). Moreover, Vitamin C has been reported to be essential in the management of SLE as it reduces oxidative stress as well as decreases inflammation and lowers the Immunoglobulin G (IgG) antibody levels. This can in turn prevent cardiovascular complications associated with SLE. The approved dose for vitamin C indicated for SLE patients is 500 mg along with vitamin E 800 IU supplement (Klack et al. 2012). Hence a high dietary regimen containing all the above natural components like in the Mediterranean diet is preferred over supplements unless prescribed by the physician for SLE as many supplements can interact with medications.

4.3 Psoriasis and Nutrition

The autoimmune skin condition psoriasis is typified by the excessive growth of keratinocytes due to an overactive immune system. It presents as raised, well-defined red plaques with irregular borders and silvery scales, primarily affecting areas like the knees, elbows, trunk, and scalp. The condition, initially defined as a distinct entity by Ferdinand von Hebra in 1841 and is known by various names such as Psoriasis Vulgaris, guttate psoriasis and erythrodermic psoriasis (Al-Shobaili and Qureshi 2013).

Understanding the complex interactions between different cells involved in psoriasis has been a subject of ongoing debate. Recent advancements in cellular immunology, biology, and genetic studies have provided valuable insights into the causes and treatment of this condition. The disease begins with the activation of T lymphocytes by an unknown antigen or gene product. This activation triggers the release of various cytokines by T cells, leading to processes such as increased keratinocyte growth, migration of neutrophils, enhanced immune response, formation of new blood vessels, and increased cell production in the outermost layer of the skin. Symptoms and classification of psoriasis vary depending on its type. Various aspects, including morphological features of elementary lesions, degree of inflammation, distribution pattern, extent, time of onset, and rate of spread, are considered in classifying psoriasis (Tripathi and Bhardwaj 2020).

In the management of psoriasis, there is currently no universally accepted standard of care for patients with moderate to severe symptoms. The choice of systemic therapy should be carefully considered, weighing the benefits and risks for each patient (Martin et al. 2019). Phototherapy is often the first line of treatment, especially for those with extensive psoriasis or severe symptoms, however, the most effective treatment plan is the one that the patient is most likely to adhere to. If phototherapy is not feasible or ineffective, systemic treatments with oral medications or biologics may be necessary. Psoriasis requires long-term treatment due to its high prevalence and as it is affecting the quality of life (Arrieta Valero 2019). The function of nutrients, especially vitamin D, in the onset and treatment of inflammatory autoimmune diseases such as psoriasis has drawn a lot of attention in recent years. Extensive research has been conducted to investigate the role of vitamin D in the proliferation and differentiation of keratinocytes, the predominant cells in the epidermis, after it was discovered that these cells have vitamin D receptors. It has been discovered that vitamin D inhibits keratinocytes' excessive growth and encourages their appropriate differentiation. As a result, research into the role of vitamin D in the pathophysiology of psoriasis has been conducted, indicating that elevated keratinocyte proliferation and cutaneous inflammation may be facilitated by a drop in vitamin D levels (Pona et al. 2019). Diverse findings from studies examining the connection between psoriasis and serum vitamin D levels have given rise to a continuing discussion (Marek-Jozefowicz et al. 2022). In 2018, a meta-analysis revealed a strong correlation between low vitamin D levels and psoriasis; however, the limitations in sample size and study heterogeneity precluded the establishment of causality. There have been mixed results from subsequent case-control studies; some have found a strong correlation between vitamin D levels and psoriasis, while others have not found a significant correlation (Stanescu et al. 2021).

A study was carried out to assess serum vitamin D levels in a larger cohort of psoriasis patients and investigate the relationship between vitamin D levels and disease severity and duration, given the contradictory evidence and the need for a more thorough understanding of the relationship between vitamin D and psoriasis. This study attempted to offer important insights into the course of psoriasis patients and the possible advantages of oral vitamin D supplementation as part of their management, while accounting for potential confounding variables. Research on the function of nutrients, especially vitamin D, in inflammatory autoimmune diseases such as psoriasis is generally an important field of study. Comprehending the intricate relationship among vitamin D, immune responses, and the etiology of psoriasis may facilitate the development of more effective therapeutic approaches and enhanced patient outcomes (Pokharel et al. 2022).

Given that psoriasis is a chronic inflammatory skin condition that is influenced by a number of factors, such as immune dysregulation and inflammation, it is reasonable to assume that the etiology of psoriasis involves the stimulation of T-helper (Th-1) and (Th-17) lymphocytes, which results in an increase in the production of inflammatory cytokines. Certain dietary components have an impact on the inflammatory response, and nutrition plays a role in modulating inflammation (Miyagawa et al. 2022).

Lipids, especially saturated fats like those in butter and red meat, can raise interleukins' concentration and aid in the development of inflammation (Saghafi-Asl et al. 2021). Overconsumption of the n-6 family of unsaturated fatty acids, which is present in refined vegetable oils and margarine, can also contribute to inflammation. On the other hand, monounsaturated and polyunsaturated fatty acids (PUFAs) from nuts, fish, and plant-based oils have anti-inflammatory qualities and are beneficial to health (Mattavelli et al. 2021).

Simple sugars in the diet, when consumed excessively, may exacerbate inflammation in psoriasis. Lowering the glycemic index (GI) has been associated with reduced inflammation. Some studies suggest that dietary interventions, including low-carbohydrate diets and polyphenol supplementation, may improve the condition of psoriasis patients, although results have been conflicting (Fu et al. 2021). Compounds with antioxidant and anti-inflammatory properties are essential for psoriasis sufferers' diet. These consist of substances like dietary fibers, polyphenols, omega-3 acids, and vitamins A, E, and C. They also include trace minerals like copper, manganese, zinc, and selenium. Probiotics, fruits, vegetables, vitamin D, fish oil, and vitamin E have all been shown to help psoriasis sufferers' condition (Pokharel et al. 2022).

Additionally, inflammatory skin diseases are significantly influenced by the gut microbiota (Polkowska-Pruszyńska et al. 2020). For those with psoriasis, probiotics and prebiotics are advised as vital components of a well-balanced diet. Psoriasis can be made worse by dysbiosis of the gut microbiota, but complex carbohydrates, which are present in fruits and vegetables, have been shown to reduce inflammation (Gioia et al. 2020).

Certain food ingredients, such as cyclitols, genistein, and selenium, have shown the potential in inhibiting psoriasis development. Cyclitols have immunomodulatory effects, while genistein, found in soybeans, may reduce inflammation, and inhibit disease progression. Selenium, as a trace element, contributes to the cellular antioxidant defense system and serves to inhibit. Inflammation-related pathways (Fu et al. 2021).

Psoriasis sufferers need to be aware of dietary triggers, which include processed foods, nightshade vegetables, gluten, dairy products, and white sugar (Garbicz et al. 2022). These substances have the capacity to upset the immune system, irritate the gut epithelium, and disrupt the gut microbiome. Antioxidants and high-fiber complex carbohydrates from fruits and vegetables, on the other hand, have been shown to have an inverse relationship with inflammatory markers. All things considered, diet affects how inflammation is modulated in psoriasis. It has been suggested that managing the illness and enhancing general health can be accomplished with a balanced diet that excludes dietary triggers and contains anti-inflammatory and antioxidant components (Kanda et al. 2020).

The inflammatory and itchy skin condition known as psoriasis can be supported by a variety of diets and nutritional regimens. Numerous studies have demonstrated the benefits of various diets in controlling the symptoms of psoriasis. Diets emphasizing lower energy intake, vegetarian diets, and weight loss programs incorporating dietary modifications are among the diets that are helpful for psoriasis patients. These diets are usually high in fruits and vegetables, which provide you plenty of essential vitamins, minerals, and other beneficial substances with less calories (Duchnik et al. 2023). Particularly the Mediterranean diet has been linked to improvements in inflammatory conditions like psoriasis (Wu and Weinberg 2019). This diet lowers animal fats and simple carbs while increasing fresh fruits and vegetables, olive oil, legumes, and seeds. It is thought that the polyphenols and antioxidants present in Mediterranean diets have anti-inflammatory and antioxidant qualities. This diet's monounsaturated fatty acids also have other health benefits, like lowering the risk of osteoporosis, altering immunological and inflammatory responses, and lowering the risk of cancer and heart disease. The Mediterranean diet's signature nutrients—beta-carotenoids, polyphenols, folic acid, and fiber—are essential for preventing oxidative stress (Guasch-Ferré and Willett 2021). Certain components of the Mediterranean diet, such as fruits, vegetables, whole grains, and seafood, have been related in studies to lower levels of inflammatory markers (Pona et al. 2019).

An alternative dietary approach for individuals with psoriasis is the ketogenic diet, which involves reducing carbohydrate intake and increasing protein and fat consumption. This diet shifts the body to ketone metabolism, using fat as the primary energy source and decreasing glucose levels. Research has indicated that adopting a low-calorie ketogenic diet can lead to a reduction in pro-inflammatory cytokines and a decrease in the severity of symptoms associated with psoriasis (Kanda et al. 2020).

There is q study by Ewa et al. didn't strongly support the idea of some literature that suggests the therapeutic role and current reports suggest limited evidence regarding the effectiveness of a gluten-free diet in managing psoriasis. Elevated levels of anti-gliadin antibodies in certain psoriasis patients might signal the presence of coexisting celiac disease, warranting a gluten-free diet exclusively for those with established gluten intolerance (Passali et al. 2020).

Another recommended dietary approach for individuals with psoriasis is a vegetarian diet, comprising legumes, grains, vegetables, fruits, nuts, and mushrooms. Variations may include the inclusion of dairy products and eggs. Because this diet excludes meat, it lowers intake of cholesterol and saturated fats, which are important for controlling uric acid levels and preventing cardiovascular diseases (Islam et al. 2021). It has been noted that eating a vegetarian diet can result in decreased blood levels of triglycerides and C-reactive protein (CRP), a decreased risk of being overweight, and a decrease in the production of pro-inflammatory chemicals (Garbicz et al. 2022; Wu and Weinberg 2019).

In summary, psoriasis is a long-term skin condition marked by keratinocyte hyperproliferation brought on by an inflammatory response. Numerous types of psoriasis are linked to the disease, such as erythrodermic psoriasis, pustular psoriasis, guttate psoriasis, inverse psoriasis, psoriasis vulgaris, and palmoplantar psoriasis.. Various diets, such as low-energy and vegetarian diets, the Mediterranean diet, and the ketogenic diet, can have positive effects on psoriasis by reducing inflammation and improving overall health. However, the therapeutic role of a gluten-free diet in psoriasis treatment remains uncertain and Should be adopted exclusively by individuals who have confirmed gluten intolerance (Bruzzese et al. 2020). Recent advancements in cellular immunology, biology, and genetic studies have provided valuable insights into the causes and treatment of psoriasis, highlighting the complex interactions between different cells. Maintaining a balanced diet that incorporates anti-inflammatory and antioxidant components while avoiding dietary triggers can help manage psoriasis and enhance overall health (Caso et al. 2020).

4.4 Multiple Sclerosis and Nutrition

Multiple sclerosis is described as an immune-mediated condition affecting the central nervous system. It is marked by persistent inflammation, demyelination, gliosis, and neuronal damage, which often result in physical or cognitive disabilities and various neurological impairments (Noyes and Weinstock-Guttman 2013). It is estimated that in 2020 a worldwide population of 2.8 million are affected by MS (Walton et al. 2020). The symptoms of MS include numbness of limbs, optic neuritis, visual loss, tremor and ataxic gait, double vision, dysarthria, and fatigue. Increased risk of stroke (Hong et al. 2019) and a decrease in life expectancy (Scalfari et al. 2013) have also been reported. Multiple sclerosis is linked to conditions such as disability, fatigue, depression, cognitive challenges, and reduced employment opportunities (Gil-González et al. 2020), hence suicide tendencies (Fredrikson et al. 2003) are more in MS patients.

Common medications used for MS as the first line are baclofen, tizanidine, and gabapentin. Other medications such as diazepam and Nabiximols, and intrathecal phenols are used as add-on therapy along with the first-line treatment to reduce symptoms associated with MS (Otero-Romero et al. 2016). The medications given to patients with MS do not cure the disease but only relieve the associated symptoms, hence many alternative therapies have been reported (Cavalli et al. 2019). It has been reported that the reasons for resorting to the alternative treatment options is dare to reduced long-term efficacy and lack of access to immunomodulators such as interferon-b and glatiramer acetate. The commonest alternative option was Oenothera seed oil and vitamin supplements (Fryze et al. 2006).

The commonest vitamin supplement in MS is Vitamin D which is observed to improve the quality of life in MS patients (Niino and Miyazaki 2017). The mechanism of action of Vitamin D on antiinflammation is its regulatory action on lymphocyte and cell proliferation (Piec et al. 2021). It has been reported that no significant benefit in Vitamin D supplements on the expanded disability status scale (EDSS) on which MS patients are assessed (Doosti-Irani et al. 2019). Though studies relating to the use of Vitamin D did not directly treat MS they are recommended as supplements due to various reasons. One such reason is that Vitamin D plays a significant role in mental health. Many studies have observed that MS patients are prone to depression and suicidal tendencies (Vellekkatt and Menon 2019; Cheng et al. 2020) Apart from mental health Vitamin D supplement has proved to have health benefits for patients with various diseases to counter osteoporosis. MS patients have been reported widely to have Vitamin D deficiency (Amrein et al. 2020; Haines and Park 2012; Feige et al. 2020).

One of the symptoms of multiple sclerosis is neuronal degeneration. Neuronal degeneration is directly linked to increased oxidative stress leading to mitochondrial dysfunction and neuronal cell death (Blagov et al. 2022). Diets such as Vitamin D, fatty acids, and curcumin have been reported to dampen oxidative stress (Stoiloudis et al. 2022). Antioxidants have been observed to reduce inflammatory factors and oxidative stresses and prevent chronic demyelination as well as axonal damage (Theodosis-Nobelos and Rekka 2023). Vitamin A has been reported among the vitamin supplements indicated for MS. Vitamin A is a fat-soluble nutrient that includes retinoids and carotenoids in diets. Dairy products such as milk, cheese, and oil have Vitamin A. A dosage of 400 IU/day is recommended for fatigue in MS patients (Bitarafan et al. 2016). Another dietary supplement that has proved to have benefits in the quality of life for MS is Omega-3. Some studies proved Omega 3 to be beneficial (Ovcharova et al. 2022) and some reported otherwise (AlAmmar et al. 2021).

Marine n-3 PUFAs have been recently reported to improve the quality of life in patients diagnosed with MS. The mechanism of action includes its antioxidant, anti-inflammatory, and neuroprotective properties by suppressing inflammatory transcription factors, reducing oxidative stresses, and blocking proinflammatory interleukins (La Rosa et al. 2023). They also possess protective action against nerve demyelination (Bagur et al. 2017). Amongst the polyunsaturated fatty acids, α-linolenic acid (ALA) when consumed is reported to have reduced chances of MS (AlAmmar et al. 2021). Omega 3 fatty acids enriched fish oil was observed by Riccio et al. To reduce inflammation and oxidative stress (Moles and Otaegui 2020).

Both pre and probiotics have been reported for their effects in the reduction of MS symptoms. Prebiotics consist of living microorganisms that benefit the host microbiota. Dairy products are a rich source of prebiotics such as Lactobacillus in curds (Lombardi et al. 2018; Jiang et al. 2021). Pre and probiotic supplements help in the reduction of inflammatory markers and balance the gut microbiota in MS (Kouchaki et al. 2017). The Commonest Dietary regiments containing the recommended supplements for MS include the Mediterranean diet (Katz Sand et al. 2023), Paleolithic diet (Wahls et al. 2019), Ketogenic diet (Lin et al. 2022), McDougall Diet (Evans et al. 2019), restricted carbohydrate diet (Fanara et al. 2021), Swank diet (Shemirani et al. 2023) as well as the MIND diet (Noormohammadi et al. 2022) and DASH diet (Rotstein et al. 2019).

4.5 *Hashimoto's Thyroiditis (HT) and Nutrition*

Hashimoto's thyroiditis (HT) is an autoimmune condition characterized by infiltration of lymphocytes, causing the damage of the thyroid gland. The primary cause of this disease is an elevated level of antibodies against thyroid peroxidase and thyroglobulin. As a result, HT patients experience various physical and psychological symptoms due to alterations in thyroid hormone levels and metabolism. The severity of symptoms is often associated with higher levels of antibodies targeting thyroid antigens. Environmental factors are required to initiate the immune system's attack on the thyroid gland, even though genetic factors are a major contributing factor to the development of hypothyroidism (HT). Stress, exposure to toxins, an imbalance in gut flora, and insufficient nutrition are a few examples of these factors (Ralli et al. 2020).

The role of diet intervention a crucial role in Hashimoto's treatment. The objective is to consume an anti-inflammatory diet in order to give the body the right nutrients and control the immune system. Studies have indicated that HT patients frequently experience nutritional deficits. The significance of potassium, selenium, copper, iodine, magnesium, zinc, and vitamins A, D, C, and B in the diet of HT patients has been shown in a number of studies. Furthermore, it has been recommended to consume dietary fiber, keep your protein intake at a suitable level, and include unsaturated fatty acids—especially those in the omega-3 family (Ihnatowicz et al. 2020).

It is recommended that HT patients avoid lactose from their diet due to lactose intolerance and potential interactions with levothyroxine, a medication commonly used to manage the condition. Furthermore, gluten elimination is recommended due to potential interactions between gliadin (a component of gluten) and thyroid antigens. This article provides an overview of the nutritional factors relevant to HT patients and offers dietary recommendations for managing the condition (Asik et al. 2014).

However, it's still not clear which strategy of diet would be the best, as diet can be a complementary treatment for Hashimoto's disease possibly by affecting thyroid function and anti-inflammatory properties (Osowiecka and Myszkowska-Ryciak 2023). Excessive and prolonged consumption of iodine can result in autoimmune thyroiditis due to the triggering of a strong immune response thyroglobulin (Tg), which has been highly iodinated. Similar results may be obtained from the recently implemented universal salt iodization program. Although it is temporary (Mikulska et al. 2022). Conversely, selenium proteins are essential for thyroid function; in particular, they shield the thyroid by removing excess hydrogen peroxide generated during Tg iodination. Genetic data indicates that a higher risk of autoimmune thyroiditis is associated with the anti-inflammatory selenoprotein S (Hu and Rayman 2017). Selenium and selenoproteins have been shown to lower thyroid peroxidase (TPO)-antibody levels and the incidence of hypothyroidism and postpartum thyroiditis, according to observational studies and randomized controlled trials. A low iron intake is linked to impaired thyroid metabolism (Rayman 2019). Heme, an iron-containing molecule, is necessary for TPO, the enzyme that produces thyroid hormone, to become active on the surface of thyrocytes. People with autoimmune thyroiditis frequently have iron deficiency because autoimmune gastritis, which impairs iron absorption, is a common comorbidity. Iron therapy raises thyroid hormone levels in anemic women with impaired thyroid function, but thyroxine plus iron works better to improve iron status. Vitamin D levels have been found to be lower in HT patients (those with autoimmune thyroiditis) than in healthy individuals. Additionally, there have been reports of an inverse relationship between TPO/Tg antibodies and serum vitamin D levels. Nevertheless, additional information and the absence of trial evidence point to the possibility that autoimmune disease processes involving vitamin D receptor dysfunction are more likely to be the cause of low vitamin D status. In order to address any deficiency, clinicians are advised to check their patients' iron levels and inquire about the menstrual cycle and vitamin D status in female patients. It is imperative to ensure adequate consumption of selenium in areas where iodine levels are either excessive or inadequate. When the body doesn't get enough selenium from food, it could be a good idea to take a 50–100 μg daily supplement (Hu and Rayman 2017).

Indeed, vitamin D supplementation can help HT patients experience less disease activity; however, more large-scale, carefully monitored studies are required to determine whether or not this treatment can be implemented in clinical settings (Chahardoli et al. 2019). When levothyroxine is being used to treat Hashimoto's thyroiditis in women with normal vitamin D levels, vitamin D preparations have been shown to reduce thyroid autoimmunity (Krysiak et al. 2017).

4.6 Inflammatory Bowel Disease (IBD) and Nutrition

Inflammatory Bowel Disease (IBD) is a chronic inflammation of the Gastrointestinal tract with an excess of immune response to normal gut flora and is grouped with Ulcerative colitis and Crohn's disease. It (Santacroce et al. 2021). Chronic inflammation causes cytokines, chemokines, eicosanoids, and other inflammatory mediators as nitric oxide to cause tissue destruction (Vezza et al. 2016). The current medications used to treat IBD cause systemic immunosuppression and cannot be used for a longer duration. The most common medications used are steroids such as prednisone which can cause side effects such as elevated blood sugar and osteoporosis. The other medications are 5-aminosalicylates, and immunomodulators such as Azathioprine, 6-Mercaptopurine, and Methotrexate. They reduce the activity of cells causing inflammation of GIT. They have adverse effects such as liver and pancreas inflammation. They are suggested to be still safer than steroids. The other group of medications used are biologics which block the effect of tumor necrosis factor-alpha thereby decreasing cell activity and inflammation. Examples are infliximab, adalimumab, and certolizumab pegol. They have adverse effects reported as lymphoma. There is another group that blocks the ability of white blood cells to cause inflammation as Natalizumab. Adverse effects have been reported as a rare chance of developing progressive multifocal encephalopathy (Triantafillidis et al. 2011). The serious adverse effects of these medications have led to the search for alternative medicine as a treatment option. In 1800 a substance salicin was extracted from willow trees and used for RA. The present-day acetylsalicylic acid was refined from the salicin preparation (Mahdi et al. 2006). Herbs has been haveorted to possess anti-inflammatory properties and were used in traditional Chinese medicine with a successful reduction in inflammation (Wang et al. 2023). Probiotics are the most frequently used alternative therapy for inflammatory bowel disease. The gut microbiota has been studied in individuals suffering from Crohn's disease and it was observed that there is an increase in levels of aerobic and facultative anaerobes which are prone to invade the gut epithelium leading to inflammation. Crohn's patients also showed a decrease in the beneficial obligate anaerobes (Liu et al. 2021). Probiotic therapy alters intestinal pH, competes for nutrients, and produces antimicrobial compounds decreasing the bacteria causing inflammation thereby reducing the symptoms (Coqueiro et al. 2019). In dietary regulation, anti-inflammatory foods such as low fermentable oligosaccharides, disaccharides, monosaccharides, and polyols (FODMAPs) are recommended. The FODMAPs consist of fruits, vegetables, nuts, and gluten-free oats and rice. This dietary regulation needs to be done under the supervision of a dietician or treating doctor. A low-fat diet is also recommended in inflammatory bowel disease patients (Jorgensen and Santer 2023; Boradyn et al. 2021). The specific carbohydrate diet used is reported to be effective in ulcerative colitis. The mechanism of action is only the consumption of monosaccharides (glucose, fructose, and galactose). It was observed from the evidence that di and polysaccharides go undigested into the colon producing bacterial and yeast growth with excessive mucous that can damage the gastrointestinal mucosa. This diet includes fruits, vegetables, honey, lentils, split peas, and lactose-free dairy products (Walsh et al. 2020). Another diet plan is the Palaeolithic diet. This consists of lean meat and plant-based food up to 100 gm daily. The limitation is that it can lead to vitamin D deficiency (Bland 2021). Another dietary regimen that was reported by Olendzki et al. was an anti-inflammatory diet which is based on dysbiosis theory. This diet limits all refined sugars and gluten-based diets that cause inflammatory bacteria in the gut. The diet consists of fibrous food such as leek and onion, fermented food, and food sources rich in Omega-3 fatty acids (Agrawal et al. 2020). It was reported also that the elimination of the inflammatory substances from diet regimens has improved both symptoms associated with IBD as well as endoscopic inflammation which is known as Autoimmune Protocol Diet (AIP) (Konijeti et al. 2017).

5 Summary of Diet Regimens for Inflammatory Autoimmune Diseases

An anti-inflammatory agent is defined as any substance that reduces inflammation symptoms. Inflammation can be acute, characterized by sudden flare-ups of inflammatory symptoms, or chronic, lasting for months or even years. Anti-inflammatory substances can be natural or synthetic in nature. This chapter focuses on natural anti-inflammatory substances and diet regimens that have been proven to alleviate most of the symptoms of inflammatory autoimmune diseases. By incorporating a diet rich in anti-inflammatory agents, patients with autoimmune diseases such as rheumatoid arthritis, multiple sclerosis, systemic lupus erythematosus and inflammatory bowel disease can experience an improved quality of life. These diet plans, such as the Mitterrandian diet, have shown effectiveness in reducing symptoms such as inflammation, swelling, tenderness, pain, and fatigue. They are rich in food substances such as flavonoids, polyphenols, fibers, herbs, vitamins, and other anti-inflammatory compounds.

Among the commonly used herbs and spices, turmeric stands out for its high content of polyphenol curcumin, which is widely recommended for all types of inflammation. Garlic contains diallyl disulfide, cinnamon contains cinnamaldehyde and cinnamic acid, and cardamom has been shown to block inflammatory pathways. Ginger is known for its ability to suppress leukotrienes, while thyme, an herb rich in antioxidants, also exhibits anti-inflammatory properties. Green tea, which contains epigallocatechin-3-gallate, caffeine, and chlorogenic acid, has been beneficial in reducing inflammatory symptoms. Pepper, specifically piperine, possesses antioxidant, antimicrobial, and anti-inflammatory properties. Cayenne, found in chili peppers, is also known for its anti-inflammatory action, according to the Arthritis Foundation. Other herbs and plants, such as willow bark, borage seed oil, and thunder god vine (used in traditional Chinese medicine), have been used to reduce inflammation. However, caution must be exercised as some of these substances can interfere with other medications and may not be suitable for individuals with specific medical conditions.

Seafood is another category of food that contains anti-inflammatory substances, particularly omega-3 fatty acids. Fatty fish like mackerel, bluefin tuna, sardines, anchovies, cod, and whitefish are rich sources of omega-3 fatty acids. Green-lipped mussels and oysters also provide substantial amounts of these beneficial fatty acids. Omega-3 fatty acids in seafood are most effective when the fish is baked, grilled, or steamed.

Fruits and vegetables are a third category of food that contains anti-inflammatory substances, mainly antioxidants. Tomatoes, for example, contain the antioxidant lycopene, while kale and spinach are rich in vitamin K. Broccoli contains phenolics and sulforaphane, and various berries are abundant in flavonols. Fruits with purple skin, such as grapes, berries, and plums, as well as red wine, peanuts, and dark chocolate, contain resveratrol, another powerful antioxidant. Pineapple contains bromelain, an enzyme known for its anti-inflammatory effects. Consuming these fruits and vegetables in their raw form provides the greatest benefit.

Nuts and seeds make up the fourth category and are good sources of PUFAs and vitamin E, both of which possess antioxidant properties. Incorporating these foods into a balanced diet can help inhibit inflammatory pathways and promote healthy gut microbiota. Therefore, a balanced diet that is composed of high-fiber, vegetables, fruits, nuts, and legumes is desired to ease symptoms inflammatory autoimmune disorders. This dietary approach should be part of a comprehensive treatment regimen that also includes medications, physical activity, stress control, and avoidance of a Western diet.

Among the various dietary studies conducted, vegetarian diets consisting of whole grains, vegetables, legumes, fruits, and nuts have been shown to reduce inflammatory markers such as C-reactive protein when followed for extended periods. The Mediterranean diet is another widely recommended approach for autoimmune disorders. This diet includes whole grains, fish, olive oil, nuts, vegetables, and fruits and is commonly consumed in Greece and Southern Italy. The MIND diet, a variation of the Mediterranean as well as DASH diets, has been recognized as the best diet by experts in a study of 39 diets conducted by US News and World Report in 2021. The DASH diet, originally developed to tackle blood pressure, cholesterol, and heart diseases, emphasizes vitamin and mineral-rich foods, low-fat and low-sodium options, whole grains, fruits, vegetables, lean meats, proteins, fish, fat-free or low-fat dairy products, herbal seasonings instead of salt, and limited consumption of processed foods. The Mediterranean diet includes vegetables, fruits, whole grains, berries, olive oil, beans, nuts, and fish. The MIND diet combines elements of both the DASH and Mediterranean diets and has shown significant effectiveness in reducing inflammatory symptoms in autoimmune disorders.

Another diet plan that is gaining attention is the macrobiotic diet. In addition to grains, vegetables, fruits, nuts, seeds, and legumes, this diet allows for the consumption of seasoned foods, soups, fermented foods, and natural sweeteners, along with mild beverages. The macrobiotic diet is relatively easier to follow as a lifestyle, with equal portions of 25% for grains, vegetables, fruits, nuts, and seeds. Other popular diet plans include vegan diets, ketogenic diets, low-carbohydrate diets, intermittent fasting, raw food diets, and paleo diets. However, it is important to note that these regimens are primarily focused on weight loss rather than specifically targeting inflammation.

The Autoimmune Protocol (AIP) diet is another recommended diet plan for individuals with autoimmune disorders. This diet involves the elimination of all foods known to cause inflammation, followed by a gradual reintroduction of foods once symptoms improve. The AIP diet consists of two phases: the elimination and reintroduction phase. During the elimination phase, foods such as grains, nuts, seeds, dairy products, nonsteroidal anti-inflammatory drugs, processed foods, oils, and certain spices are avoided. The effects of nutritional consideration to different inflammatory autoimmune diseases is provided in Table 9.1 and Fig. 9.1.

Table 9.1 Nutritional strategies for inflammatory autoimmune diseases

Autoimmune disease	Nutritional consideration	Impact of nutrients
Rheumatoid arthritis	Herbal remedies (thyme, chamomile, borage, lavender, ginger, cinnamon and curcumin)	Anti-inflammatory effects
	Vitamin D supplementation	Immune modulation, bone health promotion, and reducing inflammation
Systemic lupus erythematosus	Curcumin	Anti-inflammatory effects and manage renal manifestations
	Vitamins D, C, A, E & B supplementation	Adjunct to pharmacotherapy, reducing inflammation and managing deficiencies
	Omega-3 supplementation	Reduces ESR, CRP, inflammatory markers, and oxidative stress
Multiple sclerosis (MS)	Vitamin D supplementation	Reducing inflammation and managing deficiencies
	Omega-3 supplementation	Reducing inflammation, oxidative stress, and nerve demyelination, positively impacting the progression of MS
	Pre and probiotics supplementation	Reduction in inflammatory markers and potentially alleviating symptoms associated with MS
Psoriasis	Vitamin D levels are crucial (Pona et al. 2019)	Managing and improving the symptoms
	Omega-3 (Miyagawa et al. 2022; Saghafi-Asl et al. 2021)	Modulating inflammation
	Vitamins A, E, and C (Pokharel et al. 2022)	Anti-inflammatory and antioxidant properties
	Probiotics and prebiotics (Polkowska-Pruszyńska et al. 2020)	Maintaining gut health and mitigating inflammation
Inflammatory bowel disease (IBD)	Low-fat diet, a specific carbohydrate diet A Palaeolithic diet Autoimmune protocol diet (AIP)	Dietary regulations aim to reduce inflammation by elimination of inflammatory substances
Hashimoto's Thyroiditis (HT)	Selenium, potassium, iodine, copper, magnesium, zinc, iron, and vitamins A, C, D, and B	Reduce thyroid peroxidase (TPO)-antibody levels
	Vitamin D supplementation	Alleviate disease activity

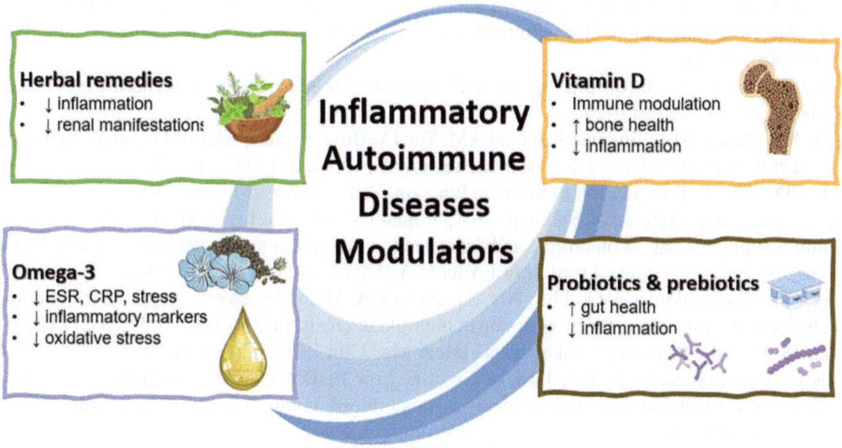

Fig. 9.1 Inflammatory autoimmune diseases modulators

References

Aaron L, Torsten M, Patricia W. Autoimmunity in celiac disease: extra-intestinal manifestations. Autoimmun Rev. 2019;18(3):241–6.

Abdelrahman W, Al-Shaarawy A, El-Zorkany B. Influence of perception of glucocorticoids on compliance of treatment in patients with rheumatoid arthritis and systemic lupus erythematosus. Egyptian Rheumatologist. 2023;45(2):159–64.

Adorini L. Intervention in autoimmunity: the potential of vitamin D receptor agonists. Cell Immunol. 2005;233(2):115–24.

Agrawal G, Clancy A, Huynh R, Borody T. Profound remission in Crohn's disease requiring no further treatment for 3–23 years: a case series. Gut Pathog. 2020;12(1):1–8.

AlAmmar WA, Albeesh FH, Ibrahim LM, Algindan YY, Yamani LZ, Khattab RY. Effect of omega-3 fatty acids and fish oil supplementation on multiple sclerosis: a systematic review. Nutr Neurosci. 2021;24(7):569–79.

Al-Shobaili HA, Qureshi MG. Pathophysiology of psoriasis: current concepts. In: Psoriasis-types, causes and medication. Rijeka: In Tech; 2013. p. 91–105.

Amrein K, Scherkl M, Hoffmann M, Neuwersch-Sommeregger S, Köstenberger M, Tmava Berisha A, et al. Vitamin D deficiency 2.0: an update on the current status worldwide. Eur J Clin Nutr. 2020;74(11):1498–513.

Ao T, Kikuta J, Ishii M. The effects of vitamin D on immune system and inflammatory diseases. Biomol Ther. 2021;11(11):1624.

Aparicio-Soto M, Sánchez-Hidalgo M, Alarcón-de-la-Lastra C. An update on diet and nutritional factors in systemic lupus erythematosus management. Nutr Res Rev. 2017;30(1):118–37.

Arpaia N, Campbell C, Fan X, Dikiy S, Van Der Veeken J, Deroos P, et al. Metabolites produced by commensal bacteria promote peripheral regulatory T-cell generation. Nature. 2013;504(7480):451–5.

Arrieta Valero I. Autonomies in interaction: dimensions of patient autonomy and non-adherence to treatment. Front Psychol. 2019;10:1857.

Aryaeian N, Shahram F, Mahmoudi M, Tavakoli H, Yousefi B, Arablou T, et al. The effect of ginger supplementation on some immunity and inflammation intermediate genes expression in patients with active rheumatoid arthritis. Gene. 2019;698:179–85.

Asik M, Gunes F, Binnetoglu E, Eroglu M, Bozkurt N, Sen H, et al. Decrease in TSH levels after lactose restriction in Hashimoto's thyroiditis patients with lactose intolerance. Endocrine. 2014;46:279–84.

Bach J-F. The effect of infections on susceptibility to autoimmune and allergic diseases. N Engl J Med. 2002;347(12):911–20.

Bagur MJ, Murcia MA, Jiménez-Monreal AM, Tur JA, Bibiloni MM, Alonso GL, et al. Influence of diet in multiple sclerosis: a systematic review. Adv Nutr. 2017;8(3):463–72.

Bikle DD. Vitamin D regulation of immune function. Vitam Horm. 2011;86:1–21.

Bitarafan S, Saboor-Yaraghi A, Sahraian M-A, Soltani D, Nafissi S, Togha M, et al. Effect of vitamin A supplementation on fatigue and depression in multiple sclerosis patients: a double-blind placebo-controlled clinical trial. Iran J Allergy Asthma Immunol. 2016;15:13–9.

Blagov AV, Sukhorukov VN, Orekhov AN, Sazonova MA, Melnichenko AA. Significance of mitochondrial dysfunction in the progression of multiple sclerosis. Int J Mol Sci. 2022;23(21):12725.

Bland JS. Why the Pegan diet makes sense. Integr Med Clin J. 2021;20(2):16.

Boradyn KM, Przybyłowicz KE, Jarocka-Cyrta E. Low FODMAP diet is not effective in children with functional abdominal pain: a randomized controlled trial. Ann Nutr Metab. 2021;76(5):334–44.

Bruzzese V, Scolieri P, Pepe J. Efficacy of gluten-free diet in patients with rheumatoid arthritis. Reumatismo. 2020;72(4):213–7.

Caso F, Navarini L, Carubbi F, Picchianti-Diamanti A, Chimenti MS, Tasso M, et al. Mediterranean diet and psoriatic arthritis activity: a multicenter cross-sectional study. Rheumatol Int. 2020;40(6):951–8.

Castelló A, Rodríguez-Barranco M, Fernández de Larrea N, Jakszyn P, Dorronsoro A, Amiano P, et al. Adherence to the Western, prudent and Mediterranean dietary patterns and colorectal cancer risk: findings from the Spanish cohort of the European Prospective Investigation into Cancer and Nutrition (EPIC-Spain). Nutrients. 2022;14(15):3085.

Cavalli E, Mammana S, Nicoletti F, Bramanti P, Mazzon E. The neuropathic pain: an overview of the current treatment and future therapeutic approaches. Int J Immunopathol Pharmacol. 2019;33:2058738419838383.

Chahardoli R, Saboor-Yaraghi A-A, Amouzegar A, Khalili D, Vakili AZ, Azizi F. Can supplementation with vitamin D modify thyroid autoantibodies (Anti-TPO Ab, Anti-Tg Ab) and thyroid profile (T3, T4, TSH) in Hashimoto's thyroiditis? A double blind, randomized clinical trial. Horm Metab Res. 2019;51(05):296–301.

Chandran B, Goel A. A randomized, pilot study to assess the efficacy and safety of curcumin in patients with active rheumatoid arthritis. Phytother Res. 2012;26(11):1719–25.

Chen S, Sims GP, Chen XX, Gu YY, Chen S, Lipsky PE. Modulatory effects of 1, 25-dihydroxyvitamin D3 on human B cell differentiation. J Immunol. 2007;179(3):1634–47.

Cheng YC, Huang YC, Huang WL. The effect of vitamin D supplement on negative emotions: a systematic review and meta-analysis. Depress Anxiety. 2020;37(6):549–64.

Conde J, Scotece M, Gómez R, López V, Gómez-Reino JJ, Lago F, et al. Adipokines: biofactors from white adipose tissue. A complex hub among inflammation, metabolism, and immunity. Biofactors. 2011;37(6):413–20.

Constantin MM, Nita IE, Olteanu R, Constantin T, Bucur S, Matei C, et al. Significance and impact of dietary factors on systemic lupus erythematosus pathogenesis. Exp Ther Med. 2019;17(2):1085–90.

Coqueiro AY, Raizel R, Bonvini A, Tirapegui J, Rogero MM. Probiotics for inflammatory bowel diseases: a promising adjuvant treatment. Int J Food Sci Nutr. 2019;70(1):20–9.

Correale J, Ysrraelit MC, Gaitán MI. Immunomodulatory effects of vitamin D in multiple sclerosis. Brain. 2009;132(5):1146–60.

Correa-Rodríguez M, Pocovi-Gerardino G, Callejas-Rubio J-L, Ríos-Fernández R, Martín-Amada M, Cruz-Caparrós M-G, et al. Vitamin D levels are associated with disease activity and damage accrual in systemic lupus erythematosus patients. Biol Res Nurs. 2021;23(3):455–63.

Cortese M, Yuan C, Chitnis T, Ascherio A, Munger KL. No association between dietary sodium intake and the risk of multiple sclerosis. Neurology. 2017;89(13):1322–9.
Cotsapas C, Hafler DA. Immune-mediated disease genetics: the shared basis of pathogenesis. Trends Immunol. 2013;34(1):22–6.
de Medeiros MCS, Medeiros JCA, de Medeiros HJ, Leitão JCGC, Knackfuss MI. Dietary intervention and health in patients with systemic lupus erythematosus: a systematic review of the evidence. Crit Rev Food Sci Nutr. 2019;59(16):2666–73.
Desai MS, Seekatz AM, Koropatkin NM, Kamada N, Hickey CA, Wolter M, et al. A dietary fiber-deprived gut microbiota degrades the colonic mucus barrier and enhances pathogen susceptibility. Cell. 2016;167(5):1339–53. e21
Doosti-Irani A, Tamtaji OR, Mansournia MA, Ghayour-Mobarhan M, Ferns G, Kakhaki RD, et al. The effects of vitamin D supplementation on expanded disability status scale in people with multiple sclerosis: a critical, systematic review and metaanalysis of randomized controlled trials. Clin Neurol Neurosurg. 2019;187:105564.
Duarte-Garcia A, Myasoedova E, Karmacharya P, Hocaoğlu M, Murad MH, Warrington KJ, et al. Effect of omega-3 fatty acids on systemic lupus erythematosus disease activity: a systematic review and meta-analysis. Autoimmun Rev. 2020;19(12):102688.
Duchnik E, Kruk J, Tuchowska A, Marchlewicz M. The impact of diet and physical activity on psoriasis: a narrative review of the current evidence. Nutrients. 2023;15(4):840.
Ebrahimzadeh A, Abbasi F, Ebrahimzadeh A, Jibril AT, Milajerdi A. Effects of curcumin supplementation on inflammatory biomarkers in patients with rheumatoid arthritis and ulcerative colitis: a systematic review and meta-analysis. Complement Ther Med. 2021;61:102773.
Evans E, Levasseur V, Cross AH, Piccio L. An overview of the current state of evidence for the role of specific diets in multiple sclerosis. Mult Scler Relat Disord. 2019;36:101393.
Fanara S, Aprile M, Iacono S, Schirò G, Bianchi A, Brighina F, et al. The role of nutritional lifestyle and physical activity in multiple sclerosis pathogenesis and management: a narrative review. Nutrients. 2021;13(11):3774.
Fazeli MS, Khaychuk V, Wittstock K, Han X, Crocket G, Lin M, et al. Rheumatoid arthritis-associated interstitial lung disease: epidemiology, risk/prognostic factors, and treatment landscape. Clin Exp Rheumatol. 2021;39(5):1108–18.
Feige J, Moser T, Bieler L, Schwenker K, Hauer L, Sellner J. Vitamin D supplementation in multiple sclerosis: a critical analysis of potentials and threats. Nutrients. 2020;12(3):783.
Fredrikson S, Cheng Q, Jiang G-X, Wasserman D. Elevated suicide risk among patients with multiple sclerosis in Sweden. Neuroepidemiology. 2003;22(2):146–52.
Fryze W, Mirowska-Guzel D, Wiszniewska M, Darda-Ledzion L, Członkowski A, Członkowska A. Alternative methods of treatment used by multiple sclerosis patients in Poland. Neurol Neurochir Pol. 2006;40(5):386–90.
Fu H, Tang Z, Wang Y, Ding X, Rinaldi G, Rahmani J, et al. Relationship between vitamin D level and mortality in adults with psoriasis: a retrospective cohort study of NHANES data. Clin Ther. 2021;43(2):e33–e8.
Garbicz J, Całyniuk B, Górski M, Buczkowska M, Piecuch M, Kulik A, et al. Nutritional therapy in persons suffering from psoriasis. Nutrients. 2022;14(1):119.
García-Montero C, Fraile-Martínez O, Gómez-Lahoz AM, Pekarek L, Castellanos AJ, Noguerales-Fraguas F, et al. Nutritional components in Western diet versus Mediterranean diet at the gut microbiota–immune system interplay. Implications for health and disease. Nutrients. 2021;13(2):699.
Gergianaki I, Garantziotis P, Adamichou C, Saridakis I, Spyrou G, Sidiropoulos P, et al. High comorbidity burden in patients with SLE: data from the community-based lupus registry of crete. J Clin Med. 2021;10(5):998.
Gil-González I, Martín-Rodríguez A, Conrad R, Pérez-San-Gregorio MÁ. Quality of life in adults with multiple sclerosis: a systematic review. BMJ Open. 2020;10(11):e041249.

Gioia C, Lucchino B, Tarsitano MG, Iannuccelli C, Di Franco M. Dietary habits and nutrition in rheumatoid arthritis: can diet influence disease development and clinical manifestations? Nutrients. 2020;12(5):1456.

Guasch-Ferré M, Willett W. The Mediterranean diet and health: a comprehensive overview. J Intern Med. 2021;290(3):549–66.

Guo B. IL-10 modulates Th17 pathogenicity during autoimmune diseases. J Clin Cell Immunol. 2016;7(2):400.

Haines ST, Park SK. Vitamin D supplementation: what's known, what to do, and what's needed. Pharmacotherapy. 2012;32(4):354–82.

Haneishi Y, Furuya Y, Hasegawa M, Picarelli A, Rossi M, Miyamoto J. Inflammatory bowel diseases and gut microbiota. Int J Mol Sci. 2023;24(4):3817.

Higgins M, Mackie S, Thalayasingam N, Bingham S, Hamilton J, Kelly C. The effect of vitamin D levels on the assessment of disease activity in rheumatoid arthritis. Clin Rheumatol. 2013;32:863–7.

Holick F, M. Vitamin D: evolutionary, physiological and health perspectives. Curr Drug Targets. 2011;12(1):4–18.

Hong Y, Tang HR, Ma M, Chen N, Xie X, He L. Multiple sclerosis and stroke: a systematic review and meta-analysis. BMC Neurol. 2019;19(1):1–11.

Hu S, Rayman MP. Multiple nutritional factors and the risk of Hashimoto's thyroiditis. Thyroid. 2017;27(5):597–610.

Ihnatowicz P, Drywień M, Wątor P, Wojsiat J. The importance of nutritional factors and dietary management of Hashimoto's thyroiditis. Ann Agric Environ Med. 2020;27(2):184–93.

Islam MA, Khandker SS, Kotyla PJ, Hassan R. Immunomodulatory effects of diet and nutrients in systemic lupus erythematosus (SLE): a systematic review. Front Immunol. 2020;11:1477.

Islam SU, Ahmed MB, Ahsan H, Lee Y-S. Recent molecular mechanisms and beneficial effects of phytochemicals and plant-based whole foods in reducing LDL-C and preventing cardiovascular disease. Antioxidants. 2021;10(5):784.

Jacob J, Amalraj A, Raj KJ, Divya C, Kunnumakkara AB, Gopi S. A novel bioavailable hydrogenated curcuminoids formulation (CuroWhite™) improves symptoms and diagnostic indicators in rheumatoid arthritis patients – a randomized, double blind and placebo controlled study. J Tradit Complement Med. 2019;9(4):346–52.

Javanbakht P, Taghizadeh F, Takabi FS, Tajik M, Iranshahi S, Pasbakhsh P, et al. Effects of calorie restriction on multiple sclerosis: a review of the preclinical and clinical studies. Neurochem Res. 2023;48(6):1597–610.

Jiang J, Chu C, Wu C, Wang C, Zhang C, Li T, et al. Efficacy of probiotics in multiple sclerosis: a systematic review of preclinical trials and meta-analysis of randomized controlled trials. Food Funct. 2021;12(6):2354–77.

Jiao H, Acar G, Robinson GA, Ciurtin C, Jury EC, Kalea AZ. Diet and systemic lupus erythematosus (SLE): from supplementation to intervention. Int J Environ Res Public Health. 2022;19(19):11895.

Jobin K, Müller DN, Jantsch J, Kurts C. Sodium and its manifold impact on our immune system. Trends Immunol. 2021;42(6):469–79.

Jorgensen M, Santer D. Promote or prevent? Gut microbial function and immune status may determine the effect of fiber in inflammatory bowel. Gastroenterology. 2023;164:182–6.

Kanda N, Hoashi T, Saeki H. Nutrition and psoriasis. Int J Mol Sci. 2020;21(15):5405.

Katz Sand I, Levy S, Fitzgerald K, Sorets T, Sumowski JF. Mediterranean diet is linked to less objective disability in multiple sclerosis. Mult Scler J. 2023;29(2):248–60.

Kiernan K, Nichols AG, Alwarawrah Y, MacIver NJ. Effects of T cell leptin signaling on systemic glucose tolerance and T cell responses in obesity. PloS One. 2023;18(6):e0286470.

Klack K, Bonfa E, Borba Neto EF. Diet and nutritional aspects in systemic lupus erythematosus. Rev Bras Reumatol. 2012;52:395–408.

Kleinewietfeld M, Manzel A, Titze J, Kvakan H, Yosef N, Linker RA, et al. Sodium chloride drives autoimmune disease by the induction of pathogenic TH17 cells. Nature. 2013;496(7446):518–22.

Kloesch B, Becker T, Dietersdorfer E, Kiener H, Steiner G. Anti-inflammatory and apoptotic effects of the polyphenol curcumin on human fibroblast-like synoviocytes. Int Immunopharmacol. 2013;15(2):400–5.

Konijeti GG, Kim N, Lewis JD, Groven S, Chandrasekaran A, Grandhe S, et al. Efficacy of the autoimmune protocol diet for inflammatory bowel disease. Inflamm Bowel Dis. 2017;23(11):2054–60.

Kostoglou-Athanassiou I, Athanassiou L, Athanassiou P. The effect of omega-3 fatty acids on rheumatoid arthritis. Mediterr J Rheumatol. 2020;31(2):190–4.

Kouchaki E, Tamtaji OR, Salami M, Bahmani F, Kakhaki RD, Akbari E, et al. Clinical and metabolic response to probiotic supplementation in patients with multiple sclerosis: a randomized, double-blind, placebo-controlled trial. Clin Nutr. 2017;36(5):1245–9.

Krysiak R, Szkróbka W, Okopień B. The effect of vitamin D on thyroid autoimmunity in levothyroxine-treated women with Hashimoto's thyroiditis and normal vitamin D status. Exp Clin Endocrinol Diabetes. 2017;125(04):229–33.

La Rosa G, Lonardo MS, Cacciapuoti N, Muscariello E, Guida B, Faraonio R, et al. Dietary polyphenols, microbiome, and multiple sclerosis: from molecular anti-inflammatory and neuroprotective mechanisms to clinical evidence. Int J Mol Sci. 2023;24(8):7247.

Lerner A, Matthias T. Changes in intestinal tight junction permeability associated with industrial food additives explain the rising incidence of autoimmune disease. Autoimmun Rev. 2015;14(6):479–89.

Letarouilly J-G, Sanchez P, Nguyen Y, Sigaux J, Czernichow S, Flipo R-M, et al. Efficacy of spice supplementation in rheumatoid arthritis: a systematic literature review. Nutrients. 2020;12(12):3800.

Li B, Leung JC, Chan LY, Yiu WH, Tang SC. A global perspective on the crosstalk between saturated fatty acids and toll-like receptor 4 in the etiology of inflammation and insulin resistance. Prog Lipid Res. 2020;77:101020.

Lin W-S, Lin S-J, Liao P-Y, Suresh D, Hsu T-R, Wang P-Y. Role of ketogenic diets in multiple sclerosis and related animal models: an updated review. Adv Nutr. 2022;13(5):2002–14.

Liu S, Zhao W, Lan P, Mou X. The microbiome in inflammatory bowel diseases: from pathogenesis to therapy. Protein Cell. 2021;12(5):331–45.

Lombardi VC, De Meirleir KL, Subramanian K, Nourani SM, Dagda RK, Delaney SL, et al. Nutritional modulation of the intestinal microbiota; future opportunities for the prevention and treatment of neuroimmune and neuroinflammatory disease. J Nutr Biochem. 2018;61:1–16.

Luo Y, Burrington CM, Graff EC, Zhang J, Judd RL, Suksaranjit P, et al. Metabolic phenotype and adipose and liver features in a high-fat Western diet-induced mouse model of obesity-linked NAFLD. Am J Physiol-Endocrinol Metabol. 2016;310(6):E418–E39.

Ma X, Nan F, Liang H, Shu P, Fan X, Song X, et al. Excessive intake of sugar: an accomplice of inflammation. Front Immunol. 2022;13:988481.

Mahdi J, Mahdi A, Mahdi A, Bowen I. The historical analysis of aspirin discovery, its relation to the willow tree and antiproliferative and anticancer potential. Cell Prolif. 2006;39(2):147–55.

Manzel A, Muller DN, Hafler DA, Erdman SE, Linker RA, Kleinewietfeld M. Role of "Western diet" in inflammatory autoimmune diseases. Curr Allergy Asthma Rep. 2014;14:1–8.

Marek-Jozefowicz L, Czajkowski R, Borkowska A, Nedoszytko B, Żmijewski MA, Cubała WJ, et al. The brain–skin axis in psoriasis—psychological, psychiatric, hormonal, and dermatological aspects. Int J Mol Sci. 2022;23(2):669.

Martin G, Young M, Aldredge L. Recommendations for initiating systemic therapy in patients with psoriasis. J Clin Aesthet Dermatol. 2019;12(4):13–26.

Mattavelli E, Catapano AL, Baragetti A. Molecular immune-inflammatory connections between dietary fats and atherosclerotic cardiovascular disease: which translation into clinics? Nutrients. 2021;13(11):3768.

Mazzucca CB, Raineri D, Cappellano G, Chiocchetti A. How to tackle the relationship between autoimmune diseases and diet: well begun is half-done. Nutrients. 2021;13(11):3956.

Merlino LA, Curtis J, Mikuls TR, Cerhan JR, Criswell LA, Saag KG. Vitamin D intake is inversely associated with rheumatoid arthritis: results from the Iowa Women's Health Study. Arthritis Rheum. 2004;50(1):72–7.

Mikulska AA, Karaźniewicz-Łada M, Filipowicz D, Ruchała M, Główka FK. Metabolic characteristics of Hashimoto's thyroiditis patients and the role of microelements and diet in the disease management—an overview. Int J Mol Sci. 2022;23(12):6580.

Milovanovic J, Arsenijevic A, Stojanovic B, Kanjevac T, Arsenijevic D, Radosavljevic G, et al. Interleukin-17 in chronic inflammatory neurological diseases. Front Immunol. 2020;11:947.

Miyagawa I, Nakayamada S, Ueno M, Miyazaki Y, Ohkubo N, Inoue Y, et al. Precision medicine based on the phenotypic differences in peripheral T helper cells in patients with psoriatic arthritis: one year follow-up outcomes. Front Med. 2022;9:9.

Mok CC, Hamijoyo L, Kasitanon N, Chen S, Yamaoka K, Oku K, et al. The Asia-Pacific League of Associations for Rheumatology consensus statements on the management of systemic lupus erythematosus. Lancet Rheumatol. 2021;3(7):e517–e31.

Moles L, Otaegui D. The impact of diet on microbiota evolution and human health. Is diet an adequate tool for microbiota modulation? Nutrients. 2020;12(6):1654.

Momtazi-Borojeni AA, Haftcheshmeh SM, Esmaeili S-A, Johnston TP, Abdollahi E, Sahebkar A. Curcumin: a natural modulator of immune cells in systemic lupus erythematosus. Autoimmun Rev. 2018;17(2):125–35.

Genetic risk and a primary role for cell-mediated immune mechanisms in multiple sclerosis. Nature. 2011;476(7359):214–9.

Nelson J, Sjöblom H, Gjertsson I, Ulven SM, Lindqvist HM, Bärebring L. Do interventions with diet or dietary supplements reduce the disease activity score in rheumatoid arthritis? A systematic review of randomized controlled trials. Nutrients. 2020;12(10):2991.

Niino M, Miyazaki Y. Role of vitamin D in multiple sclerosis pathogenesis and therapy. Nutrition and lifestyle in neurological autoimmune diseases. Elsevier; 2017. p. 71–80.

Noble PW, Bernatsky S, Clarke AE, Isenberg DA, Ramsey-Goldman R, Hansen JE. DNA-damaging autoantibodies and cancer: the lupus butterfly theory. Nat Rev Rheumatol. 2016;12(7):429–34.

Noormohammadi M, Ghorbani Z, Naser Moghadasi A, Saeedirad Z, Shahemi S, Ghanaatgar M, et al. MIND diet adherence might be associated with a reduced odds of multiple sclerosis: results from a case–control study. Neurol Ther. 2022;11(1):397–412.

Noyes K, Weinstock-Guttman B. Impact of diagnosis and early treatment on the course of multiple sclerosis. Am J Manag Care. 2013;19(17 Suppl):s321–31.

Okada H, Kuhn C, Feillet H, Bach J-F. The 'hygiene hypothesis' for autoimmune and allergic diseases: an update. Clin Exp Immunol. 2010;160(1):1–9.

Osowiecka K, Myszkowska-Ryciak J. The influence of nutritional intervention in the treatment of Hashimoto's thyroiditis—a systematic review. Nutrients. 2023;15(4):1041.

Otero-Romero S, Sastre-Garriga J, Comi G, Hartung H-P, Soelberg Sørensen P, Thompson AJ, et al. Pharmacological management of spasticity in multiple sclerosis: systematic review and consensus paper. Mult Scler J. 2016;22(11):1386–96.

Ovcharova EM, Danovska MP, Marinova DL, Pendicheva DI, Tonchev PT, Shepherd NM. Role of diet and supplementation with omega-3 polyunsaturated fatty acids for managing chronic fatigue in patients with relapsing-remitting multiple sclerosis. J Biomed Clin Res. 2022;15(2):99–104.

Passali M, Josefsen K, Frederiksen JL, Antvorskov JC. Current evidence on the efficacy of gluten-free diets in multiple sclerosis, psoriasis, type 1 diabetes and autoimmune thyroid diseases. Nutrients. 2020;12(8):2316.

Piec P-A, Pons V, Rivest S. Triggering innate immune receptors as new therapies in alzheimer's disease and multiple sclerosis. Cells. 2021;10(8):2164.

Pokharel R, Agrawal S, Pandey P, Lamsal M. Assessment of vitamin D level in patients with psoriasis and its correlation with disease severity: a case–control study. Psoriasis Targets Ther. 2022;12:251–8.

Polkowska-Pruszyńska B, Gerkowicz A, Krasowska D. The gut microbiome alterations in allergic and inflammatory skin diseases–an update. J Eur Acad Dermatol Venereol. 2020;34(3):455–64.
Pona A, Haidari W, Kolli SS, Feldman SR. Diet and psoriasis. Dermatol Online J. 2019;25:2.
Porsche CE, Delproposto JB, Geletka L, O'Rourke R, Lumeng CN. Obesity results in adipose tissue T cell exhaustion. JCI Insight. 2021;6(8):e139793.
Radu A-F, Bungau SG. Management of rheumatoid arthritis: an overview. Cells. 2021;10(11):2857.
Ralli M, Angeletti D, Fiore M, D'Aguanno V, Lambiase A, Artico M, et al. Hashimoto's thyroiditis: an update on pathogenic mechanisms, diagnostic protocols, therapeutic strategies, and potential malignant transformation. Autoimmun Rev. 2020;19(10):102649.
Rayman MP. Multiple nutritional factors and thyroid disease, with particular reference to autoimmune thyroid disease. Proc Nutr Soc. 2019;78(1):34–44.
Richards JL, Yap YA, McLeod KH, Mackay CR, Mariño E. Dietary metabolites and the gut microbiota: an alternative approach to control inflammatory and autoimmune diseases. Clin Transl Immunology. 2016;5(5):e82.
Rook GA. Hygiene hypothesis and autoimmune diseases. Clin Rev Allergy Immunol. 2012;42:5–15.
Rose NR. Prediction and prevention of autoimmune disease in the 21st century: a review and preview. Am J Epidemiol. 2016;183(5):403–6.
Rotstein DL, Cortese M, Fung TT, Chitnis T, Ascherio A, Munger KL. Diet quality and risk of multiple sclerosis in two cohorts of US women. Mult Scler J. 2019;25(13):1773–80.
Ruiz-Irastorza G, Danza A, Khamashta M. Glucocorticoid use and abuse in SLE. Rheumatology. 2012;51(7):1145–53.
Russo GL, Siani A, Fogliano V, Geleijnse JM, Giacco R, Giampaoli S, et al. The Mediterranean diet from past to future: key concepts from the second "Ancel Keys" International Seminar. Nutr Metab Cardiovasc Dis. 2021;31(3):717–32.
Saghafi-Asl M, Mirmajidi S, Asghari Jafarabadi M, Vahid F, Shivappa N, Hébert JR, et al. The association of dietary patterns with dietary inflammatory index, systemic inflammation, and insulin resistance, in apparently healthy individuals with obesity. Sci Rep. 2021;11(1):7515.
Salgado E, Bes-Rastrollo M, de Irala J, Carmona L, Gomez-Reino JJ. High sodium intake is associated with self-reported rheumatoid arthritis: a cross sectional and case control analysis within the SUN cohort. Medicine. 2015;94(37):e0924.
Santacroce L, Man A, Charitos IA, Haxhirexha K, Topi S. Current knowledge about the connection between health status and gut microbiota from birth to elderly. A narrative review. Front Biosci (Landmark Ed). 2021;26(6):135–48.
Scalfari A, Knappertz V, Cutter G, Goodin DS, Ashton R, Ebers GC. Mortality in patients with multiple sclerosis. Neurology. 2013;81(2):184–92.
Schoenfeld SR, Kasturi S, Costenbader KH. The epidemiology of atherosclerotic cardiovascular disease among patients with SLE: a systematic review. Semin Arthritis Rheum. 2013;43(1):77–95. Elsevier
Sepriano A, Kerschbaumer A, Smolen JS, Van Der Heijde D, Dougados M, Van Vollenhoven R, et al. Safety of synthetic and biological DMARDs: a systematic literature review informing the 2019 update of the EULAR recommendations for the management of rheumatoid arthritis. Ann Rheum Dis. 2020;79(6):760–70.
Sepriano A, Kerschbaumer A, Bergstra SA, Smolen JS, Van Der Heijde D, Caporali R, et al. Safety of synthetic and biological DMARDs: a systematic literature review informing the 2022 update of the EULAR recommendations for the management of rheumatoid arthritis. Ann Rheum Dis. 2023;82(1):107–18.
Shams S, Martinez JM, Dawson JR, Flores J, Gabriel M, Garcia G, et al. The therapeutic landscape of rheumatoid arthritis: current state and future directions. Front Pharmacol. 2021;12:680043.
Shemirani F, Titcomb TJ, Saxby SM, Ten Eyck P, Rubenstein LM, Hoth KF, et al. Association of serum homocysteine, folate, and vitamin B12 and mood following the Swank and Wahls elimination dietary interventions in relapsing-remitting multiple sclerosis: secondary analysis of the WAVES trial. Mult Scler Relat Disord. 2023;75:104743.

Shishehbor F, Rezaeyan Safar M, Rajaei E, Haghighizadeh MH. Cinnamon consumption improves clinical symptoms and inflammatory markers in women with rheumatoid arthritis. J Am Coll Nutr. 2018;37(8):685–90.
Song GG, Bae S-C, Lee YH. Association between vitamin D intake and the risk of rheumatoid arthritis: a meta-analysis. Clin Rheumatol. 2012;31:1733–9.
Stanescu AMA, Simionescu AA, Diaconu CC. Oral vitamin D therapy in patients with psoriasis. Nutrients. 2021;13(1):163.
Stoiloudis P, Kesidou E, Bakirtzis C, Sintila S-A, Konstantinidou N, Boziki M, et al. The role of diet and interventions on multiple sclerosis: a review. Nutrients. 2022;14(6):1150.
Terahara N. Flavonoids in foods: a review. Nat Prod Commun. 2015;10(3) https://doi.org/10.1177/1934578X1501000334.
Theodosis-Nobelos P, Rekka EA. Efforts towards repurposing of antioxidant drugs and active compounds for multiple sclerosis control. Neurochem Res. 2023;48(3):725–44.
Triantafillidis JK, Merikas E, Georgopoulos F. Current and emerging drugs for the treatment of inflammatory bowel disease. Drug Des Devel Ther. 2011;5:185–210.
Tripathi P, Bhardwaj P. Psoriasis: an autoimmune disorder. J Drug Deliv Ther. 2020;10(5):316–24.
Vellekkatt F, Menon V. Efficacy of vitamin D supplementation in major depression: a meta-analysis of randomized controlled trials. J Postgrad Med. 2019;65(2):74–80.
Venter C, Eyerich S, Sarin T, Klatt KC. Nutrition and the immune system: a complicated tango. Nutrients. 2020;12(3):818.
Vezza T, Rodríguez-Nogales A, Algieri F, Utrilla MP, Rodriguez-Cabezas ME, Galvez J. Flavonoids in inflammatory bowel disease: a review. Nutrients. 2016;8(4):211.
Vieira SM, Pagovich OE, Kriegel M. Diet, microbiota and autoimmune diseases. Lupus. 2014;23(6):518–26.
Wahls TL, Chenard CA, Snetselaar LG. Review of two popular eating plans within the multiple sclerosis community: low saturated fat and modified paleolithic. Nutrients. 2019;11(2):352.
Walsh JJ, Myette-Côté É, Neudorf H, Little JP. Potential therapeutic effects of exogenous ketone supplementation for type 2 diabetes: a review. Curr Pharm Des. 2020;26(9):958–69.
Walton C, King R, Rechtman L, Kaye W, Leray E, Marrie RA, et al. Rising prevalence of multiple sclerosis worldwide: insights from the atlas of MS. Mult Scler J. 2020;26(14):1816–21.
Wang F, Liu J, Fang Y, Li X, He M, Han Q. Traditional Chinese medicine may be associated with a reduced risk of recurrent exacerbation in patients with rheumatoid arthritis: a matched cohort study based on 1383 individuals. Heliyon. 2023;9(4):e15054.
Wu AG, Weinberg JM. The impact of diet on psoriasis. Cutis. 2019;104(2S):7–10.
Zou M-S, Song Q-J, Yin T-Y, Xu H-T, Nie G-M. Vitamin D activates miR-126a-5p to target GSK-3β and alleviates systemic lupus erythematosus in MRL/LPR mice. Curr Pharm Biotechnol. 2023;24:1803–11.

Chapter 10
Nutraceuticals and Infertility

Antoni Sureda, Margalida Monserrat-Mesquida, and Silvia Tejada

1 Introduction

Infertility is a disease affecting the reproductive system of both sexes defined as the inability to conceive after 1 year or more without barrier contraception or 6 months for women over 35 years old. The World Health Organization has also reported that one in six adults experience infertility (World Health Organization (WHO) 2023). It is a prevalent health condition that is getting higher over the time and affecting millions of couples worldwide, reaching around 15–17.5% of pairs (Boivin et al. 2007; Mascarenhas et al. 2012; World Health Organization (WHO) 2023). However, this prevalence of infertility varies among different populations, with a wider range from 10% to 17.5% globally, and observing greater prevalence amounts in African, European countries, Western Pacific and The Americas (Cox et al. 2022; World Health Organization (WHO) 2023). There are not enough studies focusing on gender, but female infertility seems to be more common than male infertility,

A. Sureda (✉) · M. Monserrat-Mesquida
Research Group in Community Nutrition and Oxidative Stress (NUCOX), University of Balearic Islands, Palma de Mallorca, Spain

Health Research Institute of Balearic Islands (IdISBa), Palma de Mallorca, Spain

CIBER Fisiopatología de la Obesidad y Nutrición (CIBEROBN), Instituto de Salud Carlos III, Palma de Mallorca, Spain
e-mail: antoni.sureda@uib.es

S. Tejada
Health Research Institute of Balearic Islands (IdISBa), Palma de Mallorca, Spain

CIBER Fisiopatología de la Obesidad y Nutrición (CIBEROBN), Instituto de Salud Carlos III, Palma de Mallorca, Spain

Laboratory of Neurophysiology, University of the Balearic Islands, Palma de Mallorca, Spain

S. M. Jafari et al. (eds.), *Nutraceuticals: A New Millennium Approach for Preventive Medicine*, Food Bioactive Ingredients,
https://doi.org/10.1007/978-3-031-76481-3_10

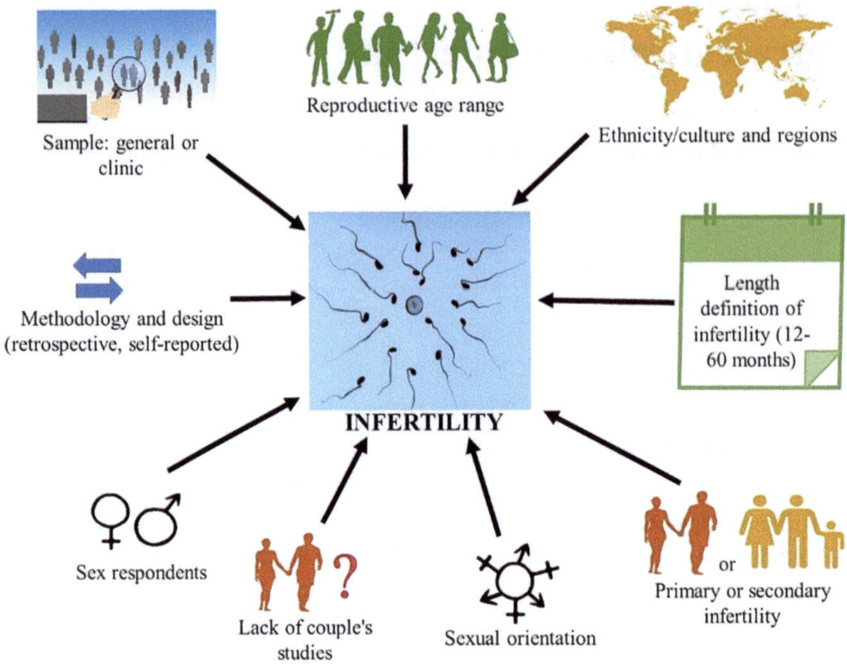

Fig. 10.1 Troubles emerging from infertility studies

accounting for approximately 40% to 50% of all cases, although more studies are focused on males. Moreover, the current lifestyle leads to seek pregnancy at older ages, observing that the highest prevalence of infertility appears in the 35–39 age groups (Cox et al. 2022), In addition, different researchers have used modified definitions of infertility or different study designs to assess this inability to conceive so the results obtained may differ and may not be comparable (Fig. 10.1). For instance, the time-to-pregnancy (TTP), considered the gold standard, is defined as the menstrual cycles or months that elapse until pregnancy (Baird et al. 1986; Joffe 1989), but it could differ depending on several factors, for instance the contraceptive methods or the frequency of sexual activity (Olsen et al. 1998; Weinberg et al. 1994). Another example is related to primary and secondary infertility which are not always described, taking into account that primary infertility denotes to the inability of an individual to attain pregnancy, whereas secondary infertility is characterized by the situation where one prior pregnancy or more has been reached before (World Health Organization (WHO) 2023).

Infertility can have significant psychological and emotional impacts on individuals and couples, including depression, anxiety, and social isolation (Gdanska et al. 2017). Therefore, efforts in identifying risk factors that can be modified, such as diet, would be crucial in the prevention and management of this inability.

2 Infertility in the Society

Around 15% of couples attempting to achieve pregnancy face fertility problems (Tamrakar and Bastakoti 2019) and a signal of this issue is the number of couples that ask for assisted reproductive technologies (de Geyter et al. 2020). The causes of infertility in couples can be both male and female, with combined factors being identified in around 40% of cases (Tamrakar and Bastakoti 2019). Infertility can result from several medical conditions, such as ovulatory disorders, tubal factor infertility, endometriosis, and uterine factors in females (Practice Committee of the American Society for Reproductive Medicine 2015; Tamrakar and Bastakoti 2019), and abnormal semen parameters, testicular disorders, obstructive azoospermia, and hormonal imbalances in males (Tamrakar and Bastakoti 2019; Winters and Walsh 2014). However, infertility could also be a combined problem of both female and male abnormalities.

In addition, there are risk factors for infertility that include age, obesity, smoking, alcohol consumption, exposure to toxins in the environment, and lifestyle factors including diet, exercise, and pressure in couples. In fact, last years, there has been growing attention in the impact of diet in the development of infertility. A number of studies have suggested that dietary factors, such as high consumption of red meat, trans fats, and processed foods, could rise the risk of infertility, while a healthy diet rich in fruits, vegetables, whole grains, and lean proteins may improve fertility (Afeiche et al. 2014; Çekici and Akdevelioğlu 2019; Gaskins and Chavarro 2018; Salas-Huetos et al. 2017) For instance, a higher chance of achieving pregnancy through *in vitro* fertilization (IVF), with a 40% increased chance of successful conception, was described in pairs following a Mediterranean diet (Karayiannis et al. 2018).

Male infertility has increased last decades, observing changes in sperm quality and an increase in testicular tumours (Sengupta et al. 2017; World Health Organization 2021); indeed, male infertility has been associated to a higher risk of cancer, malignancies, and chronic and cardiovascular diseases (Calogero et al. 2023). Genetics could play a crucial function to detect abnormality in males, currently accounting ~15%, with the Klinefelter or XXY syndrome being the most common alteration (Bojesen et al. 2003). However, this percentage could be different as new methodologies are arising and many genes are implicated in spermatogenesis (Hotaling and Carrell 2014).

Moreover, infertility in men is also correlated with the role that the reactive oxygen species (ROS) play. For a proper physiological function, ROS are important; and among these physiological functions are the development of spermatozoa in sperm and acrosome reaction and sperm-oocyte fusion during fertilisation (Ávila et al. 2022). However, an excess of ROS, exceeding the ability of seminal fluid to remove them, results in oxidative stress that initiates an assault on the polyunsaturated fatty acids (PUFAs) found in the sperm plasma membrane. These can induce the formation of reactive aldehydes (such as malondialdehyde), that contribute to mitochondrial dysfunction, triggering additional ROS production and further disruption of sperm function, such as the acrosome reaction or sperm–oocyte recognition and fusion (Aitken 2020; Takeshima et al. 2020). This could also result in DNA damage (i.e. 8-hydroxy-2′-deoxyguanosine as a mutagenic product) (Aitken and Baker 2020) and fragmentation in the sperm DNA (Agarwal et al. 2020; Cho and Agarwal 2017) which has been related to abnormality in semen analysis or to azoospermia (Aitken and Baker 2020).

Although the hormonal effects in men are not fully studied, male infertility can also be related to hormonal defects or imbalances. Various studies have associated obesity in men with hormonal changes that ultimately affect fertility. In obese men, there are alterations of the endocrine system, such as incorrect hormonal secretion, unusual circulating concentrations of hormones or alterations in their transport. This is based on the fact that obesity is linked to decreased levels of testosterone, which leads to alterations in sperm parameters. The hypothalamus-pituitary-testis axis is affected in men with increased body mass index, due to the blocking of sex hormone binding globulin because of the effect of elevated insulin concentrations, resulting in a rise in the concentrations of free androgens which are aromatized in fatty tissue (Hammoud et al. 2008). Thus, there is an increase in oestrogen levels, which by negative feedback inhibit the hypothalamic secretion of gonadotropin-releasing hormone, and in consequence, the hypothalamus-pituitary-testis axis decreases testosterone production. In fact, it has been described hypotestosteronaemia and hypogonadotropic hypogonadism in obesity, which impair fertility (Pasquali 2006). Varicocele is a disorder in which there is an enlargement of veins within the scrotum, resulting in impaired blood flow and potential negative effects on sperm production and fertility. As it is known, temperature is important for testicular physiology. In this sense, varicocele induces an increase in the testes' temperature, so that reducing the synthesis of testosterone by the Leydig cells and diminishing the functionality of the Sertoli cells (Khera and Lipshultz 2008). The alterations in these last cells include low inhibin-B secretion whose role is the inhibitory feedback in the follicle-stimulating hormone (FSH) production by the hypophysis, so that a rise of FSH levels in the circulation is also observed in patients with varicocele (Comhaire and Vermeulen 1975; Pierik et al. 2001).

Male fertility is also affected by other factors, including oncological diseases whose treatments affect the initial sperm quality and spermatogenesis phase (Vakalopoulos et al. 2015), inflammation and infections, atrophy or congenital hypoplastic testes (Ahmed et al. 2010), or other sexual dysfunctions (Lotti and Maggi 2018). Furthermore, different environmental risk factors can induce deleterious effects over spermatozoa in males and, consequently, infertility arises because those are exogenous ROS sources. Among the risk issues, it should be cited obesity together with sedentary patterns (Hammoud et al. 2008; Pasquali 2006), smoking and alcohol, exposition to radiation, and pollution. In fact, there has been cited a direct correlation between abnormal semen and smoking and sexually transmitted diseases (Garba-Alkali et al. 2018).

Smoking cigarettes are usual in reproductive individuals especially in countries of the First World, being Europe the leader regarding to smoking (Sharma et al. 2016). Both active and passive smoking have been related to infertility (Anderson et al. 2011). The many chemicals (polycyclic aromatic hydrocarbons, heavy metals, mutagenic substances) present in tobacco smoke are responsible for alterations in hormonal parameters and spermatogenesis, affecting viability and motility, morphology (such as in sperm tail) and concentration of sperm (Künzle et al. 2003; Sharma et al. 2016). Also, it has been described alterations in oocyte fertilization, being affected the acrosome reaction and capacitation in smokers (Kumosani et al. 2008; Shrivastava et al. 2014). Smokers presented an increment of inflammatory protein levels in their seminal plasma (Antoniassi et al. 2016). For this reason, smoking is related to an inflammatory condition, which can cause the alteration of the above cited spermatozoid and mitochondrial qualities, integrity of the acrosome, and the increment of DNA fragmentation. Furthermore, smoker parents affect to the health of their infants transmitting damaged DNA by oxidative stress. For instance, in male smokers, it has been related modifications in the methylation of sperm DNA to alterations in their children (Jenkins et al. 2017). Altogether, tobacco consumption is linked to a significant rise in delay (6 months) in conception in men (Practice Committee of the American Society for Reproductive Medicine American Society for Reproductive Medicine 2018). Not only tobacco is related to infertility in men, but also other consumption such as cannabis. It has been shown that infertile men who were regular cannabis users presented significantly higher sperm aneuploidy, total chromosome, and DNA fragmentation than non-cannabis users. So, sperm nuclear quality and progression of spermatogenesis have been deleteriously affected by cannabis consumption (Verhaeghe et al. 2020).

Alcohol abuse disorder is a serious condition associated with infertility. Spermatogenesis and testosterone metabolism are also affected by alcohol consumption, although there is more evidence of chronic and abuse consumption. Not only alcohol has been related to lesser seminal fluid, the morphology, motility and concentration of sperm, but also the hormone responses are altered. It has been found that alcohol caused oxidative stress and it induces a genotoxic effect on the DNA integrity and regulation of hormones, disturbing the health of the offspring (Finelli et al. 2021). Even chronic high consumption of alcohol can point to azoospermia, and it has been observed a reduction in the testosterone levels, and a rise in the FSH and luteinizing hormone (LH) in drinkers (Muthusami and Chinnaswamy 2005; Pajarinen and Karhunen 1994). However, if excessive alcohol consumption is avoided in humans, it could restore normal spermatogenesis (Jensen et al. 2014).

Drug consumers are also related to alcohol and smoking, so the studies only focused on drugs are lesser. Anyway, it has been described that the consumption of several drugs can interact with several receptors present in the pituitary gland, testicular tissues and Leydig and Sertoli cells in men. This leads to alteration in the hypothalamus-pituitary-gonadal axis that affects fertility. Consequently, hypogonadotropic hypogonadism, sperm function, spermatogenesis, sperm DNA fragmentation and apoptosis have been observed under drug abuse (Ajayi and Akhigbe 2020; Payne et al. 2019; du Plessis et al. 2015). Air pollution exposition is also related to fertility in men affecting motility, morphology and fragmentation of the DNA of sperm (Jurewicz et al. 2018). More recently, a retrospective study showed azoospermia and a reduction of the motility and volume of the sperm in men exposed to atmospheric pollution from industrial sources due to the chemical products causing endocrine disruption (Ramsay et al. 2023). Comparable results were observed in a retrospective longitudinal study in healthy and young men exposed to environmental air pollution, relating particulate matter and carbon oxygen with spermatogenesis (Qiu et al. 2020). Although it is clear that many factors can influence male fertility, the treatment for males is difficult and limited, and the trouble is normally resolved by several insemination methods or *in vitro* fertilization (Okonofua et al. 2022).

Infertility in women has also been described and studied, though the studies are retrospective and methodologically inconsistent, making the comparison among them difficult. The reproductive potential in women can be measured by the ovarian reserve, as the quality and number of oocytes (Practice Committee of the American Society for Reproductive Medicine. 2020) using transvaginal ultrasonography and together with the evaluation of the concentrations of several hormones, such as anti-mullerian hormone (AMH) or FSH.

The leading causes of female infertility are ovulatory disorders, tubal factor infertility (e.g., unilateral/bilateral blockage of fallopian tubes), endometriosis, uterine factors, and age, among others (Table 10.1).

Genetics play a complex and not well-known role in the reproductive outcomes in females; in the last decade, many common genetic risk factors have been described to modify hormone concentrations, reproductive characteristics or to be related to pathologies (Gajbhiye et al. 2018). In a review study, the variations in the hormones FSH, LH, AMH and oestradiol concentrations were related to different variants on chromosomes in females pointing out the importance that phenotypes can have in the age of menopause, ovarian reserve, and infertility (McGrath et al. 2021). Some treatments have the goal to increase the diminished levels of the hormones in females. When the pituitary gland suffers from a blockage of the negative feedback, it allows a rise of the hypothalamic gonadotropin-releasing hormone (GnRH, or also direct treatments to increase this hormone), which in turn increases the release of the FSH and LH inducing the growth and stimulation of the follicles (Adashi 1984; Christou et al. 2017; Legro et al. 2014).

There are different alterations or diseases in women that can affect conception, including anatomical alterations such as malformations in the uterus that can be corrected through surgery in order to improve fertility and achieve pregnancy (Bosteels et al. 2018). Among them, ovulatory dysfunctions are usual, being the polycystic ovary syndrome the main cause of anovulation (Sirmans and Pate 2013). In this sense, galactorrhoea is another cause of infertility due to a luteal phase defect and ovulatory cycles (El Mahgoub 1978). As in males, oxidative stress is also implicated in the oocyte quality, viability and fertilization in women. Although there is a lack of clinical trials in this line, the physiological levels of ROS allow the selection of the leading follicle (Devine et al. 2012; Shkolnik et al. 2011), and these levels are higher in a post-ovulatory oocyte since it is part of the mechanisms participating in the initiation of the apoptosis process when the fertilization does not occur (Aitken 2020; Lord et al. 2015). However, if the antioxidant capacity is diminished in the ovarian, it could induce local inflammation and fibrosis, which lead to alterations in

Table 10.1 Main causes of infertility in males and females

Male	Female
Abnormality in semen: Sperm concentration Sperm amount Vitality Motility Morphology Leukocyte concentration	Uterine abnormalities and disorders: Irregular cavity Small uterus Malformations (arcuate/septate/bicournate uterus) Endometrial polyp Uterine fibroid
Genetics (affects spermatogenesis)	Genetics
Sperm disorders: Asthenozoospermia (motility) Oligozoospermia (low sperm) Azoospermia (absence) Teratozoospermia (morphology)	Irregular menstrual pattern Oligomenorrhoea Amenorrhea Dysmenorrhoea Heavy bleeding
Hormonal imbalances: Less testosterone concentrations Hypogonadotropic hypogonadism	Hormonal imbalances: Hyperprolactinemia Hypothyroidism (thyroid abnormality) Galactorrhoea
Testicular tumours/disorders/damage	Adnexal pathology/Tumours
Sperm DNA damage/fragmentation	Ovulatory dysfunction/low ovarian reserve
Varicocele	Disease: Polycystic ovarian disease Endometriosis
Sperm transport disorders/seminal tract obstruction	Unilateral and bilateral blockage of fallopian tubes
Presence of bacterial organisms	
Another sexual dysfunction: Early ejaculation Alterations in desire, arousal and orgasm	Another sexual dysfunction: Problem in desire, arousal and orgasm
Both sexes	
Age	
Obesity	
Smoke	
Alcohol	
Drugs	
Environmental toxins	
Diet	
Idiopathic infertility	
Job situation (self-business, employee, abroad, etc.)	

the function of the tissue and fertility problems (Shi et al. 2016). It also seems that sex hormones modulate ROS levels, so that they regulate endometrium shedding. However, high levels of ROS in the uterus tissue could alter this process in some diseases, like endometriosis (Mier-Cabrera et al. 2011). Endometriosis is a chronic inflammatory disease that is oestrogen-dependent in which the proliferation of the endometrial cells occurs external to the uterine cavity; in consequence, this alteration affects an important percentage of women in reproductive age since endometrium is basic for implantation and growth of the embryo. The alterations observed included alterations of the gene profile and the expression of related proteins for implantation; for instance, the progesterone response in females with this pathology is lower which could explain, at least in part, some of the infertility problems (Zondervan et al. 2020).

Women, in contrast to men, experience a finite period of reproductive capability determined by the number of primordial follicles. Consequently, traditional chemotherapy drugs can induce an irreversible loss of the function of the ovary and the absence of menstruation due to a depletion of oocytes (Bedoschi et al. 2016). Tumour cells can disrupt the normal functioning of reproductive organs, such as the ovaries or uterus, by occupying space or causing structural changes. This can affect the release and maturation of oocytes, the ability of the uterus to support a pregnancy, or the production of related hormones (Fabiani et al. 2022; Volodarsky-Perel et al. 2019). Additionally, cancer can induce an imbalance in hormonal production, leading to irregular menstrual cycles that can impair fertility. In fact, adnexal tumours are common in infertile women growing near the uterus, principally in fallopian tubes, ovaries, or closed connective tissues. It can induce irregular menstrual periods. However, most adnexal tumours are non-cancerous but require treatment including surgery (Biggs and Marks 2016).

As in men, environmental risk factors, including obesity and sedentary patterns (Hammoud et al. 2008; Pasquali 2006) smoking and alcohol, exposition to radiation, and pollution, can also alter the reproduction function in females, so as a consequence, infertility problems appear. The overweight and obesity on women's fertility influence the ovarian and endometrial functions as consequence of the alterations in the endocrine mechanisms. The main cause derives from the functional alteration of the hypothalamic-pituitary-ovarian axis (HPO) (Marinelli et al. 2022). Higher plasma concentrations of insulin are generally observed in obese women, which induces a rise in the production ovarian androgen. These androgens are substrates for aromatase that convert them into oestrogens in adipose tissue, which generates negative feedback on the HPO axis, resulting in a longer follicular phase and a shortened luteal phase. This manifests as menstrual abnormalities and ovulatory dysfunction (Marinelli et al. 2022). Related to this, polycystic ovarian disease is a pathology normally found in obese patients. In this case, it has been observed a diminished abundance of the insulin receptor b-subunit in the adipose tissue of the women affected by that disease which reduces the sensitivity for the insulin; moreover, these patients show abnormal function of the β-cells and high levels of leptin that is related to diminished fertility (Macut et al. 2017).

Active and passive smoking also affect fertility in women (Anderson et al. 2011). The chemicals present in cigarettes induce alterations in hormonal parameters (lower oestrogens and progesterone levels), ovarian reserve and quality, ovarian dysfunction, menstrual cycle disorders, endometrial receptivity and early placentation. In fact, tobacco consumption has been related to a significant rise in delay in conception, greater than 1 year in smoking women (Anderson et al. 2011). It seems that the current smokers and passive ones (and also the women that suffered utero exposition) diminish the options of fertilization, but the former smokers recover the normal capability to get a pregnancy. According to research studies, alcohol consumption has been related to reduced female fertility and a higher risk of suffering menstrual pathologies, although the mechanism has not been determined. A possible theory for the undesirable effect of alcohol consumption on the fertility in females could be the alteration of the concentrations of endogenous hormones, a direct influence on the development of the oocyte, ovulation, the development of the early blastocyst and implantation (Skoracka et al. 2021). Similarly, addictive drugs studies in infertility are scarce in humans and mainly focused only on marijuana. Although many studies are focused on animal models, it has been described that cannabinoids alter the ovulatory function by means of the interaction with the hypothalamus so that in the HPO axis decreases the pulsatile gonadotropin-releasing hormone in a dose-dependent manner (Fonseca and Rebelo 2022).

3 Infertility and Dietary Patterns

The basis of a balanced life is based on maintaining adequate nutrition and a healthy lifestyle. Following a balanced and complete diet and having active habits are the foundations of a healthy lifestyle that supports maintaining an adequate weight and diminishing the risk of non-communicable chronic illnesses (Ng et al. 2020). The nutritional needs of an individual correspond to the sum of their basic needs and other variables depending on their age, sex, physiological state, activity and the environment in which they live. Today it is necessary to think about the importance of the social cost derived from inadequate nutritional states that lead to an increase in disorders and pathologies with a clear nutritional background.

At present, there is a consensus that changes in diet can promote health protection and prevent disorders caused by excesses, deficits and, above all, dietary imbalances. Today, perhaps the concept that best defines the message to be conveyed in relation to recommended nutrition is healthy eating. In this term are added the concepts of sufficiency both in energy and nutrients), variety to ensure the contributions of macro and micronutrients, balance in the recommended proportions and adaptation to the social and cultural characteristics of each person. All this should promote a healthy lifestyle and that is supported by the agreed nutritional recommendations for the population of our environment.

Nutritional epidemiology basically seeks to identify elements of the diet that are linked to a greater or lesser risk of disease. Scientific evidence supports the relevant role of numerous dietary and nutritional aspects in the development of various diseases of high importance in public health, including cancer, cardiovascular diseases, diabetes or mental illnesses. Although the effects of micronutrients, such as antioxidants, have been demonstrated, the complex relationships established between nutrients have led to working with global indices or dietary patterns, for instance, the Mediterranean diet pattern. The great variety of foods, with their combinations of nutrients that make up a specific dietary pattern, can interact and have both synergistic and antagonistic effects, making it necessary to evaluate their global effects.

Infertility is a common phenomenon today, increasing every year the number of couples who cannot conceive. Fertility in humankind is influenced by numerous factors, including aspects that affect the females such as uterine disorders or ovulation disturbances, and other aspects that affect the males such as abnormal sperm production and function, various non-modifiable factors such as genetics or age, and most significantly, lifestyle factors that can be modified such as nutrition, physical activity / sedentary lifestyle, overweight, alcohol, smoking, chronic stress and prolonged utilization of contraceptives (Acharya and Gowda 2017). Although it is widely acknowledged that nutrition significantly influences fertility outcomes for both women and men, there are currently no official recommendations provided for couples in the reproductive age group. (Rossi et al. 2016). Nutrition can exert positive or negative effects on fertility in both women and men, and the influence is dependent on different properties of the diet, both measurable and descriptive ones, including caloric content and profiles of fatty acids, proteins, and carbohydrate and micronutrient balance (Panth et al. 2018).

Diverse studies have revealed that nutritional patterns can play a basic part in the development of fertility disorders when considered as a whole and not as individual food components. This makes sense since people generally eat a complex combination of meals and not separate nutrients. A dietary pattern could be explained as the quantity, and how diverse foods and beverages are combined in diets, along with their consumption frequency. In addition, studying the effects of eating patterns can provide more accurate information on the association among disease and diet than the information based on individual nutrients or foods. In fact, for both sexes, a prevalent aspect in many cases of infertility is the presence of chronic inflammation and oxidative stress; these challenges can be ameliorated through adopting healthy practices, including the incorporation of an antioxidant-rich diet and bioactive compounds with anti-inflammatory activity (Barati et al. 2020; Ojo et al. 2023).

Within the healthiest dietary patterns, the traditional Mediterranean diet would be the most relevant example (Fig. 10.2). This dietary pattern has been associated with numerous helpful health effects, from reduced risk of death from any cause to numerous illnesses as it can be cancer, cognitive deterioration or cardiovascular disease (Ros et al. 2014).

In various meta-analyses, it has been shown that in recent years semen quality has decreased in men and ovulatory infertility increased in women which have been largely associated with progressive worsening in diet (Chavarro and Schlaff 2018; Chiu et al. 2018; Nassan et al. 2018a). It has been described that following the Mediterranean diet closely has the potential to enhance rates of pregnancy and live births, although the results are not entirely clear because of the heterogeneity of the works analysed (Winter et al. 2023). An interesting study has shown that following a Mediterranean pattern rose wealthy pregnancies by nearly 40% among couples who underwent in vitro fertilization (Vujkovic et al. 2010). In addition, the Mediterranean dietary model was also positively related to blood vitamin B6 and folate and in follicular fluid for vitamin B6, important for the homocysteine

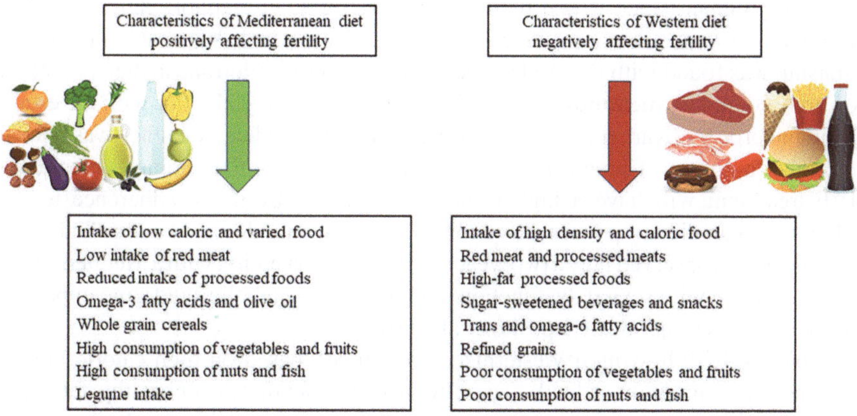

Fig. 10.2 Basic features of the Mediterranean and Western diets with potential positive and negative effects of fertility

pathway. In the case of men, various studies have shown improvements in sperm parameters associated with the Mediterranean diet pattern.

In a cross-sectional work, 225 men joining a fertility health centre were classified into tertiles based on their degree of adherence to the Mediterranean diet (Karayiannis et al. 2017). The results revealed that men in the highest adherent tertile had lesser probabilities of presenting non-standard total sperm count and motility and sperm concentration, with respect to men in the lowest tertile. The authors conclude, however, whether these observed changes translate into increased male fertility is still incomplete. In a second study, it has been investigated whether there is a relation between semen quality factors and following the Mediterranean dietary model in 106 healthy and young men (Salas-Huetos et al. 2019). Using a multivariate linear regression model applied to evaluate the association between the parameters of the sperm quality and tertiles of adherence to the Mediterranean diet, higher total sperm motility was observed in the tertile with the highest adherence.

Regarding women, in a prospective study (Nurses Health Cohort Study II) 17,544 women without previous infertility problems were included, it was evidenced that those women with lesser consumption of food related to ovulatory disorder infertility presented an inferior risk of alterations in fertility than those who had a higher consumption after 8 years of follow-up (Chavarro et al. 2007). The work concluded that most cases of infertility because of ovulation alterations could be avoided or significantly reduced by diet and lifestyle modifications. However, in this same cohort of women, no relationship was observed concerning a higher adherence to the Mediterranean pattern and the risk of suffering incident spontaneous abortions and stillbirths (Gaskins et al. 2014b). In a nested work involving 485 young women that had difficulties to get pregnant and 1669 controls, lesser difficulties to conceive were observed in women in the highest quartile of adherence to the Mediterranean pattern when compared with the lowest one (Toledo et al. 2011). The authors conclude the need to create nutritional strategies for women who wish to become pregnant since it will reduce infertility rates. In a cohort study with 244 women who were non-obese, undertook prospectively a first IVF treatment, no relationship was found with the grade of adherence to the Mediterranean dietary pattern on embryo quality measures, fertilization rate, and oocyte yield, nor with the establishment process (Karayiannis et al. 2018). In contrast, higher adherence was related to higher amounts of pregnancy and live birth suggesting that women undergoing IVF treatment will have a higher successful pregnancy if the adherence to the Mediterranean dietary pattern was stronger. Also, in a randomized controlled clinical trial, it was observed how a two-year nutritional intervention with a Mediterranean diet in women with sexual dysfunction together with metabolic syndrome improved the sexual function index (Esposito et al. 2007).

Conversely, following a Westernized diet model, based on high caloric intake, red meat consumption, added sugars, refined carbohydrates, unhealthy fats, and processed foods, is linked to obesity and negative effects on fertility. In this sense, the Western diet affects parameters such as the amounts of hormones, and function and concentration of sperm, and gamete composition (Meldrum et al. 2017). When comparing different parameters of sperm quality among 209 young and healthy subjects who followed the Mediterranean diet with respect to others who adhered to a more Western diet, it has been observed that those who followed more to the Mediterranean dietary pattern had not only higher sperm levels, but also better body mass index and healthier lifestyle (Cutillas-Toĺn et al. 2015). Nevertheless, it remains to be clarified if the described differences in the parameters of the semen translate into differences in fertility rates between pairs who wish to conceive. In a retrospective study carried out on 5598 pregnant women, it was described that higher consumption of fast food and a lesser consumption of fruit were related to increases in time to pregnancy, revealing how a more Western diet can make conception difficult (Grieger et al. 2018). The importance of changes in some elements of the diet is evident in a study in which healthy men following a Western diet were supplemented with nuts (60 grams during 14 weeks), where improvements were achieved in overall sperm count, motility and morphology when compared to the control individuals (Salas-Huetos et al. 2018). In addition, a decrease in the degree of damage and fragmentation of sperm DNA was observed, which could justify the improvements observed.

An important factor related to diet, in addition to a sedentary lifestyle which can lead to infertility is overweight and obesity. These terms included an abnormal or excessive build-up of fat, are a global burden due to their effects on health and at the socio-sanitary level. In both women and men, overweight/obesity are related to a longer term for conception, in addition to being a risk issue for the development of chronic diseases that affect the evolution of pregnancy (Baheerati and Devi 2018). In this sense, in a study where a total of 159 infertile subjects (men and women) and 143 fertile subjects were analysed, it was shown that physical inactivity and having values of fat mass above the reference values were positively associated with infertility (Foucaut et al. 2019). From a fertility point of view, the influence of overweight and obesity is mainly due to endocrine mechanisms mostly derived from the functional alteration of the hypothalamic-pituitary-gonadal axis (HPG) (Broughton and Moley 2017; Hammoud et al. 2008). In the case of women, ovulatory dysfunction is the main reason of female infertility, which in many cases is a consequence of PCOS that usually occurs with obesity. In addition, obesity is related to a rise in the degree of oxidative stress and a state of subclinical chronic inflammation that also contributes to reduced fertility (Manna and Jain 2015). In this sense, it has been shown that following a Mediterranean diet significantly reduces the risk of insulin resistance and weight gain, which can favour fertility and the probability of pregnancy. Furthermore, the Mediterranean diet is characterized by being rich in bioactive compounds that can help reduce oxidative stress and the proinflammatory state. Thus, it should be emphasized that weight loss should be established as a strategy for obese people with infertility troubles, since it allows normalizing the hormonal balance and reducing the pro-oxidative state (McGrice and Porter 2017; Ricci et al. 2018). A meta-analysis examining weight loss interventions in infertile and overweight women found that participants assigned to a food regime and exercise intervention were more prone to achieving pregnancy and a live birth than women in control groups (Hunter et al. 2021). In another work, the combined effects of a 4-month nutritional and physical activity intervention were investigated in a randomized controlled trial including 263 healthy young men (Montano et al. 2022). After the intervention, an increase in the concentration of the sperm and progressive motility was observed, while the proportion of spermatozoa with abnormal morphology was reduced in the individuals of the intervention group. The authors conclude that lifestyle interventions are recommendable strategies to protect and improve semen quality in young men. Conversely, it has also been shown that low-weight women (body mass index <19 kg/m^2) take longer to become pregnant than normal-weight women, probably due to inhibition of the HPG axis due to chronic energy deficiency (Fontana and Della Torre 2016).

However, sometimes following a presumably healthy diet can hide risks that should be taken into consideration. In this sense, healthy diets are usually associated with an increase in foods of plant origin, which may contain chemical substances derived from pest control. Many of the pesticides used to act as endocrine disruptors can have effects even in minute concentrations. Results proposes that intake of foods from plants with considerable amounts of pesticides affects fertility in both men (testicular atrophy, reduced spermatogenesis, lower total sperm concentration and motility, etc.) and women (oestrous cycle defects, altered folliculogenesis, follicular atresia, etc.), in addition to being related to implantation defects, endometriosis, spontaneous abortions, foetal, and other birth defects (Bhardwaj et al. 2020; Mehrpour et al. 2014). Another factor to consider is the recommendation to increase the consumption of fish, especially oily fish with heart-healthy fats. However, these fish are usually large pelagic animals that accumulate methylmercury in their tissues, a compound with the capacity to produce neurotoxicity and reproductive disorders if consumed in excess (Mínguez-Alarcón et al. 2018). Alcohol consumption is another element to consider in relation to fertility. Although the Mediterranean dietary pattern comprises a modest amount of alcohol consumption (low-proof and non-distilled), its consumption is related to a decline in sperm quality and menstrual dysfunction, abortion, and ovulation disorders (Gabrielsen and Tanrikut 2016;

María del Mendoza-López et al. 2015). In this sense, it has been shown that patients with azoospermia associated with chronic alcohol consumption recovered normal semen parameters in less than 3 months (Sermondade et al. 2010).

4 Nutrients and Infertility

Nutrients play a pivotal role in fertility for both men and women (González-Rodríguez et al. 2018). Proper nutrition plays a crucial role in ensuring the optimal operation of the reproductive system and the overall health of individuals. Additionally, a balanced and varied diet, along with a healthy lifestyle, can help ensure an adequate intake of nutrients for overall reproductive health (Silva et al. 2019). While certain nutrients are important for reproductive health, it is essential to note that nutrient deficiencies have also been related to a higher risk of infertility (Bindari et al. 2013). Therefore, adhering to a nutritious and diverse diet holds significant promise for the reproductive well-being of both men and women. Nevertheless, beyond achieving a balance in macronutrients, the specific sources and quantities of individual micronutrients are crucial, as they dictate the impact of diet on fertility.

4.1 Macronutrients

When dealing with macronutrients (carbohydrates, proteins and lipids) not only is important their caloric content, but for a good state of health dietary macronutrient balance and quality must be considered (Senior et al. 2019). In this sense, Western diets also frequently comprise excessively refined sugars, with negative impacts on health by promoting insulin resistance and type II diabetes. Typically, Western diets exhibit an abundance of saturated fats (SFAs) and trans fatty acids, along with an unfavourable ratio of omega-6 to omega-3 fatty acids. Furthermore, maintaining a good balance of amino acids is vital for development, reproductive health, as well as cardiometabolic well-being and aging.

4.1.1 Fats

The prevention of fertility-related issues is significantly influenced by both the quantity and quality of consumed fatty acids. SFAs are abundant in foods derived from animals including processed meat, red meat, and whole-fat dairy products, may have adverse effects on fertility. On the contrary, monounsaturated and PUFAs are fats mainly present in vegetable foods and fish with positive effects on fertility (Benatta et al. 2020). However, among the fats, trans fatty acids, present mainly in fast and processed food, snacks and ready-made confectionery, tend to exert the most detrimental influence on fertility.

Trans fatty acids have proinflammatory characteristics and also induce insulin resistance, rising the probability of evolving type 2 diabetes or other metabolic disorders, including polycystic ovary syndrome (Çekici and Akdevelioğlu 2019) directly related to infertility. In this sense, in a study where 2416 healthy women were followed for a minimum of 1 year or until they got pregnant, a relationship was observed between an elevated consumption of trans fatty acids and a low omega-3 consumption with reduced fertility (Wise et al. 2018). Moreover, there is evidence indicating that the intake of trans fatty acids and SFAs is linked to a high risk of anovulatory infertility and endometriosis. Conversely, the consumption of PUFAs is related to elevated levels of luteal progesterone and estrogen, along with a reduced probability of anovulation (Chiu et al. 2018; Jurkiewicz-Przondziono et al. 2017). In the case on men, it has been reported negative effects of dietary trans and saturated fat on sperm parameters, specifically, it has been related to decreased total amounts and concentration of the sperm, and poor quality of the semen (Attaman et al. 2012; Jensen et al. 2013). On the one hand, the consumption of trans-saturated fat has been linked to reduced testosterone levels and decreased testicular volume, both indicative of compromised testicular function (Chavarro et al. 2014; Mínguez-Alarcón et al. 2017). Additionally, omega-3 PUFAs were related to greater seminal parameters and testicular volume (Attaman et al. 2012; Chavarro et al. 2014; Jensen et al. 2013; Mínguez-Alarcón et al. 2017). Specifically, it has been indicated that palmitic acid (C16:0) concentration was greater in infertile men in front of decreased concentrations of PUFAs (Gulaya et al. 2001; Khosrowbeygi and Zarghami 2007; Tavilani et al. 2006). As well as, researchers have shown that a main PUFA in human spermatozoa is docosahexaenoic acid (DHA, 22:6n-3). So, a lack of it in sperm cells is a common indication of infertility in men (Aksoy et al. 2006; Martínez-Soto et al. 2013). The importance of the fatty acid profile in men is due to the structure of the membrane of the sperm cell, a critical component for its proper functioning. This membrane has a fundamental part in the fertilization steps, including uptake, the reaction of the acrosome, and the fusion of sperm and oocytes. Long-chain PUFAs in the human sperm membrane cannot be synthesized endogenously, so they require to be obtained from the diet (vegetable oils, nuts, seeds, shellfish), which is why the consumption of such Food in men has been linked to a greater percentage of sperm with normal morphology and higher sperm parameters (Nassan et al. 2018a).

4.1.2 Proteins

Another important component of the diet for fertility is the origin and quality of the protein ingested. A deficiency in protein can play a role in infertility by affecting the necessary structural components. This is evident in reported reductions in the weight of the epididymis, testicles, and seminal vesicles, as well as a decrease in the circulating concentration of testosterone in men (Ferramosca and Zara 2022). However, the main problem usually comes from an excess protein intake, especially of animal origin, while that from vegetables is reduced. In accordance, it has been shown that elevated intake of animal proteins is associated with a greater risk of ovulatory disorders, while if proteins of vegetable origin predominate, fertility increases in women aged over 32 years. (Chavarro et al. 2008a). In another study carried out in healthy women without dietary restrictions, it was observed that those located in the tertile with the lowest intake of plant-based proteins presented lower luteal phase progesterone, higher follicle-stimulating hormone, and a higher risk of anovulation (Kim et al. 2021). In relation to men, an inverse relationship has been observed between higher consumption of animal protein and a poor quality of the sperm (Chavarro et al. 2011). Similarly, in a work where men with reduced semen quality and healthy men were analysed, it was observed that those with infertility ingested more protein of animal origin (Mendiola et al. 2009). One of the possible causes may be the differential response that proteins induce on insulin production. Thus, the insulin reaction is less pronounced when consuming proteins from vegetables compared to animal ones (Chen et al. 2020). In addition, the greater consumption of animal protein is parallel to the greater consumption of SFAs, which can favour infertility in a combined way.

Within protein consumption, the importance of red meat, and mainly the processed one, in relation to infertility should be highlighted. A research study conducted in the United States (US) explored the connection between meat consumption and semen quality parameters, revealing a robust inverse correlation between the intake of processed red meat and both total sperm count and ejaculate volume (Afeiche et al. 2014). Another study in China revealed that a higher consumption of processed meat may rise the risk of suffering asthenozoospermia (Liu et al. 2022), which is a primary factor contributing to the infertility in males, with absent or diminished sperm motility. (Tu et al. 2020). Nevertheless, consumption of untreated meat and raw vegetable might lessen asthenozoospermia risk (Liu et al. 2022). A recent study associated diets abundant in sweets, potatoes, red and processed meats and sweetened drinks with reduced semen quality, while the opposite effect had been seen with higher consumption of grains, poultry, fruits and vegetables and seafood (Gaskins and Chavarro 2018). Another study reported that women who intake >2 servings per day of red meat increased the risk to suffer endometriosis about 56% in comparison with those who consume ≤1 serving per week (Yamamoto et al. 2018).

Also highlight the benefits of fish protein as long as large pelagic are avoided due to the accumulation of toxins, especially heavy metals. In this sense, it has been observed that changing a meat service for a fish service increases fertility in assisted reproductive technology (Nassan et al. 2018b).

4.1.3 Carbohidrates

As previously mentioned, it seems that insulin plays a relevant role in infertility. Thus, lipid metabolism and above all the glycaemic index and load are dietary factors to consider (Skoracka et al. 2021). Continued consumption of high glycaemic index foods is associated with hyperinsulinemia, higher fasting glucose levels, and insulin resistance, which also leads to increased insulin-like growth-factor I (IGF-I) concentrations and androgens (Willis et al. 2020). All of this can contribute to female endocrine disturbances and, therefore, could be linked to a poor development of the oocyte and reduced fertility. In a prospective study, 18,555 fertile women who wished to become pregnant were followed for 8 years (Chavarro et al. 2009). During the development of the study, a total of 438 women presented ovulatory infertility. When analysing the results, a direct correlation was found between the overall consumption of carbohydrates and the dietary glycemic load with issues related to ovulatory function. In a meta-analysis focused on women with PCOS, low-glycaemic diets decreased insulin resistance, improved lipid profile, and reduced testosterone levels, indicating potential beneficial effects on infertility (Kazemi et al. 2021). In the case of men, it appears that an raised glycaemic load might be related to elevated rates of infertility due to increased oxidative stress and insulin resistance and decreased levels of testosterone. Insulin resistance can reduce glucose uptake and utilization by reducing ATP concentration, which can affect sperm motility (Dias et al. 2014). The rise of the levels in the oxidative stress biomarkers and the reduction in testosterone can affect the mitochondria, reducing their respiratory efficiency and, consequently, contributing to the decline in sperm motility (Ferramosca et al. 2013).

Within carbohydrates, various studies have shown the negative effects on both female and male fertility of the high consumption of sweetened drinks. In a study with 189 healthy young men, it was observed how the consumption of sugar-sweetened beverages was related to lesser motility of the sperm (Chiu et al. 2014). Similarly, in a prospective study of 3628 women planning pregnancy, those who consumed three or more sugar-sweetened soft drinks reported a pregnancy rate of 52% that was lower than women who did not consume these drinks (Hatch et al. 2012).

On the other hand, the increase in the intake of indigestible carbohydrates, such as fibre, seems to be associated with an increased probability of conception, although the results in this regard are few and further study is required. Its beneficial effects would come from its ability to reduce the rate of carbohydrate uptake, reducing blood glucose (Willis et al. 2020). In addition, the fermentation of the fibre can improve the quality of the microbiota and provide benefits for the health of the host (Makki et al. 2018). However, in one study it was shown that a high fibre intake reduced levels of sex hormones and increased anovulatory cycles (Gaskins et al. 2009).

4.2 Micronutrients

The existence of relations of specific foods and nutrients to fertility can give a significant idea of the probable mechanisms associated with diet quality to reproductive health. Micronutrients play an important role in different physiological

Table 10.2 Role of essential micronutrients on the healthy reproduction of both male and female

Micronutrient	Effects on male	Effects on female
Selenium	Vital for spermatogenesis, semen function and semen quality	Potential effect on the placenta Adequate development of foetus' nervous system
Zinc	Testosterone synthesis Improve androgen	Importance in oocyte maturation and development and on oocyte quality.
Iron	Necessary for spermatogenesis and spermatozoa metabolism Reduces sperm quality and sperm morphology and motility	Overload levels: Provoke reproductive disorders, such as endometriosis, preeclampsia and pregnancy complication
Folic acid	Essential role in spermatogenesis	Reduces risk of ovulatory infertility; improves chances of pregnancy
Vitamin E	Deficiency: Oligozoospermia, azoospermia, asthenozoospermia	Placenta development; the implantation of the fertilized egg
Vitamin A	Spermatogenesis; maturation of spermatozoa	Ovarian production; oocyte maturation; folliculogenesis
Vitamin C	Sperm motility; sperm morphology; sperm count; sperm quality	Improve ovaries fecundity function; prevent endometriosis
Vitamin D	Semen quality; increase of testosterone levels	Steroidogenesis of oestradiol and progesterone
Vitamin B12	Improve sperm quality	Essential for the placenta (development and functionality) Prevent spontaneous abortion

processes for the human being, including reproductive health. These are basic nutrients necessary in minimal amounts for development, typical growth, and overall well-being. They include vitamins (vitamins A, C, E, folate, B12), and minerals (iron, selenium, and zinc). Micronutrients contribute to the correct function of the reproductive system in both males and females, especially in pregnant women (Table 10.2) (Hosseini and Eslamian 2015; Tvrdá et al. 2021). However, the intake of dietary supplements should be approached cautiously, as an excessive consumption can have adverse effects on overall human health.

In males, certain micronutrients are crucial for maintaining sperm health, quality, and motility. Zinc, for instance, is necessary for the testosterone production and spermatogenesis. Deficiencies in zinc have been related to decreased sperm count and injured sperm function (Osadchuk et al. 2021). Other antioxidants, like vitamin C, selenium, and vitamin E, help protect sperm from oxidative damage, which can affect their viability and fertilizing capacity (Dimitriadis et al. 2023).

In females, micronutrients are essential for optimal hormonal balance, healthy ovulation, and a receptive environment for fertilization and implantation (Hosseini and Eslamian 2015). In the Nurses Health Cohort Study II, multivitamin supplementation is contrariwise associated with anovulatory dysfunction (Chavarro et al. 2008b). Iron, for example, has a pivotal role in the production of haemoglobin, which carries oxygen to the reproductive organs. Insufficient iron levels can lead to anaemia, which may negatively impact fertility (Kadhum and Al-Shammaree 2021). Folate, a B-vitamin, is vital for cell division and DNA synthesis, and its deficiency has been linked to fertility problems and neural tube defects in early pregnancy (Ebisch et al. 2007).

4.2.1 Selenium

Selenium (Se) is a crucial micronutrient, which deficiency or excess is related to gamete dysfunction and gonadal insufficiency in both females and males, developing to unsuccessful implantation, altered embryonic growth, and infertility (Dahlen et al. 2022). As well as, during pregnancy, Se deficiency or overload is related to gestational diabetes, miscarriage, foetal growth restriction, preeclampsia, and preterm birth (Dahlen et al. 2022). In women, Se is a part of selenoproteins displaying antioxidant features and is vital for spermatogenesis, DNA protection and the male fertility's preservation, since oxidative stress plays a vital role in male infertility. Se has antioxidative features and is vital for spermatogenesis and the preservation of fertility in man. Decreased levels of selenium have negative effects on human reproductive health, sperm function, semen quality and, consequently, fertility potency in men (Mirnamniha et al. 2019). Also, adequate ranks of Se are necessary for normal spermatogenesis and sperm maturation. Dietary selenium ought to be in optimum measure to avoid infertility and preserve reproductive function in males (Ahsan et al. 2014). Se is also essential due to its potential to decrease oxidative stress in the placenta, along with the proper development of the foetal nervous system (Habibi et al. 2020).

4.2.2 Zinc

Zinc is one of the metals that plays extensive physiological functions in hundreds of proteins and in enzymes for adequate development. Among its functions, zinc is a cofactor for several proteins participating in the antioxidant balance; therefore, it has an important function against oxidative stress. In women, zinc deficiency has been linked to decreased pregnancy rates and abnormal menstruation, as well as to require more time to get pregnant (Kohil et al. 2022). Moreover, it has been evidenced the significance of zinc in oocyte development and maturation, as well as, the effects of zinc on oocyte quality protecting against ROS. A deficiency in zinc results in dysfunctional antioxidant enzymes and an increment of ROS accumulation causing infertility diseases, such as endometriosis (Camp et al. 2023). In men, zinc has a diverse impact on both fertility and sperm function. It is essential for the adequate development of the testicles, normal spermatogenesis, and protection of the Leydig cells from damage through its antioxidant capabilities (Beigi Harchegani et al. 2020). Zinc insufficiency is also described to be related to oxidative damage and apoptosis of Leydig cells, a decrease in the production and secretion of testosterone and impaired spermatogenesis (Te et al. 2023).

4.2.3 Iron

Iron is a crucial micronutrient for maintaining human well-being, although deficient or excessive iron levels have been associated with dysregulated reproductive health. It is necessary for the formation of sperms and spermatozoa metabolism, so iron deficiency blocks spermatogenesis. Moreover, an excess of iron might stimulate male infertility by rising oxidative damage to sperm via the Fenton reaction. Increased levels of iron in seminal plasma have been related to low sperm quality and negative sperm morphology and motility, which are mediated by oxidative damage. Deficiency in iron together with the anaemia that can induce are widespread among women, in particular of reproductive age. In women, insufficient iron levels are associated with negative health outcomes that can impact both physical and emotional well-being. Furthermore, low iron levels can lead to impaired fertility and pathological alterations in vital reproductive tissues or organs, including the ovaries and endometrium. Among women experiencing infertility attributed to ovulatory disorders, the intake of iron supplements was associated with a markedly reduced risk of ovulatory infertility compared to those who did not take supplements (Chavarro et al. 2006). However, an iron overload in women may also provoke reproductive disorders, such as endometriosis, preeclampsia, and pregnancy complication (Li et al. 2023).

4.2.4 Folic Acid (Vitamin B9)

Water-soluble B9 vitamins are vital, and among them, there are the folate. Their importance is because of one-carbon synthesis and epigenetic processes. Moreover, folate is crucial for the synthesis of DNA, RNA, and proteins, making it essential during stages of quick cell development and proliferation, including the maturation of the germ cell and the pregnancy growth. Folates can be acquired in diet, especially in beans, legumes, whole grains, dark green leafy vegetables, juices, citrus fruits, and milk. Folic acid, as a synthetic form of folate, could be ingested as fortified foods or nutritional supplements. A low-quality diet or malabsorption of folate can be the origin of its deficiency in humans. Inadequate levels of key micronutrients crucial for folate metabolism, such as vitamins B2, B6, and B12, along with iron and zinc, can lead to the occurrence of functional folate deficiency (Laanpere et al. 2010). Folate plays an essential part in spermatogenesis, thus, folate deficiency decrease significantly the gene expression involved in spermatogenesis, resulting in a reduced sperm concentration (Yuan et al. 2017). Especially, a folic acid deficiency increases the methylation of significant genes causing sperm DNA damage (Wang et al. 2022). Folic acid also decreases the risk of ovulatory infertility and increases the chances of pregnancy in females. In this sense, elevated folate intake in women undergoing in vitro fertilisation was related to an augmented ratio of implantation and live births (Gaskins et al. 2014a).

4.2.5 Vitamin E

Vitamin E, alpha-tocopherol, is an important antioxidant present in semen that plays an essential role to prevent the functional competence of sperm cells exposed to an oxidative attack (Thérond et al. 1996). In fact, the concentration of vitamin E in seminal plasma is related to sperm motility and, consequently, insufficiency of vitamin E was observed in the semen of interfile men, evidencing oligozoospermia, azoospermia, and asthenozoospermia (Omu et al. 1999). Moreover, an association between decreased vitamin E in dysfunctional sperm and higher production of ROS was reported. A randomized controlled study conducted across multiple centres provided prospective evidence that vitamin E can enhance the percentage of motile sperm, sperm concentration, and the rate of natural pregnancy (Chen et al. 2012). A cross-sectional study conducted on females evidenced that an appropriate quantity of vitamin E in the follicle improves the possibility of oocyte maturation, which translates in a better reproductive outcome after intra-cytoplasmic sperm injection in infertile females (Ashraf et al. 2020). In females, vitamin E also participates in placenta development and the implantation of the fertilized egg (Mohd Mutalip et al. 2018).

4.2.6 Vitamin A

Vitamin A (retinol) has been known to be indispensable for spermatogenesis and for the spermatozoa maturation in the epididymis (Chung and Wolgemuth 2004). Specifically, retinol acts on both Sertoli cells and germ cells to induce in the undifferentiated spermatogonia their differentiation. All-trans-retinoic acid, which is the active form of vitamin A, binds to nuclear retinoic acid receptors (RARs), and retinoid X receptors (RXRs), that modulate the transcriptional activity of goal genes regulating continuous spermatogenesis (Wolgemuth and Chung 2007). Previous results reported that the total motility and number of sperm were positively correlated with serum levels of retinol. In females, vitamin A is important for oocyte maturation and ovarian steroid production. Also, retinoic acid may play an essential role in folliculogenesis (ovarian follicular growth) and the growth of oocyte competency, contributing to an effective fertilization (Best et al. 2015). Specifically, retinoic acid is necessary for the antral follicles for granulosa cell differentiation modulated by FSH and the reproductive capacity of the ovaries.

4.2.7 Vitamin C

Vitamin C, also known as ascorbic acid, is a non-enzymatic antioxidant that detoxifies free radicals in the aqueous phase. Vitamin C is regarded the main antioxidant in seminal plasma, eliminating ROS but also serving to recycle oxidized vitamin E (Luck et al. 1995). There are some results that evidence that supplementation of vitamin C in infertile men may help sperm motility, sperm morphology, sperm count, and as well as semen quality (Akmal et al. 2006). However, other studies have described slight or no outcome of vitamin C on seminal parameters or as protection against DNA fragmentation, however when it is used together with other antioxidants like selenium or vitamin E, do obvious recoveries occur. In women, it has been shown that treatment with vitamin C had a significant effect on improving the fecundity function of ovaries, as well as, prevent to endometriosis (Hoorsan et al. 2022). Also, a cohort study reported prospectively a positive relationship between vitamin C consumption and fertilization taxes in couples experiencing assisted reproductive technology (Li et al. 2019).

4.2.8 Vitamin D

Vitamin D is an important micronutrient whose function is promoting bone mineralization and preserving calcium and phosphorus homeostasis (Gil et al. 2018). Moreover, it has been evidenced that vitamin D also acts as a sex steroid hormone, the typical regulator of reproduction for humankind which can regulate reproductive processes in men and women. In fact, the vitamin D receptor (VDR) has been found in the reproductive tissues of women and men. In women, this vitamin may impact in the steroidogenesis of sex hormones (progesterone and oestradiol) but increased levels could be related to endometriosis. However, various cohort studies have not found any clear association between vitamin D plasma levels with the risk of ovulatory infertility, chances of conceiving and risk of spontaneous abortion (Gaskins and Chavarro 2018). Instead, supplementation of vitamin D may enhance metabolic disturbances and menstrual frequency in women suffering from PCOS and also increase pregnancy rates after in vitro fertilisation. It is reported a direct effect on the endometrium by the vitamin D on in vitro fertilisation. Regarding men, vitamin D is found to be positively associated with semen quality, androgen status and an increment of testosterone levels (Lerchbaum and Obermayer-Pietsch 2012).

4.2.9 Vitamin B12

Vitamin B12, cobalamin, is essential for human growth, development acting as a cofactor in fatty acid synthesis and DNA. Moreover, this vitamin can protect DNA from the damage induced by the free radicals and decrease homocysteine toxicity (Hosseinabadi et al. 2020). In males, it has been shown that supplements with vitamin B12 could enhance sperm quality, increasing sperm amount and sperm motility and decreasing the damage in the sperm DNA (Hosseinabadi et al. 2020). A vitamin B12 supplement is capable to induce the activity of sperm via rising mitotic activity in Sertoli cells. In females, vitamin B12 prevents spontaneous abortion and congenital defects due to its essential role in the growth and functionality of the placenta. For this reason, an insufficient intake of vitamin B12 may cause reduced foetal growth and, consequently, vegetarian women may need supplementation of vitamin B12 (Hovdenak and Haram 2012).

4.3 *Caffeine*

Caffeine is a natural chemical located in drinks made from coffee, but also many manufacturers add it to their sodas and energy drinks. Caffeine has the power to stimulate the brain and the central nervous system, avoiding being sleepy in order to keep active. Caffeine consumption might have adverse effects on the reproduction in males; nevertheless, the studied parameters related to fertility and semen are unreliable and unsatisfying. However, it has been suggested that consuming coffee (<6 cups per day) can to reduce fertility in pairs (Hassan and Killick 2004). Altogether, it is necessary design well works, in which the selection of the participants, semen analysis and lifestyle habits are predefined to discover reliable effects of the caffeine on male fertility and semen parameters (Ricci et al. 2017). In women, the results are not clear either, although there seems to be an increased risk of spontaneous abortion if the women take caffeine previous to the pregnancy more than 400 mg per day (Gaskins et al. 2018).

5 Conclusions

Infertility is a complex and multifactorial issue that can affect both men and women. Fertility disorders have become one of the main reproductive health problems, especially in developed countries, where the voluntary delay in the reproductive plan has led to a progressive increase in maternity age, as well as an increase in the incidence of reproductive dysfunctions associated with germinal aging. While clinical conditions such as ovulatory, testicular and hormonal alterations are the primary causes of infertility, lifestyle factors like avoiding tobacco smoking, drugs, alcohol consumption, sedentarism and diet can also impact fertility. In fact, diverse dietary and lifestyle components might help diminishing the risk of fertility alterations in the reproductive-age population and might also be an effective approach for women and men already experiencing infertility. Considering that prevention is the most effective tactic to face up the fertility disorders in human couples; consuming a diet high in vegetables, fruits, whole grains, and healthy fats and avoiding processed and saturated fats may improve fertility outcomes in both men and women. Moreover, maintaining body weight through adequate nutrition and also promoting physical activity is key to improving fertility in men and women. More standardized integration of nutrition advice into treatment delivery for infertility is required, since there are solid described results that following eating habits that are healthy and adopted before conception in both men and women have beneficial effects on fertility.

Acknowledgements The authors are supported by the Programme of Promotion of Biomedical Research and Health Sciences, Instituto de Salud Carlos III (CIBEROBN CB12/03/30038).

Glossary

Assisted reproductive technology Medical procedures and interventions used to assist couples or individuals in achieving pregnancy, such as in vitro fertilization (IVF), intracytoplasmic sperm injection (ICSI), and embryo transfer.

Asthenozoospermia Asthenozoospermia refers to a condition characterized by reduced sperm motility or poor sperm movement. It is a common form of male infertility where the majority of sperm present in the semen exhibit decreased motility, making it difficult for them to swim effectively and reach the egg for fertilization.

Azoospermia Azoospermia is a condition characterized by the absence of sperm in the ejaculated semen is known as azoospermia. This can occur due to either the man's inability to produce sperm or the obstruction preventing the sperm from entering the semen.

Endocrine disruptors Substances, often found in certain chemicals or environmental pollutants, that can interfere with the normal functioning of the endocrine system, leading to adverse effects on hormone regulation and various aspects of health and development.

Endometriosis Chronic condition characterized by the presence of endometrial-like tissue outside the uterus, causing pain and potential fertility issues.

Folliculogenesis Folliculogenesis refers to the process of the development and maturation of ovarian follicles in the female reproductive system. It involves a series of complex cellular and hormonal interactions that occur in the ovary.

Mediterranean diet It is a pattern of healthy eating originating in the Mediterranean area that is characterized by a great diversity of foods and a predominance of foods of vegetable origin and consumption of fresh foods or with minimal transformations.

Oligozoospermia Oligozoospermia or oligospermia is a sperm alteration which consist of a decrease in the number of spermatozoa in the ejaculated semen count.

Polycystic ovarian disease Hormonal disorder characterized by multiple cysts on the ovaries, menstrual irregularities, excess androgen production, and various symptoms including infertility, acne, and hirsutism.

Polyunsaturated fatty acids (PUFAs) Dietary fat that contains more than one double bond in their chemical structure, and are essential for various physiological functions in the body, including cell membrane integrity, hormone production, and cardiovascular health.

Processed food Products modified by the addition or introduction of substances (salt, sugar, oil, preservatives and/or additives) that change the nature of the original foods, in order to prolong their duration, make them more pleasant or attractive.

Reactive oxygen species (ROS) Chemically reactive molecules containing oxygen that can be generated in cells as by-products of normal cellular metabolism, and at high levels, can cause oxidative stress and damage to cells, proteins, and DNA, potentially contributing to various diseases and aging processes.

Steroidogenesis Steroidogenesis refers to the process of synthesis and production of steroid hormones in the body. Steroid hormones are a class of lipid molecules derived from cholesterol and include hormones such as cortisol, aldosterone, progesterone, oestrogen, and testosterone.

Trans fatty acids Unhealthy fatty acids containing at least one double bond in the trans configuration that are produced primarily when vegetable oil is hydrogenated to increase its plasticity and chemical stability for further food processing.

Varicocele Condition where the veins in the scrotum become enlarged and twisted, potentially leading to testicular discomfort and impaired fertility.

References

Acharya S, Gowda CR. Lifestyle factors associated with infertility in a rural area: a cross-sectional study. Int J Med Sci Public Health. 2017;6(3):502–7.

Adashi EY. Clomiphene citrate: mechanism(s) and site(s) of action--a hypothesis revisited. Fertil Steril. 1984;42(3):331–47.

Afeiche MC, Williams PL, Gaskins AJ, Mendiola J, Jørgensen N, Swan SH, et al. Meat intake and reproductive parameters among young men. Epidemiology. 2014;25(3):323–30.

Agarwal A, Majzoub A, Baskaran S, Selvam MKP, Cho CL, Henkel R, et al. Sperm DNA fragmentation: A new guideline for clinicians. World J Mens Health. 2020;38(4):412–71.

Ahmed A, Bello A, Mbibu N, Maitama H, Kalayi G. Epidemiological and aetiological factors of male infertility in northern Nigeria. Niger J Clin Pr. 2010;13(2):205–9.

Ahsan U, Kamran Z, Raza I, Ahmad S, Babar W, Riaz MH, et al. Role of selenium in male reproduction – a review. Anim Reprod Sci. 2014;146(1–2):55–62.

Aitken RJ. Impact of oxidative stress on male and female germ cells: implications for fertility. Reproduction. 2020;159(4):R189–201.

Aitken RJ, Baker MA. The role of genetics and oxidative stress in the etiology of male infertility – a unifying hypothesis? Front Endocrinol (Lausanne). 2020;11:581838.

Ajayi AF, Akhigbe RE. The physiology of male reproduction: impact of drugs and their abuse on male fertility. Andrologia. 2020;52(9):e13672.

Akmal M, Qadri JQ, Al-Waili NS, Thangal S, Haq A, Saloom KY. Improvement in human semen quality after oral supplementation of vitamin C. J Med Food. 2006;9(3):440–2.

Aksoy Y, Aksoy H, Altinkaynak K, Aydin HR, Özkan A. Sperm fatty acid composition in subfertile men. Prostaglandins Leukot Essent Fatty Acids. 2006;75(2):75–9.

Anderson K, Norman RJ, Middleton P. Preconception lifestyle advice for people with subfertility. Cochrane Database Syst Rev. 2011;4(4):CD008189.

Antoniassi MP, Intasqui P, Camargo M, Zylbersztejn DS, Carvalho VM, Cardozo KHM, et al. Analysis of the functional aspects and seminal plasma proteomic profile of sperm from smokers. BJU Int. 2016;118(5):814–22.

Ashraf M, Mustansir F, Baqir SM, Alam F, Rehman R. Changes in vitamin E levels as a marker of female infertility. J Pak Med Assoc. 2020;70(10):1762–6.

Attaman JA, Toth TL, Furtado J, Campos H, Hauser R, Chavarro JE. Dietary fat and semen quality among men attending a fertility clinic. Hum Reprod. 2012;27(5):1466–74.

Ávila C, Vinay JI, Arese M, Saso L, Rodrigo R. Antioxidant intervention against male infertility: time to design novel strategies. Biomedicine. 2022;10(12):3058.

Baheerati MM, Devi RG. Obesity in relation to infertility. Res J Pharm Technol. 2018;11(7):3183–5. A & V Publications

Baird DD, Wilcox AJ, Weinberg CR. Use of time to pregnancy to study environmental exposures. Am J Epidemiol. 1986;124(3):470–80.

Barati E, Nikzad H, Karimian M. Oxidative stress and male infertility: current knowledge of pathophysiology and role of antioxidant therapy in disease management. Cell Mol Life Sci. 2020;77(1):93–113.

Bedoschi G, Navarro PA, Oktay K. Chemotherapy-induced damage to ovary: mechanisms and clinical impact. Future Oncol. 2016;12(20):2333–44.

Beigi Harchegani A, Dahan H, Tahmasbpour E, Bakhtiari Kaboutaraki H, Shahriary A. Effects of zinc deficiency on impaired spermatogenesis and male infertility: the role of oxidative stress, inflammation and apoptosis. Hum Fertil (Camb). 2020;23(1):5–16.

Benatta M, Kettache R, Buchholz N, Trinchieri A. The impact of nutrition and lifestyle on male fertility. Arch Ital Urol Androl. 2020;92(2):121–31.

Best MW, Wu J, Pauli SA, Kane MA, Pierzchalski K, Session DR, et al. A role for retinoids in human oocyte fertilization: regulation of connexin 43 by retinoic acid in cumulus granulosa cells. Mol Hum Reprod. 2015;21(6):527–34.

Bhardwaj JK, Mittal M, Saraf P, Kumari P. Pesticides induced oxidative stress and female infertility: a review. Toxin Rev. 2020;39(1):1–13.

Biggs WS, Marks ST. Diagnosis and management of adnexal masses. Am Fam Physician. 2016;93(8):676–81.

Bindari YR, Shrestha S, Shrestha N, Gaire TN. Effects of nutrition on reproduction: a review. Adv Appl Sci Res. 2013;4(1):421–9.

Boivin J, Bunting L, Collins JA, Nygren KG. International estimates of infertility prevalence and treatment-seeking: potential need and demand for infertility medical care. Hum Reprod. 2007;22(6):1506–12.

Bojesen A, Juul S, Gravholt CH. Prenatal and postnatal prevalence of Klinefelter syndrome: a national registry study. J Clin Endocrinol Metab. 2003;88(2):622–6.

Bosteels J, van Wessel S, Weyers S, Broekmans FJ, D'Hooghe TM, Bongers MY, et al. Hysteroscopy for treating subfertility associated with suspected major uterine cavity abnormalities. Cochrane Database Syst Rev. 2018;12(12):CD009461.

Broughton DE, Moley KH. Obesity and female infertility: potential mediators of obesity's impact. Fertil Steril. 2017;107(4):840–7.

Calogero AE, Cannarella R, Agarwal A, Hamoda TAA, Rambhatla A, Saleh R, et al. The renaissance of male infertility management in the golden age of andrology. World J Mens Health. 2023;41(2):237.

Camp OG, Bembenek JN, Goud PT, Awonuga AO, Abu-Soud HM. The implications of insufficient zinc on the generation of oxidative stress leading to decreased oocyte quality. Reprod Sci. 2023;30(7):2069–78.

Çekici H, Akdevelioğlu Y. The association between trans fatty acids, infertility and fetal life: a review. Hum Fertil (Camb). 2019;22(3):154–63.

Chavarro JE, Schlaff WD. Introduction: impact of nutrition on reproduction: an overview. Fertil Steril. 2018;110(4):557–9.

Chavarro JE, Rich-Edwards JW, Rosner BA, Willett WC. Iron intake and risk of ovulatory infertility. Obstet Gynecol. 2006;108(5):1145–52.

Chavarro JE, Rich-Edwards JW, Rosner BA, Willett WC. Diet and lifestyle in the prevention of ovulatory disorder infertility. Obstet Gynecol. 2007;110(5):1050–8.

Chavarro JE, Rich-Edwards JW, Rosner BA, Willett WC. Protein intake and ovulatory infertility. Am J Obstet Gynecol. 2008a;198(2):210.e1–7.

Chavarro JE, Rich-Edwards JW, Rosner BA, Willett WC. Use of multivitamins, intake of B vitamins, and risk of ovulatory infertility. Fertil Steril. 2008b;89(3):668–76.

Chavarro JE, Rich-Edwards JW, Rosner BA, Willett WC. A prospective study of dietary carbohydrate quantity and quality in relation to risk of ovulatory infertility. Eur J Clin Nutr. 2009;63(1):78–86.

Chavarro JE, Furtado J, Toth TL, Ford J, Keller M, Campos H, et al. Trans-fatty acid levels in sperm are associated with sperm concentration among men from an infertility clinic. Fertil Steril. 2011;95(5):1794–7.

Chavarro JE, Mínguez-Alarcón L, Mendiola J, Cutillas-Tolín A, López-Espín JJ, Torres-Cantero AM. Trans fatty acid intake is inversely related to total sperm count in young healthy men. Hum Reprod. 2014;29(3):429–40.

Chen X-F, Li Z, Ping P, Dai J-C, Zhang F-B, Shang X-J. Efficacy of natural vitamin E on oligospermia and asthenospermia: a prospective multi-centered randomized controlled study of 106 cases. Zhonghua Nan Ke Xue. 2012;18(5):428–31.

Chen Z, Franco OH, Lamballais S, Ikram MA, Schoufour JD, Muka T, et al. Associations of specific dietary protein with longitudinal insulin resistance, prediabetes and type 2 diabetes: the Rotterdam study. Clin Nutr. 2020;39(1):242–9.

Chiu YH, Afeiche MC, Gaskins AJ, Williams PL, Mendiola J, Jorgensen N, et al. Sugar-sweetened beverage intake in relation to semen quality and reproductive hormone levels in young men. Hum Reprod. 2014;29(7):1575–84.

Chiu YH, Chavarro JE, Souter I. Diet and female fertility: doctor, what should I eat? Fertil Steril. 2018;110(4):560–9.

Cho CL, Agarwal A. Role of sperm DNA fragmentation in male factor infertility: a systematic review. Arab J Urol. 2017;16(1):21–34.

Christou F, Pitteloud N, Gomez F. The induction of ovulation by pulsatile administration of GnRH: an appropriate method in hypothalamic amenorrhea. Gynecol Endocrinol. 2017;33(8):598–601.

Chung SSW, Wolgemuth DJ. Role of retinoid signaling in the regulation of spermatogenesis. Cytogenet Genome Res. 2004;105(2–4):189–202.

Comhaire F, Vermeulen A. Plasma testosterone in patients with varicocele and sexual inadequacy. J Clin Endocrinol Metab. 1975;40(5):824–9.

Cox CM, Thoma ME, Tchangalova N, Mburu G, Bornstein MJ, Johnson CL, et al. Infertility prevalence and the methods of estimation from 1990 to 2021: a systematic review and meta-analysis. Hum Reprod Open. 2022;2022(4):hoac051.

Cutillas-Toĺn A, Manguez-Alarcón L, Mendiola J, López-Espn JJ, Jørgensen N, Navarrete-Munoz EM, et al. Mediterranean and western dietary patterns are related to markers of testicular function among healthy men. Hum Reprod. 2015;30(12):2945–55.

Dahlen CR, Reynolds LP, Caton JS. Selenium supplementation and pregnancy outcomes. Front Nutr. 2022;9:1011850.

de Geyter C, Wyns C, Calhaz-Jorge C, de Mouzon J, Ferraretti AP, Kupka M, et al. 20 years of the European IVF-monitoring consortium registry: what have we learned? A comparison with registries from two other regions. Hum Reprod. 2020;35(12):2832–49.

Devine PJ, Perreault SD, Luderer U. Roles of reactive oxygen species and antioxidants in ovarian toxicity. Biol Reprod. 2012;86(2):27.

Dias TR, Alves MG, Silva BM, Oliveira PF. Sperm glucose transport and metabolism in diabetic individuals. Mol Cell Endocrinol. 2014;396(1–2):37–45.

Dimitriadis F, Borgmann H, Struck JP, Salem J, Kuru TH. Antioxidant supplementation on male fertility—a systematic review. Antioxidants. 2023;12(4):836.

du Plessis SS, Agarwal A, Syriac A. Marijuana, phytocannabinoids, the endocannabinoid system, and male fertility. J Assist Reprod Genet. 2015;32(11):1575–88.

Ebisch IMW, Thomas CMG, Peters WHM, Braat DDM, Steegers-Theunissen RPM. The importance of folate, zinc and antioxidants in the pathogenesis and prevention of subfertility. Hum Reprod Update. 2007;13(2):163–74.

El Mahgoub S. Galactorrhea and the defective luteal phase of the menstrual cycle. Int J Gynaecol Obstet. 1978;16(2):124–7.

Esposito K, Ciotola M, Giugliano F, Schisano B, Autorino R, Iuliano S, et al. Mediterranean diet improves sexual function in women with the metabolic syndrome. Int J Impot Res. 2007;19(5):486–91.

Fabiani C, Ferrante MG, Meneghini C, Licata E, Paciotti G, Gallo M, et al. Female fertility preservation: impact of cancer on ovarian function and oocyte quality. Int J Gynaecol Obstet. 2022;156(1):166–71.

Ferramosca A, Zara V. Diet and male fertility: the impact of nutrients and antioxidants on sperm energetic metabolism. Int J Mol Sci. 2022;23(5):2542.

Ferramosca A, Pinto Provenzano S, Montagna DD, Coppola L, Zara V. Oxidative stress negatively affects human sperm mitochondrial respiration. Urology. 2013;82(1):78–83.

Finelli R, Mottola F, Agarwal A. Impact of alcohol consumption on male fertility potential: A narrative review. Int J Environ Res Public Health. 2021;19(1):328.

Fonseca BM, Rebelo I. Cannabis and cannabinoids in reproduction and fertility: where we stand. Reprod Sci. 2022;29(9):2429–39.

Fontana R, Della Torre S. The deep correlation between energy metabolism and reproduction: a view on the effects of nutrition for women fertility. Nutrients. 2016;8(2):87.

Foucaut AM, Faure C, Julia C, Czernichow S, Levy R, Dupont C. Sedentary behavior, physical inactivity and body composition in relation to idiopathic infertility among men and women. PloS One. 2019;14(4):e0210770.

Gabrielsen JS, Tanrikut C. Chronic exposures and male fertility: the impacts of environment, diet, and drug use on spermatogenesis. Andrology. 2016;4(4):648–61.

Gajbhiye R, Fung JN, Montgomery GW. Complex genetics of female fertility. NPJ Genom Med. 2018;3(1):29.

Garba-Alkali AE, Adesiyun AG, Randawa AJ. Semen profile of male partners of women attending infertility clinic in Zaria, Nigeria. Trop J Obstet Gynaecol. 2018;35(3):256–60.

Gaskins AJ, Chavarro JE. Diet and fertility: a review. Am J Obstet Gynecol. 2018;218(4):379–89.

Gaskins AJ, Mumford SL, Zhang C, Wactawski-Wende J, Hovey KM, Whitcomb BW, et al. Effect of daily fiber intake on reproductive function: the BioCycle study. Am J Clin Nutr. 2009;90(4):1061–9.

Gaskins AJ, Afeiche MC, Wright DL, Toth TL, Williams PL, Gillman MW, et al. Dietary folate and reproductive success among women undergoing assisted reproduction. Obstet Gynecol. 2014a;124(4):801–9.

Gaskins AJ, Rich-Edwards JW, Hauser R, Williams PL, Gillman MW, Penzias A, et al. Prepregnancy dietary patterns and risk of pregnancy loss. Am J Clin Nutr. 2014b;100(4):1166–72.

Gaskins AJ, Rich-Edwards JW, Williams PL, Toth TL, Missmer SA, Chavarro JE. Pre-pregnancy caffeine and caffeinated beverage intake and risk of spontaneous abortion. Eur J Nutr. 2018;57(1):107–17.

Gdanska P, Drozdowicz-Jastrzebska E, Grzechocinska B, Radziwon-Zaleska M, Wegrzyn P, Wielgos M. Anxiety and depression in women undergoing infertility treatment. Ginekol Pol. 2017;88(2):109–12.

Gil Á, Plaza-Diaz J, Mesa MD. Vitamin D: classic and novel actions. Ann Nutr Metab. 2018;72(2):87–95.

González-Rodríguez LG, López-Sobaler AM, Sánchez JMP, Ortega RM. [Nutrition and fertility]. Nutr Hosp. 2018;35(Spec No 6):7–10.

Grieger JA, Grzeskowiak LE, Bianco-Miotto T, Jankovic-Karasoulos T, Moran LJ, Wilson RL, et al. Pre-pregnancy fast food and fruit intake is associated with time to pregnancy. Hum Reprod. 2018;33(6):1063–70.

Gulaya NM, Margitich VM, Govseeva NM, Klimashevsky VM, Gorpynchenko II, Boyko MI. Phospholipid composition of human sperm and seminal plasma in relation to sperm fertility. Arch Androl. 2001;46(3):169–75.

Habibi N, Grieger JA, Bianco-Miotto T. A review of the potential interaction of selenium and iodine on placental and child health. Nutrients. 2020;12(9):1–22.

Hammoud AO, Gibson M, Peterson CM, Meikle AW, Carrell DT. Impact of male obesity on infertility: a critical review of the current literature. Fertil Steril. 2008;90(4):897–904.

Hassan MAM, Killick SR. Negative lifestyle is associated with a significant reduction in fecundity. Fertil Steril. 2004;81(2):384–92.

Hatch EE, Wise LA, Mikkelsen EM, Christensen T, Riis AH, Sørensen HT, et al. Caffeinated beverage and soda consumption and time to pregnancy. Epidemiology. 2012;23(3):393–401.

Hoorsan H, Simbar M, Tehrani FR, Fathi F, Mosaffa N, Riazi H, et al. The effectiveness of antioxidant therapy (vitamin C) in an experimentally induced mouse model of ovarian endometriosis. Womens Health (Lond). 2022:18.

Hosseinabadi F, Jenabi M, Ghafarizadeh AA, Yazdanikhah S. The effect of vitamin B12 supplement on post-thaw motility, viability and DNA damage of human sperm. Andrologia. 2020;52(11)

Hosseini B, Eslamian G. Association of micronutrient intakes with female infertility: review of recent evidence. Thrita Brieflands. 2015;4(1)

Hotaling J, Carrell DT. Clinical genetic testing for male factor infertility: current applications and future directions. Andrology. 2014;2(3):339–50.

Hovdenak N, Haram K. Influence of mineral and vitamin supplements on pregnancy outcome. Eur J Obstet Gynecol Reprod Biol. 2012;164(2):127–32.

Hunter E, Avenell A, Maheshwari A, Stadler G, Best D. The effectiveness of weight-loss lifestyle interventions for improving fertility in women and men with overweight or obesity and infertility: a systematic review update of evidence from randomized controlled trials. Obes Rev. 2021;22(12)

Jenkins TG, James ER, Alonso DF, Hoidal JR, Murphy PJ, Hotaling JM, et al. Cigarette smoking significantly alters sperm DNA methylation patterns. Andrology. 2017;5(6):1089–99.

Jensen TK, Heitmann BL, Jensen MB, Halldorsson TI, Andersson AM, Skakkebœk NE, et al. High dietary intake of saturated fat is associated with reduced semen quality among 701 young Danish men from the general population. Am J Clin Nutr. 2013;97(2):411–8.

Jensen TK, Gottschau M, Madsen JOB, Andersson AM, Lassen TH, Skakkebæk NE, et al. Habitual alcohol consumption associated with reduced semen quality and changes in reproductive hormones; a cross-sectional study among 1221 young Danish men. BMJ Open. 2014;4(9)

Joffe M. Male- and female-mediated reproductive effects of occupation: the use of questionnaire methods. J Occup Med. 1989;31(12):974–9.

Jurewicz J, Dziewirska E, Radwan M, Hanke W. Air pollution from natural and anthropic sources and male fertility. Reprod Biol Endocrinol. 2018;16(1)

Jurkiewicz-Przondziono J, Lemm M, Kwiatkowska-Pamula A, Ziółłko E, Wójtowicz MK. Influence of diet on the risk of developing endometriosis. Ginekol Pol. 2017;88(2):96–102.

Kadhum BS, Al-Shammaree SAW. Serum iron status in association with pregnancy outcomes in infertile women undergoing IVF/ICSI. J Phys Conf Ser. 2021;1853(1):12025.

Karayiannis D, Kontogianni MD, Mendorou C, Douka L, Mastrominas M, Yiannakouris N. Association between adherence to the Mediterranean diet and semen quality parameters in male partners of couples attempting fertility. Hum Reprod. 2017;32(1):215–22.

Karayiannis D, Kontogianni MD, Mendorou C, Mastrominas M, Yiannakouris N. Adherence to the Mediterranean diet and IVF success rate among non-obese women attempting fertility. Hum Reprod. 2018;33(3):494–502.

Kazemi M, Hadi A, Pierson RA, Lujan ME, Zello GA, Chilibeck PD. Effects of dietary glycemic index and glycemic load on cardiometabolic and reproductive profiles in women with polycystic ovary syndrome: a systematic review and meta-analysis of randomized controlled trials. Adv Nutr. 2021;12(1):161–78.

Khera M, Lipshultz LI. Evolving approach to the varicocele. Urol Clin North Am. 2008;35(2):183–9.

Khosrowbeygi A, Zarghami N. Fatty acid composition of human spermatozoa and seminal plasma levels of oxidative stress biomarkers in subfertile males. Prostaglandins Leukot Essent Fatty Acids. 2007;77(2):117–21.

Kim K, Yisahak SF, Nobles CJ, Andriessen VC, Devilbiss EA, Sjaarda LA, et al. Low intake of vegetable protein is associated with altered ovulatory function among healthy women of reproductive age. J Clin Endocrinol Metab. 2021;106(7):E2600–12.

Kohil A, Chouliaras S, Alabduljabbar S, Lakshmanan AP, Ahmed SH, Awwad J, et al. Female infertility and diet, is there a role for a personalized nutritional approach in assisted reproductive technologies? A narrative review. Front Nutr. 2022;9:927972.

Kumosani TA, Elshal MF, Al-Jonaid AA, Abduljabar HS. The influence of smoking on semen quality, seminal microelements and Ca2+-ATPase activity among infertile and fertile men. Clin Biochem. 2008;41(14–15):1199–203.

Künzle R, Mueller MD, Hänggi W, Birkhäuser MH, Drescher H, Bersinger NA. Semen quality of male smokers and nonsmokers in infertile couples. Fertil Steril. 2003;79(2):287–91.
Laanpere M, Altmäe S, Stavreus-Evers A, Nilsson TK, Yngve A, Salumets A. Folate-mediated one-carbon metabolism and its effect on female fertility and pregnancy viability. Nutr Rev. 2010;68(2):99–113.
Legro RS, Brzyski RG, Diamond MP, Coutifaris C, Schlaff WD, Casson P, et al. Letrozole versus clomiphene for infertility in the polycystic ovary syndrome. N Engl J Med. 2014;371(2):119–29.
Lerchbaum E, Obermayer-Pietsch B. Vitamin D and fertility: a systematic review. Eur J Endocrinol. 2012;166(5):765–78.
Li MC, Chiu YH, Gaskins AJ, Mínguez-Alarcón L, Nassan FL, Williams PL, et al. Men's intake of vitamin C and β-carotene is positively related to fertilization rate but not to live birth rate in couples undergoing infertility treatment. J Nutr. 2019;149(11):1977–84.
Li X, Duan X, Tan D, Zhang B, Xu A, Qiu N, et al. Iron deficiency and overload in men and woman of reproductive age, and pregnant women. Reprod Toxicol. 2023;118:108381.
Liu YS, Zhang YX, Bin WX, Wu QJ, Liu FH, Pan BC, et al. Associations between meat and vegetable intake, cooking methods, and asthenozoospermia: a hospital-based case-control study in China. Nutrients. 2022;14(9):1956.
Lord T, Martin JH, Aitken RJ. Accumulation of electrophilic aldehydes during postovulatory aging of mouse oocytes causes reduced fertility, oxidative stress, and apoptosis. Biol Reprod. 2015;92(2):33.
Lotti F, Maggi M. Sexual dysfunction and male infertility. Nat Rev Urol. 2018;15(5):287–307.
Luck MR, Jeyaseelan I, Scholes RA. Ascorbic acid and fertility. Biol Reprod. 1995;52(2):262–6.
Macut D, Bjekić-Macut J, Rahelić D, Doknić M. Insulin and the polycystic ovary syndrome. Diabetes Res Clin Pract. 2017;130:163–70.
Makki K, Deehan EC, Walter J, Bäckhed F. The impact of dietary fiber on gut microbiota in host health and disease. Cell Host Microbe. 2018;23(6):705–15.
Manna P, Jain SK. Obesity, oxidative stress, adipose tissue dysfunction, and the associated health risks: causes and therapeutic strategies. Metab Syndr Relat Disord. 2015;13(10):423–44.
María del Mendoza-López M, Reyes-Martin KI, Gutiérrez-Gómez YY. Dietary intake and infertility: a review. FASEB J. 2015;29:512–90.
Marinelli S, Napoletano G, Straccamore M, Basile G. Female obesity and infertility: outcomes and regulatory guidance. Acta Biomed. 2022;93(4):e2022278.
Martínez-Soto JC, Landeras J, Gadea J. Spermatozoa and seminal plasma fatty acids as predictors of cryopreservation success. Andrology. 2013;1(3):365–75.
Mascarenhas MN, Flaxman SR, Boerma T, Vanderpoel S, Stevens GA. National, regional, and global trends in infertility prevalence since 1990: a systematic analysis of 277 health surveys. PLoS Med. 2012;9(12):e1001356.
McGrath IM, Mortlock S, Montgomery GW. Genetic regulation of physiological reproductive lifespan and female fertility. Int J Mol Sci. 2021;22(5):1–13.
McGrice M, Porter J. The effect of low carbohydrate diets on fertility hormones and outcomes in overweight and obese women: a systematic review. Nutrients. 2017;9(3):204.
Mehrpour O, Karrari P, Zamani N, Tsatsakis AM, Abdollahi M. Occupational exposure to pesticides and consequences on male semen and fertility: a review. Toxicol Lett. 2014;230(2):146–56.
Meldrum DR, Morris MA, Gambone JC. Obesity pandemic: causes, consequences, and solutions-but do we have the will? Fertil Steril. 2017;107(4):833–9.
Mendiola J, Torres-Cantero AM, Moreno-Grau JM, Ten J, Roca M, Moreno-Grau S, et al. Food intake and its relationship with semen quality: a case-control study. Fertil Steril. 2009;91(3):812–8.
Mier-Cabrera J, Jiménez-Zamudio L, García-Latorre E, Cruz-Orozco O, Hernández-Guerrero C. Quantitative and qualitative peritoneal immune profiles, T-cell apoptosis and oxidative stress-associated characteristics in women with minimal and mild endometriosis. BJOG. 2011;118(1):6–16.

Mínguez-Alarcón L, Chavarro JE, Mendiola J, Roca M, Tanrikut C, Vioque J, et al. Fatty acid intake in relation to reproductive hormones and testicular volume among young healthy men. Asian J Androl. 2017;19(2):184–90.

Mínguez-Alarcón L, Afeiche MC, Williams PL, Arvizu M, Tanrikut C, Amarasiriwardena CJ, et al. Hair mercury (Hg) levels, fish consumption and semen parameters among men attending a fertility center. Int J Hyg Environ Health. 2018;221(2):174–82.

Mirnamniha M, Faroughi F, Tahmasbpour E, Ebrahimi P, Beigi Harchegani A. An overview on role of some trace elements in human reproductive health, sperm function and fertilization process. Rev Environ Health. 2019;34(4):339-348.

Mohd Mutalip SS, Ab-Rahim S, Rajikin MH. Vitamin E as an antioxidant in female reproductive health. Antioxidants (Basel, Switzerland). 2018;7(2):22.

Montano L, Ceretti E, Donato F, Bergamo P, Zani C, Viola GCV, et al. Effects of a lifestyle change intervention on semen quality in healthy young men living in highly polluted areas in Italy: the FASt randomized controlled trial. Eur Urol Focus. 2022;8(1):351–9.

Muthusami KR, Chinnaswamy P. Effect of chronic alcoholism on male fertility hormones and semen quality. Fertil Steril. 2005;84(4):919–24.

Nassan FL, Chavarro JE, Tanrikut C. Diet and men's fertility: does diet affect sperm quality? Fertil Steril. 2018a;110(4):570–7.

Nassan FL, Chiu YH, Vanegas JC, Gaskins AJ, Williams PL, Ford JB, et al. Intake of protein-rich foods in relation to outcomes of infertility treatment with assisted reproductive technologies. Am J Clin Nutr. 2018b;108(5):1104–12.

Ng R, Sutradhar R, Yao Z, Wodchis WP, Rosella LC. Smoking, drinking, diet and physical activity-modifiable lifestyle risk factors and their associations with age to first chronic disease. Int J Epidemiol. 2020;49(1):113–30.

Ojo OA, Nwafor-Ezeh PI, Rotimi DE, Iyobhebhe M, Ogunlakin AD, Ojo AB. Apoptosis, inflammation, and oxidative stress in infertility: a mini review. Toxicol Rep. 2023;10:448–62.

Okonofua FE, Ntoimo LFC, Ayodeji AOO, Olafusi C, Unuabonah E, Ohenhen V. Causes and risk factors for male infertility: a scoping review of published studies. Int J Gen Med. 2022;15:5985–97.

Olsen J, Juul S, Basso O. Measuring time to pregnancy. Methodological issues to consider. Hum Reprod. 1998;13(7):1751–3.

Omu AE, Fatinikun T, Mannazhath N, Abraham S. Significance of simultaneous determination of serum and seminal plasma alpha-tocopherol and retinol in infertile men by high-performance liquid chromatography. Andrologia. 1999;31(6):347–54.

Osadchuk L, Kleshchev M, Danilenko A, Osadchuk A. Impact of seminal and serum zinc on semen quality and hormonal status: a population-based cohort study of Russian young men. J Trace Elem Med Biol. 2021;68:126855.

Pajarinen JT, Karhunen PJ. Spermatogenic arrest and "Sertoli cell-only" syndrome--common alcohol-induced disorders of the human testis. Int J Androl. 1994;17(6):292–299.

Panth N, Gavarkovs A, Tamez M, Mattei J. The influence of diet on fertility and the implications for public health nutrition in the United States. Front Public Health. 2018:6.

Pasquali R. Obesity, fat distribution and infertility. Maturitas. 2006;54(4):363–71.

Payne KS, Mazur DJ, Hotaling JM, Pastuszak AW. Cannabis and male fertility: a systematic review. J Urol. 2019;202(4):674–81.

Pierik FH, Abdesselam SA, Vreeburg JTM, Dohle GR, De Jong FH, Weber RFA. Increased serum inhibin B levels after varicocele treatment. Clin Endocrinol (Oxf). 2001;54(6):775–80.

Practice Committee of the American Society for Reproductive Medicine. Diagnostic evaluation of the infertile female: a committee opinion. Fertil Steril. 2015;103(6):e44–50.

Practice Committee of the American Society for Reproductive Medicine. Testing and interpreting measures of ovarian reserve: a committee opinion. Fertil Steril. 2020;114(6):1151–7.

Practice Committee of the American Society for Reproductive Medicine American Society for Reproductive Medicine. Smoking and infertility: a committee opinion. Fertil Steril. 2018;110(4):611–8.

Qiu Y, Yang T, Seyler BC, Wang X, Wang Y, Jiang M, et al. Ambient air pollution and male fecundity: a retrospective analysis of longitudinal data from a Chinese human sperm bank (2013-2018). Environ Res. 2020;186:109528.

Ramsay JM, Fendereski K, Horns JJ, VanDerslice JA, Hanson HA, Emery BR, et al. Environmental exposure to industrial air pollution is associated with decreased male fertility. Fertil Steril. 2023;120:637–47.

Ricci E, Viganò P, Cipriani S, Somigliana E, Chiaffarino F, Bulfoni A, et al. Coffee and caffeine intake and male infertility: a systematic review. Nutr J. 2017;16(1):37.

Ricci E, Al-Beitawi S, Cipriani S, Alteri A, Chiaffarino F, Candiani M, et al. Dietary habits and semen parameters: a systematic narrative review. Andrology. 2018;6(1):104–16.

Ros E, Martínez-González MA, Estruch R, Salas-Salvadó J, Fitó M, Martínez JA, et al. Mediterranean diet and cardiovascular health: teachings of the PREDIMED study. Adv Nutr. 2014;5(3):330S–6S.

Rossi BV, Bressler LH, Correia KF, Lipskind S, Hornstein MD, Missmer SA. Lifestyle and in vitro fertilization: what do patients believe? Fertil Res Pract. 2016;2(1):11.

Salas-Huetos A, Bulló M, Salas-Salvadó J. Dietary patterns, foods and nutrients in male fertility parameters and fecundability: a systematic review of observational studies. Hum Reprod Update. 2017;23(4):371–89.

Salas-Huetos A, Moraleda R, Giardina S, Anton E, Blanco J, Salas-Salvadó J, et al. Effect of nut consumption on semen quality and functionality in healthy men consuming a Western-style diet: a randomized controlled trial. Am J Clin Nutr. 2018;108(5):953–62.

Salas-Huetos A, Babio N, Carrell DT, Bulló M, Salas-Salvadó J. Adherence to the Mediterranean diet is positively associated with sperm motility: a cross-sectional analysis. Sci Rep. 2019;9(1):3389.

Sengupta P, Dutta S, Krajewska-Kulak E. The disappearing sperms: analysis of reports published between 1980 and 2015. Am J Mens Health. 2017;11(4):1279–304.

Senior AM, Solon-Biet SM, Cogger VC, Le Couteur DG, Nakagawa S, Raubenheimer D, et al. Dietary macronutrient content, age-specific mortality and lifespan. Proc R Soc B Biol Sci. 2019;286(1902):20190393.

Sermondade N, Elloumi H, Berthaut I, Mathieu E, Delarouzire V, Ravel C, et al. Progressive alcohol-induced sperm alterations leading to spermatogenic arrest, which was reversed after alcohol withdrawal. Reprod Biomed Online. 2010;20(3):324–7.

Sharma R, Harlev A, Agarwal A, Esteves SC. Cigarette smoking and semen quality: A new meta-analysis examining the effect of the 2010 World Health Organization Laboratory methods for the examination of human semen. Eur Urol. 2016;70(4):635–45.

Shi L, Zhang J, Lai Z, Tian Y, Fang L, Wu M, et al. Long-term moderate oxidative stress decreased ovarian reproductive function by reducing follicle quality and progesterone production. PloS One. 2016;11(9):e0162194.

Shkolnik K, Tadmor A, Ben-Dor S, Nevo N, Galiani D, Dekel N. Reactive oxygen species are indispensable in ovulation. Proc Natl Acad Sci U S A. 2011;108(4):1462–7.

Shrivastava V, Marmor H, Chernyak S, Goldstein M, Feliciano M, Vigodner M. Cigarette smoke affects posttranslational modifications and inhibits capacitation-induced changes in human sperm proteins. Reprod Toxicol. 2014;43:125–9.

Silva T, Jesus M, Cagigal C, Silva C. Food with influence in the sexual and reproductive health. Curr Pharm Biotechnol. 2019;20(2):114–22.

Sirmans SM, Pate KA. Epidemiology, diagnosis, and management of polycystic ovary syndrome. Clin Epidemiol. 2013;6(1):1–13.

Skoracka K, Ratajczak AE, Rychter AM, Dobrowolska A, Krela-Kaźmierczak I. Female fertility and the nutritional approach: the Most essential aspects. Adv Nutr; 2021;12(6):2372–2386.

Takeshima T, Usui K, Mori K, Asai T, Yasuda K, Kuroda S, et al. Oxidative stress and male infertility. Reprod Med Biol. 2020;20(1):41–52.

Tamrakar SR, Bastakoti R. Determinants of infertility in couples. J Nepal Health Res Counc. 2019;17(1):85–9.

Tavilani H, Doosti M, Abdi K, Vaisiraygani A, Joshaghani HR. Decreased polyunsaturated and increased saturated fatty acid concentration in spermatozoa from asthenozoospermic males as compared with normozoospermic males. Andrologia. 2006;38(5):173–8.
Te L, Liu J, Ma J. Wang S, vol. 127124. Correlation between serum zinc and testosterone: a systematic review: J Trace Elem Med Biol; 2023. p. 76.
Thérond P, Auger J, Legrand A, Jouannet P. alpha-Tocopherol in human spermatozoa and seminal plasma: relationships with motility, antioxidant enzymes and leukocytes. Mol Hum Reprod. 1996;2(10):739–44.
Toledo E, Lopez-Del Burgo C, Ruiz-Zambrana A, Donazar M, Navarro-Blasco Í, Martínez-González MA, et al. Dietary patterns and difficulty conceiving: a nested case-control study. Fertil Steril. 2011;96(5):1149–53.
Tu C, Wang W, Hu T, Lu G, Lin G, Tan YQ. Genetic underpinnings of asthenozoospermia. Best Pract Res Clin Endocrinol Metab. 2020;34(6):101472.
Tvrdá E, Sikeli P, Lukáčová J, Massányi P, Lukáč N. Mineral nutrients and male fertility. J Microbiol Biotechnol Food Sci. 2021;2021:1–14.
Vakalopoulos I, Dimou P, Anagnostou I, Zeginiadou T. Impact of cancer and cancer treatment on male fertility. Hormones (Athens). 2015;14(4):579–89.
Verhaeghe F, Di Pizio P, Bichara C, Berby B, Rives A, Jumeau F, et al. Cannabis consumption might exert deleterious effects on sperm nuclear quality in infertile men. Reprod Biomed Online. 2020;40(2):270–80.
Volodarsky-Perel A, Cohen Y, Arab S, Son WY, Suarthana E, Dahan MH, et al. Effects of cancer stage and grade on fertility preservation outcome and ovarian stimulation response. Hum Reprod. 2019;34(3):530–8.
Vujkovic M, De Vries JH, Lindemans J, MacKlon NS, Van Der Spek PJ, Steegers EAP, et al. The preconception Mediterranean dietary pattern in couples undergoing in vitro fertilization/intracytoplasmic sperm injection treatment increases the chance of pregnancy. Fertil Steril. 2010;94(6):2096–101.
Wang W, Peng M, Yuan H, Liu C, Zhang Y, Fang Y, et al. Studying the mechanism of sperm DNA damage caused by folate deficiency. J Cell Mol Med. 2022;26(3):776–88.
Weinberg CR, Baird DD, Wilcox AJ. Sources of bias in studies of time to pregnancy. Stat Med. 1994;13(5–7):671–81.
Willis SK, Wise LA, Wesselink AK, Rothman KJ, Mikkelsen EM, Tucker KL, et al. Glycemic load, dietary fiber, and added sugar and fecundability in 2 preconception cohorts. Am J Clin Nutr. 2020;112(1):27–38.
Winter HG, Rolnik DL, Mol BWJ, Torkel S, Alesi S, Mousa A, et al. Can dietary patterns impact fertility outcomes? A systematic review and meta-analysis. Nutrients. 2023;15(11):2589.
Winters BR, Walsh TJ. The epidemiology of male infertility. Urol Clin North Am. 2014;41(1):195–204.
Wise LA, Wesselink AK, Tucker KL, Saklani S, Mikkelsen EM, Cueto H, et al. Dietary fat intake and Fecundability in 2 preconception cohort studies. Am J Epidemiol. 2018;187(1):60–74.
Wolgemuth DJ, Chung SS. Retinoid signaling during spermatogenesis as revealed by genetic and metabolic manipulations of retinoic acid receptor alpha. Soc Reprod Fertil Suppl. 2007;63:11.
World Health Organization. WHO laboratory manual for the examination and processing of human semen. 6th ed. WHO Press. World Health Organization, Department of Reproductive Health and Research; 2021. p. 276.
World Health Organization (WHO). Infertility Prevalence Estimates, 1990–2021. 2023.
Yamamoto A, Harris HR, Vitonis AF, Chavarro JE, Missmer SA. A prospective cohort study of meat and fish consumption and endometriosis risk. Am J Obstet Gynecol. 2018;219(2):178.e1–10.
Yuan HF, Zhao K, Zang Y, Liu CY, Hu ZY, Wei JJ, et al. Effect of folate deficiency on promoter methylation and gene expression of Esr1, Cav1, and Elavl1, and its influence on spermatogenesis. Oncotarget. 2017;8(15):24130–41.
Zondervan KT, Becker CM, Missmer SA. Endometriosis. Longo DL, editor. N Engl J Med. 2020;382(13):1244–56.

[illegible]

Chapter 11
Nutraceuticals and Suboptimal Health: The Relationship and Significance

Ashish Thakur, Ram Kumar, Kamal Shah, Nagendra Singh Chauhan, and Ashish Baldi

1 Introduction

Nutrition is the study of how food affects the body and is a key component of human growth. To sustain life, cells and organisms need to be adequately supplied with vitamins, minerals, fibre, water, carbohydrates, proteins, and other micronutrients. Some basic functions of essential nutrients are mentioned in Table 11.2. Malnutrition is defined by the World Health Organization as the cellular mismatch between the body's supply of nutrients and energy and its need for them to support development, maintenance, and certain functions. It is crucial to comprehend the significance that diet plays in suboptimal health (Yassine et al. 2021). Nutrition is complex and seeming getting more complicated. Most of consumers are familiar with only "essential nutrients" e.g., minerals and vitamins and more recently protein and important amino acids. These essential nutrients have reference values, referred to as dietary reference intakes developed by consensus committees of scientific experts convened by the Institute of Medicine of the National Academy of Sciences, Medicine and Engineering and carried out by the Food and Nutrition Board (Lupton et al. 2016).

The maintenance of health and well-being, lowering the risk of sickness and injury and improving performance all depend on an adequate diet. A variety of variables, including age, gender, body mass, height, and demands for growth and development, have an impact on the dietary requirements of each person. Suboptimal

A. Thakur · R. Kumar · A. Baldi (✉)
Pharma Innovation Lab, Department of Pharmaceutical Sciences and Technology, Maharaja Ranjit Singh Punjab Technical University, Bathinda, Punjab, India

K. Shah
Institute of Pharmaceutical Research, GLA University, Mathura, UP, India

N. S. Chauhan (✉)
Drugs Testing Laboratory Avam Anusandhan Kendra, Raipur, CG, India

S. M. Jafari et al. (eds.), *Nutraceuticals: A New Millennium Approach for Preventive Medicine*, Food Bioactive Ingredients,
https://doi.org/10.1007/978-3-031-76481-3_11

health status (SHS), also known as subhealth, is a condition that exhibits certain alterations in psychological behaviors, physical traits, or some medical examination indices but lacks the typical pathologic signs. It is regarded as a therapeutic working idea that identifies a stage between health and sickness. Definitions of some related terms used in this chapter are mentioned in Table 11.1. The early identification of

Table 11.1 Definition of some related terms

Terms	Definition	Reference
Diet	Diet is defined as the total quantity of food a human or different creature consumes. The term "diet" frequently denotes the utilization of a certain intake of nutrients for weight or health control determinations.	Goswami et al. (2017)
Dietary nutrients	The nutrients are defined as the chemical composites in food which are used by the body for the maintenance of function and health properly. The dietary nutrients are important as it protects you against many non-communicable diseases which are chronic, such as cancer, diabetes and also disease related to heart. The major nutrients are proteins, vitamins, carbohydrates, fats, fiber, minerals, and water.	Beal et al. (2021)
Dietary supplements	These are the manufactured good meant to be occupied as a tablet, capsule, pill, liquid, or powder in order to supplement diet. The supplement can provide nutrients that are either synthetic or extracted from various food sources in order to increase their consumption.	Nesheim (1999)
Balanced diet	Every essential vitamin that the body requires is part of a balanced diet. Among other things, a well-balanced diet should contain vitamins, carbohydrates, minerals, fats, proteins, water, and fibre. In essence, a balanced diet lowers the risk of disease and enhances general health.	Goswami et al. (2017)
Suboptimal health status	SHS, also known as subhealth, is a condition that exhibits certain alterations in psychological behaviors, physical traits, or some medical examination indices but lacks the typical pathologic signs. It is regarded as a therapeutic working idea that identifies a stage between health and sickness.	Li et al. (2013)
Nutraceuticals	Nutraceuticals are "any substance that is a food or a component of a food that delivers medicinal and/or health advantages, including disease prevention and treatment".	Roberfroid (2000)
Functional foods	Functional foods are defined as "food that seems similar to traditional food and is meant to be taken as part of a regular diet, but has been changed to perform physiological tasks beyond the provision of simple nutritional requirements".	Spence (2006)
Nutritional assessment	Nutritional assessment is a diagnostic procedure carried out by the competent nutrition-trained healthcare professionals such as deititians, specialized nutrition nurses, and medical personnel. Nutritional assessment determines nutritional state and needs of patient as well as potential nutritional hazards including malnutrition.	Ferrie (2020)
Biomarkers	These are the "cellular, biochemical or molecular alterations that are measurable in biological media such as human tissues, cells, or fluids". Some basic types of biomarkers are shown in Fig. 11.2.	Mayeux (2004)

SHS creates a window opportunity for the predictive, preventive, and personalized medicine (PPPM) in chronic diseases (Chen et al. 2014). The understanding about the relationship between dietary nutrients and supplements with respect to suboptimal health i.e., physical, mental as well as spiritual health is evolving.

2 Related Terms

Various terms viz. dietary nutrients/supplements, functional foods, nutraceuticals etc. are used intermittently in relevant literature. Therefore, definitions of all such terms are mentioned in Table 11.1 for proper understanding.

3 Assessment of Dietary Nutrients

A healthy diet can maintain and improve optimal health. A high consumption of the legumes, seeds or nuts, fruits, vegetables, fish, cereals and a low intake of meat and dairy goods characterize a healthy quality diet, while sugary drinks, a high salt intake, and red meat are signs of a morbid quality of diet. Classification and some food sources of nutrient is shown in Fig. 11.1. Optimal health can include aspects of mental, physical as well as spiritual health (Echouffo-Tcheugui and Ahima 2019).

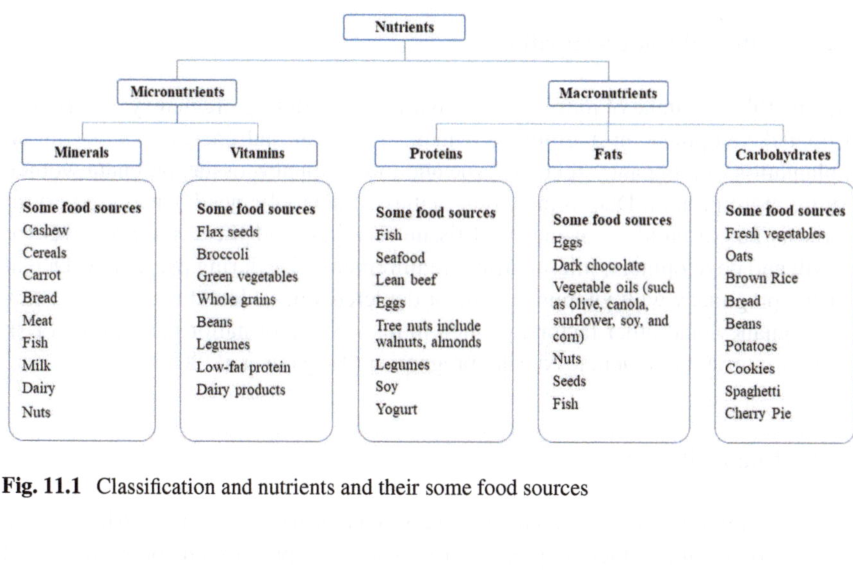

Fig. 11.1 Classification and nutrients and their some food sources

3.1 Factors Associated with the Intake of Nutrients

3.1.1 Declined Food Intake with Age

It is also known as "Anorexia of Aging" It is an important health problem and highly prevalent among older adults. It leads to negative health outcomes such as malnutrition, weight loss, sarcopenia, physical frailty and changes in composition. The onset of cachexia with muscular wasting, which is linked to high rates of mortality and morbidity, is also influenced by it. A reduction in food consumption is an uncommon and difficult component of ageing. Along with physiological changes brought on by ageing, age-related diseases, and other causes, emotional, social, and environmental factors may also have a big influence on eating habits, food consumption, and functional status (Plotkin and Taani 2020).

3.1.2 Tooth Loss

Diet is a crucial part to living a healthy life since it plays a part in the genesis and, therefore, prevention of many chronic illnesses, including cardiovascular disease, cancer, diabetes, and obesity. Nutrient consumption and tooth loss are closely related (Arora et al. 2019a). The major purpose of teeth is mastication, which takes place mostly in the oral cavity. Loss of teeth affects masticatory function and chewing abilities, which might diminish the range and number of foods one can choose from. Due to these factors, food consumption been seen as a link between a number of chronic diet-related disorders and tooth retention (Gaewkhiew et al. 2017).

3.1.3 Maternal Under-Nutrition

The insufficient intake of nutrients, which are crucial during pregnancy, contributes to poor development in addition to a lack of protein and energy. For the initial implantation of pregnancy-related tissue and to sustain new tissue, pregnant women require more protein. Due to increased maternal metabolic needs, increased erythropoiesis has, a building of maternal tissue reserves, and foetal requirements for growth and development, micronutrient requirements rise during pregnancy. Due to starting pregnancy with vitamin deficits or depleted stocks, low dietary intakes during pregnancy, and other factors, women in resource-poor nations are more at risk for micronutrient deficiencies during pregnancy (Tsegaye et al. 2017).

3.1.4 Tongue Pressure

The suboptimal pressure of tongue is related to the hazard of the malnutrition in the community dwelling elder people. Older persons with poor tongue pressure should have their nutritional condition and swallowing function evaluated thoroughly to detect subclinical malnutrition and dysphagia early. The tongue is vital in swallowing, and its failure might hypothetically lead to insufficient oral intake and eventual malnutrition (Chang et al. 2021).

4 Essential Nutrients for Maintenance of Suboptimal Health

The physiological and biochemical progression through which an organism takes food to sustain its existence is termed as nutrition. It gives nutrients to organisms, which can be digested to produce chemical structures and energy. The deficiency of nutrients, also known as malnutrition, it occurs when an organism does not have the quantity of nutrients that it needs for good health. This may be caused as the insufficient nutrients are absorbed or there is a sudden loose of nutrients. An organism will adapt in this situation by reducing its energy intake and output in order to maximize the use of nutrients that have been conserved. It will initially use stored energy before destroying its own body mass to get additional energy. Thus, nutrients are essential for the maintenance of suboptimal health (Bender 2002).

Nutrients are complex organic or inorganic compounds found in the diet. There are around 50 distinct nutrients that are generally given by the food which we intake. Some basic functions of essential nutrients are mentioned in Table 11.2. Each nutrient has a distinct purpose in the body. Most natural foods include many nutrients (Arora and Baldi 2017a). Essential nutrients are the chemicals that the body doesn't or can't produce in suitable quantities. The requirements of these nutrient are fulfilled by the other sources of food as per the WHO, and they are essential for disease prevention, growth, and well health (Goswami et al. 2017).

Table 11.2 Basic functions of some essential nutrients

Nutrients	Function	Reference
Protein	It is necessary for cell reparation and hormone, enzyme production, and formation of tissue. Protein is vital for building a health of the immune system and strong muscles.	Erickson and Kim (1990)
Carbohydrate	Carbohydrate provides good energy source for human body as well as basic elements for the creation of cells.	Stephen et al. (2012)
Fat	It provides reservoir of energy for the functioning of body as basic component of cells, and signals molecules for the appropriate communication of cells. It works to maintain body temperature and provides insulation to vital organs.	Ananthakrishnan et al. (2014)
Vitamin	It promotes or regulates the normal functions of body systems and functions the body processes.	Maqbool et al. (2018)
Mineral	It regulates the body processes, which are vital for the proper cellular function, and comprised of tissues.	Majewski et al. (2016)
Water	It transports essential nutrients to every point of body, aids with the maintenance of body temperature and remove waste products for disposal.	Boyd (2019)

4.1 Micronutrients

These are vitamin and mineral supplements. These are called as micronutrients as they are essential in tiny amounts ranging from a fraction of a milligrams to several grams (Alloway 2008).

Example: iron, cobalt, chromium, iodine, copper, zinc and molybdenum etc.

4.2 Macronutrients

These are also known as "proximate principles" of food as they make up the majority of diet. They contribute to total calorie consumption in the following percentage in the Indian diet (Macdonald 1999).

Examples are carbohydrates (65–80%), protein (7–15%) and fat (10–30%).

The essential nutrients and their function for the maintenance of suboptimal health are discussed below:

4.3 Vitamins

Vitamins are the chemicals, organic in nature which are classified in two classes on the bases of their solubility nature such as soluble in fat or either soluble in water. Vitamins which are fat soluble like as vitamin-A, vitamin-D, vitamin-E and vitamin-K get dissolved in fat tends to gather in the body. Vitamins which are water soluble like as vit-B and C complex vitamins such as vit-B_{12}, vit-B_6 and folate must get dissolved into the water before they get absorbed by the body, and therefore cannot be stored. Basic function of vitamins is given in Table 11.2. Any water-soluble vitamin which is not utilized by the body is lost primarily through the urine (Maqbool et al. 2018).

4.4 Minerals

Minerals are the in-organic elements extant in water and soil, and consumed by animals or absorbed by plants (Arora et al. 2013). Basic function of minerals is mentioned in Table 11.2. Some minerals like as Ca, Na and K including number of other trace minerals (like as Zn, Cu and iodine) needed in very minor amounts for the maintenance of health (Majewski et al. 2016).

4.5 *Protein*

Proteins are the large macromolecules and biomolecules that includes one or more long chains of including catalyzing metabolic related reactions, DNA replication, transportation of molecules from one location to another, responding to stimuli, and providing structure to organisms and cells. Proteins are differed from one another in their amino-acid sequence, which is known by the nucleotide sequence of their genes. Basic function of proteins is mentioned in Table 11.2. This causes proteins to typically fold into a certain three-dimensional structure, which controls the function of the protein (Erickson and Kim 1990).

4.6 *Fats*

Along with amino acid residues, fats are one of the three major macronutrient categories in the human diet. These are the primary building blocks of everyday foods including cooking oils milk, tallow, butter, salt pork and lard together with proteins and carbohydrate. Fats carry out a wide range of tasks inside organisms and serve vital metabolic and structural roles in most of the living things including thermal insulation, waterproofing. These are the primary source of energy for several animals. Except for a few necessary fatty acids that must consume by the human is able to synthesize the fat, which it needs from other dietary elements. Basic function of fats is mentioned in Table 11.2. Several water-insoluble vitamins, compounds of fragrance, and flavorings are transported by dietary fats (Ananthakrishnan et al. 2014).

4.7 *Water*

Water is an inorganic chemical, that is clear, flavorless, odorless, and almost colorless. It is the primary component of the hydrosphere of the earth. Despite not supplying food, organic micronutrients, or energy. For all currently understood forms of life, it is necessary. Basic function of water is mentioned in Table 11.2. Its molecules having chemical formula H_2O and are composed of two hydrogen atoms bound together by covalent bonds and the H and O atoms are linked at an angle of 104.45° (Boyd 2019).

4.8 *Carbohydrates*

Carbohydrates are the moieties of sugar. It is also one of the major nutrients which is found in foods and drinks. The carbohydrates were breaks down into the glucose by human body, which is the major source of energy for tissues, organs, and cells of human body. Basic function of carbohydrates is mentioned in Table 11.2. Glucose can be stored or used immediately in the muscles and liver for its later use (Stephen et al. 2012).

5 Biomarkers

It is well known by the name of biological markers. Free radical biologists have used the word "biomarker"from molecular epidemiology to characterize a molecular alteration in a biological molecule caused by reactive halide species, nitrogen, or oxygen. It is equally applicable to DNA, lipids, proteins, and antioxidant consumption products, where the kind of reaction may be direct addition, abstraction of proton, or transfer of electron (Griffiths et al. 2002). Biomarkers are the "cellular, biochemical or molecular alterations that are measurable in biological media such as human tissues, cells, or fluids". The concept has lately been expanded to encompass biological traits that may be empirically examined and assessed as a sign of healthy pharmacological, pathogenic, and biological reactions to the therapeutic intervention (Mayeux 2004).

Clinical biomarkers come under predictive bio-markers and have the advantages of being repeatable and able to be studied over a shorter time period. These are also less expensive and easier to monitor than last clinical end-points. Some basic types of biomarkers are shown in Fig. 11.2. For example, collecting a patient's blood pressure is far easier than using echocardiogram to test left ventricular function, and echocardiography is much easier than measuring morbidity and death from hypertension in the long run. Whereas blood pressure may be monitored instantly and often, collecting data on morbidity and death takes years (Aronson and Ferner 2017).

Currently, biomarkers are used in nutrition, and health care & research aim to quantify disease onset and progression. However, both nutrition and health care should focus on the maintenance of optimal health, where the underlying biology differs significantly from the biomedical science. Health is defined by the ability to continuously adapt in different conditions involving many mechanisms of system flexibility. A novel generation of biomarkers that quantify all aspects of system flexibility is required, it opens the way for true lifestyle-related optimization of health, self-empowerment, and related products and services (Zeisel and da Costa 2009).

5.1 Alliance Between biomarkers and Supplement Intake

Objectively measured nutritional biomarkers can help to confirm self-reported nutrient consumption by indicating the internal dosage or absorption. Biological markers may be affected by a several factors, however, with regard to the supplements as a source for intake of nutrients. Firstly, a larger spectrum of nutrients may be consumed, and various dose-response relationships can be found when

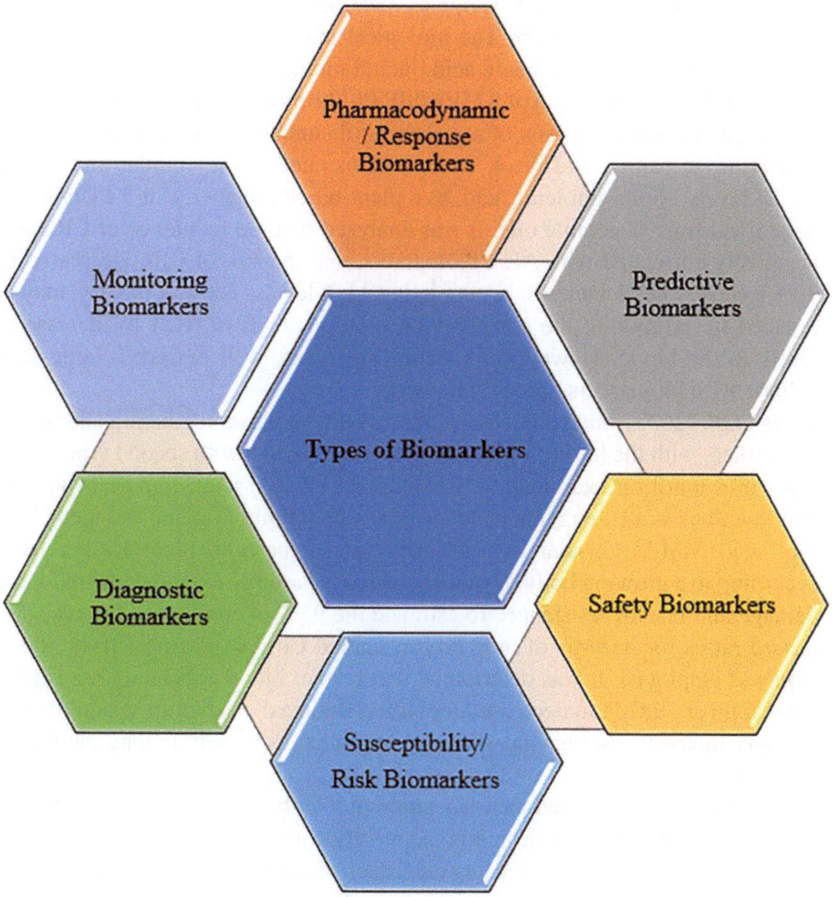

Fig. 11.2 Types of biomarkers

comparing total nutrient intake (TNI) to food-sourced intake alone. Secondly, although a dose-response connection is different and some of these results may be contradictory with reference to the "internal dosage". At last, collinearity in supplement nutrient consumption exists, just as meals include many nutrients that may interact (for example fat-soluble vitamins as antioxidants in high-fat diets) (e.g., use of multivitamin/multimineral-type supplements). Consequently, biomarkers outside of the nutrients investigated may be impacted (Lentjes 2019).

Some of the major studies related to the relationship between supplement intake and biomarkers is discusses below:

- ***Dai et al.*** (2022) studied the association of biomarker and intake of α- linolenic acid with incident Colorectal Cancer (CRC). CRC is now the fourth most deadly disease in the world, causing around 900,000 deaths annually. It is the third most common in males and the second most frequent cancer diagnosed in women

(Dekker et al. 2019). Dietary variables have been proven to have a significant influence in CRC prevention. The biological mechanisms underlying the anti-cancer effects of alpha-linolenic acid, including the suppression of nuclear factor-B (NF-B), the activation of AMPK/SIRT1, the modulation of cyclooxygenase activity, and the upregulation of novel anti-inflammatory lipid mediators such as protectins, maresins, and resol, were demonstrated at the cellular and animal model levels. Alpha-linolenic acid, as a plant-based member of n-3 PUFAs, can be derived from vegetable oils. It was analyzed that the incidence of CRC was negatively correlated with ALA biomarkers, and the risk of CRC decreased by 10% for every 0.1% increase in circulating ALA levels. The risk of CRC may be reduced by promoting the consumption of foods high in ALA to increase its levels in the blood. However, for better outcomes, well-designed, large-scale cohorts with biomarkers are still required.

- ***Wang et al.*** (2022) studied the association of the intake of cobalamin and related biomarkers with the hazard of mortality amongst adults with second type of diabetes in a nutrition examination. Cobalamin (Vit-B_{12}) is an important vitamin that functions as a co-enzyme in the metabolism of homocysteine and methylmalonic acid (MMA). Cobalamin insufficiency and diabetes are biologically related, according to a growing body of studies, which is partially explained by oxidative damage and insulin resistance. To estimate the 95% of confidence intervals and hazard ratios for hazards of mortality, weighted Cox proportional risks regression was employed. It was determined that neither blood cobalamin nor dietary cobalamin are linked to decreased mortality. Reduced cobalamin sensitivity was strongly linked to cardiac and all-cause mortality, especially among metformin nonusers.
- ***Fallah et al.*** (2022) studied the association of liver biomarkers in adults with the consumption of green tea, which is the mostly consumes beverage in the world after the water and formed by using the leaves of plant known as Camellia sinensis, and its supplements. Green tea includes beneficial compounds called polyphenols, primarily catechins, which account for not less than 15% of the dried green tea leaves. Green tea extracts (GTEs) have been developed and added to dietary supplements, foods, and beverages due to the association in between the usage of green tea and advantageous health effects, such as weight loss, improved lipid metabolism, neuro-protective effects, and the improvements in the chronic conditions. Green tea drinking was shown to be related with a lower risk of having abnormal liver biomarkers, but no meaningful connection with the usage of supplements was found.
- ***Conley et al.*** (2022) studied the relation between biomarkers and dietary phosphate intake of the bone and material metabolism. An estimated 10% of individuals globally are thought to have a chronic kidney disease (CKD), and its incidence is steadily increasing. People with CKD are more likely to get heart problems (CVD). The development of the CVD risk linked to chronic kidney disease (CKD) may be significantly influenced by the abnormalities of mineral metabolism and bone, including the hyper-phosphatemia and hyper-parathyroidism (CKD-MBD). According to dietary food records, total dietary phosphate con-

sumption levels do not substantially correspond with either biochemical markers of the bone and intermediate cardio-vascular or mineral metabolism indicators.

- ***Pitts et al.*** (2022) studied the pressure-mediated reflection spectroscopy criterion validity as a biological marker of vegetable and fruit intake. According to the US Dietary Guidelines for 2020–2025, 2 and 2.5 cups of fruits and vegetables respectively daily-focusing on orange, red, and green vegetables are necessary for good health. An inadequate intake of vegetables & fruits is linked to an increase the hazard of metabolic syndrome, diabetes, heart related disease, obesity and cancer. Overall, Americans drink fewer FVs than they should, and this is especially true in underprivileged communities. It was concluded that skin carotenoids which are assessed by spectroscopy may be rational alternative of the measurement of plasma carotenoids.
- ***Mates et al.*** (2022) reviewed the walnut intake targeting biomarkers of the metabolic syndrome and inflammation in over the age of adults. The walnut is regarded as a nutraceutical's food source due to their high level of healthy fatty acids, like as omega-3 PUFA & MUFA, nutritional value, strong antioxidant phytochemical content, and health benefits. MetS (Age-related metabolic syndrome) conditions, inflammation, chronic, and the oxidative stress are all important risk factors for morbidity and mortality. These pathophysiological factors raise the risk of age-related illnesses like cancer, dementia, cardiovascular disease, type-2 diabetes (T2D), and cognitive decline. It was discovered that walnut intake significantly improved the lipid profiles in middle-aged and older persons when compared to various control diets.

6 Role of Various Microorganisms in Improving Suboptimal Health

The role of various microorganisms in the maintenance of suboptimal health status is discussed below:

6.1 Microbiota and Probiotics

Microbiota has an essential role in multiple body functions, including physiological, immunological, and metabolic processes, via various mechanisms like as the regulated growth and functioning of several categories of the immune cells in gut. Bacteria over the mucosal surface or within the mucus layer interact with the human immune system; hence, an healthy gut microbiota is required for the growth of mucosal immunity (D'Angelo et al. 2017).

Probiotics are a mixture of live bacteria or yeast that lives in your body in order to maintain the human body health (Baldi 2018a). The intake of the probiotics can lower the hazard of ailments related with the dysfunction in intestinal barrier dysfunction (Baldi 2018b; Arora et al. 2019b). The methods by which the probiotics alter barrier function have revealed that probiotics boost barrier function by increasing anti-microbial peptides, mucus and production of secretory IgA, as well as enhancing viable adherence for pathogens and epithelial cell tight junction (TJ) integrity (Arora and Baldi 2017b). Certain lactobacilli are known to stick to mucosal surfaces, limiting harmful bacteria adhesion and increasing mucin production (Rao and Samak 2013).

6.2 Gut Microbiota and Vitamin

Obesity progression and maintenance are influenced by a variety of variables, including systemic substandard inflammation, adipose tissue malfunction, and gut microbial dysbiosis. Recently, there has been a surge of interest in the function of vitamins in the obesity and associated illnesses, both at gut and host bacterial levels. Vitamins are the mostly obtained from food, although some, particularly those in the K and B families, can be produced by the gut bacterial environment and absorbed in the colon (Arora et al. 2019c). It is critical to closely monitor the vitamin status of patients with the obesity and perhaps pre-existing comorbidities, and also to scrutinize the dysbiotic gut microbiota and hence potentially changed vitamin bacterial metabolism. This is due to the fact that vitamin shortage can change several critical cellular activities, resulting in major health problems (Voland et al. 2022).

6.3 Acidifiers

Organic acids, often known as acidifiers, play a crucial function in animal intestinal health. Acidifiers can be utilized to modify the intestinal microbial populations and boost response of immune, acting similarly to the antibiotics in the food of animals by combating harmful bacteria. The acidifiers are also promoting mineral absorption and improve nutrient digestion. The inclusion of organic acids also causes the intestinal lining to thin, allowing for greater nutrient absorption and utilization. Their impact will not be the same for all organic acids since their mode of action is determined by their pKa value. Acidifiers in feed decrease pathogenic bacterial development and reduce microbial competition for host resources by affecting the pH (Pearlin et al. 2020).

6.3.1 Function of the Acidifiers in the Nutrition

- Effect of the organic acid on mineral utilization and nutrient digestion.
- Improved production and performance of nutrition.

- Positive impact of acidifiers on immunity.
- Assistance in intermediary metabolism.
- Preservation of feedstuffs.
- pH maintenance.
- Control on pathogenic microbes.
- Facilitative absorption of nutrients.

6.3.2 Factors Influencing the Efficacy of Acidifiers

Despite of our growing understanding of the various ways that acidifiers work in animal nutrition, there are still a few things that need to be deduced. Some of the attributes that are connected to dietary, animal, or acid factors have led to some contradictory results. The inclusion level, the molecular weight, the nature of chemical (salt, acid, uncoated or coated), the minimum inhibitory concentration of the acid, the pKa values of acids, the types of species, microorganisms and their concentration, the site of action, the buffering ability of feed, the dietary composition and the species of animal must be included (Pearlin et al. 2020).

6.4 Microbial Metabolites

The microbial population that exists in the synergistic relationship with the host and is widely distributed throughout the human gastro-intestinal tract is essential to preserving cellular homeostasis. They collectively make up the microbiome. A host may develop a variety of illnesses as a result of changes in the gut microbiota's composition and disturbances in the synergy between the host and microbiome. These involves inflammatory, metabolic and neurological illnesses (Singh 2022).

6.5 Live Bifidobacterium, Lactobacillus and Enterococcus

Immunodeficiency is a fairly frequent condition in which people with poor health and numerous illnesses are being developed or treated. Probiotics have recently emerged as a crucial tool for immunological control (Arora et al. 2019c; Voland et al. 2022). The mechanisms of interactions between immune deficiency and the gut microbiota are receiving more and more attention. The combination of *live Bifidobacterium, Lactobacillus* and *Enterococcus* has an immunomodulatory effect (Lv et al. 2021).

6.6 Feed Enzymes

The health of the gut is based on the effective evolution of the intestine's absorptive capacity, which is regulated by the co-evolution of the intestinal immune systems and the microbiota. In this process, availability of nutrient is important, from the standpoint of the microbiota, this alters with age as the intestines and upper gastrointestinal tract (GIT) microbes become more effective at removing nutrients. The feeding of the enzymes is important in this process. The presence of the nutrients in the lower region of the intestines for microbiota can be decreased by the use of phytases, which can enhance the digestion of the energy, minerals, and amino acids. With regard to amino acid delivery, protease can have a similar impact. Non-starch polysaccharidases play a special role in which they release fibre fragments from the insoluble matrix into the minor and more fermentable fractions of the carbohydrates. By improving diet digestibility from the host's perspective, these agents limits the amount of nutrients available to the microbiota (Bedford and Apajalahti 2022).

7 Role of Dietary Nutrients in Various Health Conditions for Maintaining Suboptimal Health Status

It is critical to understand the influence of nutrients, not only in the context of a dietary pattern but also with the separation, which is a complex mixture of the nutrients mainly in the health conditions. Some basic functions of essential nutrients are mentioned in Table 11.2. The role of dietary nutrients in several health conditions for the maintenance of the suboptimal health is discussed below:

7.1 Osteoporosis and Related Nutrient

Osteoporosis is the chronic disease that causes a reduction in the mineral density of bone, increased hazard of fragility fractures and a decreased strength of bone. Fragility fractures cause significant mortality, disability and morbidity, and are a key public health issue around the world. Calcium and Vit-D intake status are considered essential for the bone metabolism homeostasis and maintenance of suboptimal health. Promoting healthy habits is critical for lowering the risk of osteoporosis. The dietary intake of protein, calcium, and vit-D into adequate quantity in the diet, as well as getting regular exercise with weight and giving up bad behaviors like smoking and drinking, all help to improve bone quality (Muñoz-garach et al. 2020).

7.1.1 Dairy Products, Calcium and Vit-D

Vitamin D and calcium form part of the mineral matrix as calcium phosphate are required for the strength of bone. The adherence to the healthy diet is the finest way to achieve the adequate calcium intake. The most important sources of the calcium in the diet are dairy products (cheese, yogurt and milk), fish (mainly sardines with the bones), pulses and a some of the fruits and vegetables (particularly seeds and nuts) (Chiodini and Bolland 2018).

The main regulator of calcium homeostasis is vitamin D. A small variety of foods, including fatty fish, mushrooms, as well as some fortified dairy products, provide 10–20% of the vitamin D that is consumed, with the remaining 80–90% coming through cutaneous synthesis following exposure to sunshine. Vitamin D is necessary to preserve bone health, strength, and the muscular mass (Hayes et al. 2016).

Calcium, phosphorus, and magnesium, which have morphological roles in the proper development of bones, are minerals that are mostly found in dairy products. Protein, potassium, zinc, riboflavin and vitamin B_{12} are additional nutrients found in dairy products. Dairy products are a fantastic source of vitamin D. The vitamin D deficiency may be prevented and treated by getting enough sun exposure. The shortage vitamin D has negative effects on health outcomes, and crucial for maintaining health of bone. Osteoporosis may worsen in older or post-menopausal women who are vitamin D deficient (Adami et al. 2009).

7.1.2 Minerals

Magnesium and potassium are two minerals that are crucial for healthy bones. Potassium in the diet may lessen the strain and, consequently, the loss of calcium from the bones. Along with maintaining the body's alkaline condition, potassium can also lead to an increase in the body's calcium resorption (Kong et al. 2017). The metabolism of calcium also depends on magnesium. With an average body concentration of 10–30 mm, magnesium is the second most predominant intracellular cation after potassium. The majority of entire foods, including legumes, nuts, and green leafy vegetables, contain magnesium. For males and women, the recommended daily intakes of magnesium are 400–420 mg and 310–360 mcg, respectively (Arora and Baldi 2015a). Based on their age, sex, and past nutritional condition, different people have different needs. Magnesium is necessary for muscular contractions, brain function, and the exchange of potassium and calcium ions throughout the membranes of cells (Institute of Medicine (US) Standing Committee on the Scientific Evaluation of Dietary Reference Intakes 1997).

7.1.3 Protein

The consumption of protein is essential for the maintenance of health of bone. Proteins make up roughly one-third of bone mass and nearly half of bone volume. As the mineralization takes place, these bio-blocks become a component of the collagen structure and are merged into the organic matrix of bone. Furthermore, dietary proteins have an impact on the production and activity of IGF-I, an orthotropic hormone vital for bone development. The IGF-I hormone increases the rate of the reabsorption of phosphate from the kidney and enhances absorption of phosphorus and calcium in the gut. Protein is also involved in the conception of calcitriol. Eggs, fish, meat, poultry, and dairy goods are the key protein sources in healthy diets. For the synthesis and preservation of bone matrix, enough protein consumption is essential (Rizzoli et al. 2014).

Fruits and vegetables provide abundant sources of the nutrients needed for strong bones, including magnesium and potassium as well as vitamins C, carotenoids, vitamin K, and folate (Ahmadieh and Arabi 2011).

7.1.4 Vitamin C and K

The mineralization process during which bone matrix is formed involves vitamin K. It affects other vitamin K-dependent proteins as well as the carboxylase-microsomal, which helps with the post-translational conversion of glutamyl into carboxy-glutamyl residues in the osteocalcin. Vitamin K is made up of a family of various molecular forms, with vitamin K_1 being a single form which is produced by plants and vitamin K_2 being a family of many forms, mostly produced by the bacteria. The main form of vitamin K_1 consumed by humans is type-I. The presence of vitamin K_2 in the cheese is a unique characteristic (Veronese et al. 2015).

Because of its characteristics of antioxidant, vitamin C can help with the bone health. It has the capacity to reduce osteoclast activity. Also, it contributes to the production of collagen and serves as a co-factor for osteoblast differentiation. Vitamin C is a sign of a balanced diet high in vegetables and fruits (Finck et al. 2014).

7.1.5 Fatty Acids (Omega-3 Polyunsaturated) and Some Other Nutrients

The influence of the omega-3 polyunsaturated on the metabolism of bones is a subject of contradictory research. Docosahexaenoic acid and eicosatetraenoic acid ingestion may have an impact on bone development and remodeling in the humans by preventing resorption of bone as well as by promoting formation of bone (Griel et al. 2007). Although the exact method by which the PUFAs may alter turnover of bone is unknown, it is speculated that both EPA and DHA may benefit by controlling the ratio of receptor activator with a balance toward bone production (Sun et al. 2003).

7.2 *Pulmonary Tuberculosis*

Pulmonary tuberculosis termed as TB is a serious contamination which is caused by the *Mycobacterium tuberculosis* (MTB) that includes the respiratory system, lungs and may spread to other organs. TB can also cause malnutrition. Micronutrient deficiencies and protein-energy malnutrition leads to immuno-deficiency as a result of poor nutrition. This secondary immunodeficiency makes the host more susceptible to the infection and, as a result, increases the hazard of emerging tuberculosis which origins the nutrient malabsorption, appetite loss, altered metabolism, and micronutrient malabsorption, resulting in poor nutritional status. During antituberculosis treatment, the nutritional status of patients and dietary intake improve, which helps to maintain their nutritional or suboptimal health status (Kant et al. 2015). Nutritional status may influence relapse of active tuberculosis. Along with the nutritional depletion and weight loss are common in tuberculosis patients at the period of diagnosis. The role of several dietary nutrients in pulmonary tuberculosis for maintaining suboptimal health status is discussed below:

7.2.1 Micro-nutrients

Because of their various metabolic properties and functions, micronutrients have been recognized as an essential component for optimal human health. The most common causes of secondary immunodeficiency are micronutrient deficiencies and morbidity from infection, including tuberculosis. All of the vitamins that is vitamin-A, vitamin-B_6, vitamin-E, vitamin-C, and folic acid as well as the other minerals like as zinc, selenium, iron and copper plays an important function in cellular function, metabolic pathways, and immune competence (Karyadi et al. 2000).

7.2.2 Vitamin A

In tuberculosis, vitamin A has an immunocompetent activity as it prevents pathogenic bacilli from multiplying in cultured human macrophages (Chandra 1991). It is often measured using serum retinol and also plays a significant role in antibody response production, mucosal surface maintenance, epithelial function, and lymphocyte proliferation (Semba 1998). It is essential for the normal functioning of macrophage activity, B and T lymphocytes and generation of antibody response (Mugusi et al. 2003).

7.2.3 Vitamin D

Vitamin D is essential for activation macrophages, which is required to keep tuberculosis dormant. Furthermore, vitamin D inhibits the transcription of a substance required for the intracellular survival of tuberculosis bacillus in macrophages, which contain the bacilli. The deficiency of vitamin D allows the progress of disease to the active form (Ustianowski et al. 2005).

7.2.4 Vitamin E

The level of vitamin E in the blood is significantly lower in tuberculosis patients. Non-enzymic antioxidants such as reduced glutathione and vitamin E were significantly depleted in the plasma of pulmonary tuberculosis patients (Bakaev and Duntau 2004).

7.2.5 Polyunsaturated Fatty Acid

Tuberculosis is linked to omega-3. The eicosanoid production in the macrophages was investigated, and the conclusions revealed that dietary supplementation with eicosapentaenoic acid, n-3 fatty acids, and docosahexaenoic acid can affect the resistance to the *Mycobacterium tuberculosis*, while n-6 fatty acids did not (Zent and Smith 1995).

7.3 *Heart Disease, Diabetes, Influenza and Pneumonia*

The world population is undergoing a demographic change, with the share of elderly individuals (aged 65 and up) increasing faster than individual of any other age. This age group is more likely to acquire chronic illnesses related to nutrition such diabetes, cardiovascular disease, influenza, and pneumonia. Inappropriately, due to systemic healthcare, biological, and environmental complications, insufficient nutritional insufficiency, often known as hidden hunger which disproportionately affects older persons. Hidden hunger refers to people, who consume enough calories but receive insufficient micronutrients, placing them at the risk for diseases related to nutrition. Over time, inadequate dietary consumption of macro-nutrients and micro-nutrients can have a detrimental impact on the health outcomes, potentially leading to a cycle of suboptimal health (Eggersdorfer et al. 2018).

The lifespan of individual is determined by both intrinsic (such as genetic) and extrinsic (such as lifestyle, nutrition, and environment) factors. Physiological dysregulation, particularly in the immune system, is a key alteration that happens in older persons (Bailey and Arab 2012). Aging can cause inflammation, diminish innate immune response, and decrease T cell-mediated activity. Low level of nutrition is the vital hazard factor for the emergence of chronic and infectious illnesses, as well as impaired cognitive function, all of which have serious consequences for quality of life and survival. Increased intake of particular nutrients, dietary patterns, and food that promote healthy ageing, on the other hand, may diminish the hazard or postpone the onset of a several varieties of the chronic illnesses, as well as having a positive impact on health span (Who and Consultation 2003).

7.3.1 Protein

Maintaining lean mass (LM) is crucial since muscle mass normally declines with ageing and increases the risk of falls in older persons. The highest protein intake quintile of community-dwelling women and men (70–80-year-old) lost 40% less lean mass over time than those in the lowest quintile, according to the aging, health, and body study, which came to the conclusion that the dietary protein intake may be an adjustable risk associated factor for the sarcopenia. The Current Institute of Medicine (CIM) protein consumption guidelines apply to adults of all ages. However, it has been indicated that older persons may require an essential protein consumption to maintain the nitrogen balance and counteract the reduced insulin action and lower protein synthesis efficiency that occur with age and avoid muscle mass loss (Nowson and O'Connell 2015).

7.3.2 Omega-3

Dietary fibre is essential for proper digestive health. Obesity, bowel inflammatory disease, auto-immune arthritis, and the metabolic syndrome have all been linked to an unhealthy gut microbiota and infectious and degenerative illnesses (Zapata and Quagliarello 2015). Due to the extremely low consumption of fibre in diets, fibre has been designated as the "nutrient of public health concern" (Goldszmid and Trinchieri 2012).

7.3.3 Vitamin B: Folate, B_6 and B_{12}

Vitamin B are necessary for a variety of biological functions, such as healthy nerve and immune systems and bones. Folate inhibits hyperhomocysteinemia, which is linked to endothelial dysfunction, cardiovascular disease, inflammation, and is required for DNA methylation. With ageing, vitamin B_{12} absorption becomes less effective, increasing the risk of deficiencies, which might have serious effects given the vitamin's involvement in cell growth and mental function. Animal protein consumption affects how much vitamin B_{12} is consumed, and since inadequate stomach acid affects how much of it is absorbed, a deficit may also arise with the extended usage of acid hindering drugs. The immune system and homocysteine concentration are linked by vitamin B_6, another B vitamin that is crucial for cognition and depression. Because it is easily lost during processing of food and is not typically added back into the food supply, vitamin B_6 is crucial for the metabolism of protein (Eggersdorfer et al. 2018).

7.3.4 Vitamin D

Vitamin D is vital for proper absorption of the calcium, which aids in bone mineralization and the prevention of weakness in muscle. It is a particularly significant micronutrient for lowering the risk of fractures and osteoporosis caused by falls as a vitamin D can only be created by UV-exposure and is scarce in food sources, elderly people are more likely to be deficient in this vitamin because they spend more time indoors (Eggersdorfer et al. 2018).

7.3.5 Dietary Patterns

The best way to reduce your risk of developing chronic illnesses is to consume a balanced diet of whole foods, limiting your intake of processed meat, sugar, salt, and refined grains. The examples of healthy dietary patterns like as DASH (Dietary approaches to stop the hypertension) and the Mediterranean diet are widely used because they maintain a decent balance of nutrients, which is necessary for maximum health. It has been demonstrated that the DASH diet, can help decrease blood pressure (Appel et al. 1997).

7.4 Ageing

It has been suggested that vitamin B and folate with a metabolic connection may have preventive functions in the ageing disease (vitamins B_6, B_{12}, and riboflavin). Riboflavin and folate deficiencies in elder persons are typically brought on by poor intake of diet, whereas the low level of B_{12} status is typically linked to food bounded malabsorption and subpart vit-B_6 deficiency is explained by aging-related increases in requirements. Because of the constantly growing elderly population, it is anticipated that the prevalence of various age-related disorders would significantly rise. Additionally, these degenerative illnesses lead to numerous co-morbidities in elderly individuals, which has significant social and economic repercussions. Thus health in elder age become a top public health concern (Porter et al. 2016).

7.4.1 Health of Brain in Ageing: Cognitive Dysfunction

The range of cognitive abilities may gradually disappear, ranging from (MCI) to dementia (Hedden and Gabrieli 2004). According to estimates, 50% of elderly adults with moderate cognitive impairment will go on to acquire dementia within the 5 years of diagnosis. The pace of cognitive loss differs across individuals. It has been demonstrated that cognitive impairment and cognitive decline are caused by low vitamin B_6 levels (Tucker et al. 2005).

7.4.2 Health of Bone in Ageing

The possible significance of B-vitamins and folate in bone health MTHFR 677TT polymorphism's possible impact on bone health may be mediated by impaired methylation of DNA as a result of decreased DNA structure and stability. It was suggested that individuals with the low B-vitamin status and MTHFR 677TT genotype may be most susceptible to deprived bone health in the later life (Wolffe and Matzke 1999).

7.4.3 Cardiovascular Health in Ageing

Homocysteine is the major problem associated with the dietary nutrients. The plasma homocysteine concentrations and/or low folate with an increased risk of cognitive dysfunction. It was concluded that the risk of heart disease would reduce by 11–16% by lowering homocysteine (Who and Consultation 2003; Nowson and O'Connell 2015). However, homocysteine may be a marker of poor B-vitamin status and so just represent a disruption in the metabolism of single carbon instead of the causative function in cardiovascular disease, but this is not commonly accepted (McNulty et al. 2012).

7.5 Serum Cortisol, Psychological Stress, and Suboptimal Symptoms of Emotional and Physical Health

The core cause of the multiple negative consequences of persistent psychological stress is systemic inflammation and oxidative damage. All stress reactions stem from the disturbance of some homeostatic set point, whether physical, psychological or physiological (McEwen and Wingfield 2010). Stress is linked to the activation of the autonomic and the endocrine systems, including limbic, hypothalamus, and the brainstem circuits (Chrousos 2000). Chronic activation of these systems can lead to stress-related diseases like depression and other anxiety disorders (Mitra and Sapolsky 2008). The effects of the macular carotenoids on cortisol, stress, and signs of poor mental and physical health are all dramatically reduced (Stringham et al. 2018).

7.6 Intake of Salt

Excessive consumption of dietary salt i.e., sodium chloride is linked to an increased hazard of the increase in blood pressure, which is the main hazardous factor for the stroke and the other heart related disorders, along with renal disease. Besides, High intake of salt or a liking for salty foods has been linked to an amplified hazard of stomach cancer and, an enhanced hazard of the obesity. Reduced dietary salt consumption, on the other hand, leads to a significant drop in blood pressure, particularly in individuals with hypertension, albeit to a smaller extent (Rust and Ekmekcioglu 2017). There are several studies in the context of impact of salt intake on the treatment and pathogenesis of increased blood pressure i.e., hypertension. For example:

O'Donnell et al. (2014) suggested that a diet with lower level of salt may bears negative health consequences, particularly in terms of cardiovascular disease and death and;

Mente et al. (2016) for instance reviewed an overtone between the low level of sodium intake and increased hazard of cardio-vascular disease and mortality in the both normotensive and hypertensive personalities.

8 Nutrient Assessment for Suboptimal Health

Nutritional status/Nutrient is the current status of body, of a population group or a person, related to consumption and utilization of nutrients i.e., their state of nourishment. Some basic functions of essential nutrients are mentioned in Table 11.2. The status of nutrition is determined by a complex interaction between external environment factors and internal/constitutional factors:

(a) Constitutional/internal factors like physical activity sex, age, nutrition, behavior, and ailments.
(b) External environmental factors like safety of food, economic, cultural, and social circumstances.

"Nutrient/Nutritional assessment, defined as the interpretation from laboratory, dietary, anthropometric, and clinical studies, is used to determine the nutritional status of individual or population groups as influenced by utilization and intake of nutrients". Definitions of some related terms used in this chapter are mentioned in Table 11.1. The major goals of nutritional assessment are to recognize the presence and type of malnutrition, to plan suitable diets as the prophylaxis in contrast to ailment later in life and to define health-threatening obesity (Kumar et al. 2019). Malnutrition caused by illness is an imbalance between nutrient intake, energy and needs. It causes the body to undergo a number of metabolic and functional changes (Sumit et al. 2012). There are several limits to traditional nutrition evaluation metrics such as weight loss, mass index of body, consumption of food, or common laboratory indicators such as lymphocytes or albumin (García Almeida et al. 2021).

8.1 Assessment of Vitamin Status

8.1.1 Vitamin D

8.1.1.1 Function of Vitamin D

Vitamin D is necessary for the maintenance of normal blood Ca and K level, which are required for appropriate bone mineralization, nerve transmission, muscular contraction, and general cellular function in all the cells of the human body.

8.1.1.2 Sources of Vitamin D

Vitamin D is found primarily in two forms: D_2 (ergocalciferol) and D_3 (cholecalciferol). In addition to the D_2 and D_3 and forms of vitamin D, 25-hydroxy vitamin D plays an important role in dietary vitamin D consumption. It is found in products derived from animals. Fortified food contains D_2 and D_3 forms or vitamin D metabolite 25-hydroxy vitamin D (Benedik 2021). The exposure to sunlight is and essential source for the vitamin D. During the winter, when sunshine exposure is limited, vitamin D stored in the adipose tissue is accessible.

8.1.1.3 Dietary Sources of Vitamin D

Vitamin D is only found in trace levels in a select few foods. Fish is a good source, particularly fish with a lot of fat, such sardines, salmon, herring, and mackerel. Other significant sources include beef, milk, eggs, and fortified foods like margarine (Nowson and Margerison 2002).

8.1.1.4 Metabolism of Vitamin D

Vitamin D insufficiency is a widespread concern in many people across the world (Lips 2010). Aside from its involvement in calcium and phosphate metabolism, current research has shown that vitamin D has a title role in vascular, inflammatory, neoplastic, and neurological illnesses (Holick 2007). D_3 (cholecalciferol) and D_2 (ergocalciferol) are the two primary forms of vitamin D, which different in the assembly of their side chains. The presence of a methyl group on carbon number 24 and double bond between the carbons (22 and 23) distinguishes side chain of vitamin D_2 from the D_3 (Zhao et al. 2015).

8.1.1.5 Deficiency of Vitamin D

Numerous disorders, including cardiovascular disease, osteoporosis, cancer, rickets, multiple sclerosis, and autoimmune disease, are linked to vitamin D and both their occurrence and development (Herrmann et al. 2017).

8.1.2 Vitamin B_{12} (Cobalamin)

8.1.2.1 Function of Vitamin B_{12}

The synthesis of red blood cells and deoxyribonucleic acid both require vitamin B_{12}. The formation and operation of nerve and brain cells are also significantly impacted by it. Protein in our food forms a binding site for vitamin B_{12}. Hydrochloric acids and digestive enzymes in the stomach converts vitamin B_{12} to its free form.

8.1.2.2 Sources of Vitamin B_{12}

Plant foods like as dried purple and green lavers, have significant levels of vitamin B_{12}, but other edible algae contained only or none traces. The majority of cyanobacteria used in human supplements contain pseudo-vitamin B_{12}, which is inert in humans. For older persons and vegetarians, fortified breakfast cereals are a very beneficial source of vitamin B_{12} (Watanabe 2007).

8.1.2.3 Dietary Source and Availability

Vitamin B_{12} is synthesized in microbes enters the human food chain by integration into animal-derived foods. Gastrointestinal fermentation promotes the development of these vitamin B_{12}-producing bacteria in several animals, which are then absorbed and integrated into the tissue of animal. This is especially true for the liver, which stores enormous amounts of vitamin B_{12}. Unless the animal lives in one of the numerous places known to be geochemically poor in cobalt, products from animals which are herbivorous, such as meat, eggs, and milk are essential dietary sources of the nutrient (Zent and Smith 1995).

8.1.2.4 Absorption of Vitamin B_{12}

The absorption pattern of vitamin B_{12} is complicated. Vitamin B_{12} in the food is liberated from the proteins and is bound to proteins by the action of high concentration of the hydrochloric acid in the stomach (Shils et al. 1994).

8.1.2.5 Deficiency of Vitamin B_{12}

Vitamin B_{12} (cobalamin) is vital for DNA synthesis and neurological function. Deficiency of vitamin B_{12} can cause a variety of hematologic and neuropsychiatric diseases, many of which can be reversed with early detection and therapy (Oh and Brown 2003).

8.1.3 Vitamin C

8.1.3.1 Function of Vitamin C

Vitamin C is an antioxidant or electron giver, and this role accounts for the majority of its biochemical and molecular actions.

8.1.3.2 Sources of Vitamin C

Vitamin C is plentiful in many natural sources, such as fresh vegetables and fruits. Citrus fruits such as oranges, limes, Tomatoes, lemons, papaya, potatoes, red & green peppers, strawberries and kiwifruits, Indian gooseberry, leafy green vegetables like as fortified cereals & broccoli all are rich in vitamin C (Devaki and Raveendran 2017).

8.1.3.3 Dietary Sources of Vitamin C

Numerous plants and fruits contain ascorbate. Citrus juices and fruits are particularly abundant in vitamin C, but other fruits with varying vitamin C content also include cherries, honeydew melon, cantaloupe, papaya, kiwi fruits, strawberries, watermelon, tomatoes and tangelo (Madhavi et al. 2019). The best sources of vitamin C may be vegetables rather than fruits, including brussels & bean sprouts, broccoli, cabbage, kale, cauliflower, mustard greens, peas, green & red peppers, tomatoes and potatoes (Haytowitz 1995).

8.1.3.4 Deficiency of Vitamin C

In humans, the deficiency of the vitamin C causes a major disease which is known as scurvy. The majority of symptoms related to scurvy manifest in mesenchymal tissues. It causes poor wound healing, bleeding (due to insufficient intercellular material creation) in skin, oedema, mucosal membranes, internal muscles & organs and weakness in collagenous structures in connective tissues bone, teeth, and cartilage (Devaki and Raveendran 2017).

8.1.4 Vitamin A

8.1.4.1 Function of Vitamin A

Retinol, often known as vitamin A, is a critical ingredient that supports proper growth and development, immune system health, reproduction, and the operation of the visual system in humans.

8.1.4.2 Sources of Vitamin A

Vitamin A is present in animal sources as retinol, which is the active form of vitamin A. In particular, fish liver is an excellent source. Dairy products and egg yolk (which is not white in color) and such as milk (which includes human breast milk), butter and cheese are other animal sources. Meat, which is derived from the animal's muscles, is not an excellent source. Plant sources include vitamin A in the form of carotenoids, which must be transformed into retinol during digestion before the body can utilize it. Carotenoids are the pigments that give plants their color which is green as well as certain vegetables and fruits their particular color (Gilbert 2013).

8.1.4.3 Dietary Sources of Vitamin A

Glandular meats, human milk, fish liver and liver oils (particularly), whole dairy & milk products and egg yolk, are practically entirely sources of preformed vitamin A. Preformed form of vitamin A is also added to processed foods such as sauces, sugar, oils and fats. Green leafy vegetables (e.g., amaranth spinach), yellow & orange non-citrus fruits (e.g., papaya, mangoes, and apricots) and yellow vegetables (e.g., carrots, squash, and pumpkins) contain pro-vitamin A carotenoids. Red palm oil, which is produced in numerous places across the world, is particularly high in precursor of vitamin A (Rodriguez-Amaya 1997).

8.1.4.4 Deficiency of Vitamin A

The deficiency of the vitamin A is not merely known. It is defined by world health organization as tissue of vitamin A concentrations low enough for negative health repercussions even if clinical xerophthalmia is not present. Some of particular symptoms and signs associated with the risk of irreversible blindness and xerophthalmia, non-specific symptoms of xerophthalmia also include poor reproductive health, higher morbidity and mortality, contributing to slower development and growth and an increased hazard of anaemia (World Health Organization 2009).

8.1.5 Vitamin E

8.1.5.1 Function of Vitamin E

The key biological function of vitamin E is to prevent free radicals from oxidizing PUFAs, and low-density lipoprotein and other components of cell membranes (LDL) (Duthie 1993).

8.1.5.2 Sources of Vitamin E

Vitamin E may be found in a variety of foods. Vitamin E is found in abundance in seeds and fruits. Green vegetables are another good source. It is synthesized by photosynthetic organisms and plants. Dietary supplements are also an essential source of vitamin E. The vitamin E in pills is significantly more than the vitamin E in meals (Niki and Abe 2019).

8.1.5.3 Dietary Sources of Vitamin E

Both plant-based diets and animal products naturally contain vitamin E, and producers frequently add it to processed meals and vegetable oils.

8.1.5.4 Deficiency of Vitamin E

The deficiency of vitamin E leads to issues with the muscles and nervous system (Jeet and Baldi 2015). The leakage of the muscle enzymes like pyruvate and creatine kinase into the plasma, elevated levels of lipid peroxidation products in the plasma and enhanced erythrocytes haemolysis are some of the early diagnostic symptoms of deficiency (Sokol 1988).

8.2 Assessment of Protein Status

8.2.1 Function of Protein

It is significant in the diet as the source of amino acids, that are vital for protein synthesis in body and some other important compounds consist of nitrogen. The latter include nucleotides, creatine, nucleic acids and polypeptide hormones and some neurotransmitters and other polypeptide hormones (Hoffman and Falvo 2004).

8.2.2 Sources of Protein

Eating nuts, whole grains, legumes (peas and beans), seeds, and other plant-based sources of protein is good for the health and environment.

8.2.3 Sources and Dietary Intakes

The food of animal origin is the major source of protein. Cereal grains can supply up to 23% of the total intake of the protein in these diets. The adequacy of protein depends on quality as well as quantity (Madrigal et al. 2021).

8.2.4 Absorption and Metabolism

A number of proteolytic enzymes hydrolyze dietary protein in the gastro-intestinal tract. The final products in the digestion of free amino acids, protein, and tiny peptides, are took up by the mucosal cells in small intestine by an energy-requiring process involving several carrier systems (Bilsborough and Mann 2006).

8.2.5 Deficiency of Protein

During temporary changes in dietary intake, the key loss of protein occurs from the visceral protein pool (De Blaauw et al. 1996). In chronic absence states, the largest single contributor to protein loss is the skeletal muscle.

8.2.6 Effect of High Intake

High intakes or habitual of protein can increase the risk for diabetes, renal cell cancer and prostate cancer. Additional adverse effects on bone health and immune competence were also reported (Wolfe et al. 2008).

8.3 Assessment of Mineral Status

Mineral elements recognized to be vital in human nutrition include macro minerals (calcium, potassium, magnesium, Sulphur, sodium, and chlorine) and trace minerals (iron, zinc, copper, nickel, carbon, selenium, chromium, iodine and more). Naturally, the bulk of these substances enter the body through the gastrointestinal tract on a regular basis. As a result, the bioavailability, or the portion of an eaten substance that is absorbed, is a complicated relationship between a number of dietary conditions, involves the source of protein (Arora and Baldi 2015b).

The bioavailability of minerals in ingested meals is also influenced by the chemical makeup of the element in question as well as the metabolic and nutritional status of the host (Janghorbani and Young 1980).

8.3.1 Function of Minerals

Minerals are necessary for the wellness of body. Minerals are used by the human body for a variety of purposes, including the normal functioning of bones, muscles, the heart, and the brain. They are also required for the production of enzymes and hormones. (Basic function of minerals is mentioned in Table 11.2). They are classified into two major types such as macrominerals and trace minerals (Majewski et al. 2016).

8.3.2 Sources of Minerals

Meat, cereals, fish, dairy, bread, nuts, milk, fruit (mainly dried fruit), and vegetables are very rich in minerals.

8.3.3 Deficiency of Minerals

Minerals are certain types of nutrients that the body need to function effectively. The deficiency of mineral occurs when the body does not absorb or receive enough of the mineral. To maintain health, the human body requires varying quantities of each mineral (Arora et al. 2013).

8.4 Assessment of Iron

It is crucial to evaluate the population's iron status since iron insufficiency is the most common micronutrient deficit in developed and low-income nations.

8.4.1 Function of Iron

Seventy percent of iron is present in the hemoglobin, which is the oxygen carrying pigment of the red blood cells (RBCs) that plays a critical role in transferring the oxygen from lungs to tissues. About 4% of body iron is present in myoglobin, the oxygen binding storage protein found in muscle. The structure of myoglobin is similar to hemoglobin, except that it contains only one globin chain and one heme unit. Trace quantities of iron are also associated with transport of several electrons and enzymes (Gupta 2014).

8.4.2 Sources of Iron

Food is the only natural source of iron.

8.4.3 Dietary Sources of Iron

Many dietary supplements include iron. The best supplements are those that have readily absorbable iron with little adverse effects. It is best to take iron under the guidance of a doctor. Multivitamin/multimineral iron supplements, particularly those geared toward women, often contain 18 mg of iron (Moustarah and Mohiuddin 2019).

8.4.4 Absorption and Metabolism

There are three main factors in the body operates to maintain the balance and prevent iron deficiency and iron overload. These are in-take, storage and loss of iron. Iron intake is measured by the bioavailability and quantity of iron in the diet and the capacity to absorb the iron. The quantity of dietary iron get absorbed is measured by the nutritional needs of the specific and by the factors influencing the bioavailability of the ingested iron (Anderson et al. 2009).

8.4.5 Deficiency of Iron

The deficiency of iron is prevalent in young offspring, infants and pregnant women. It may arise from adequate intakes of poor absorption, excessive losses, dietary iron or combination of these factors (Camaschella 2019).

8.5 Assessment of Magnesium

8.5.1 Function of Magnesium

Soft tissue magnesium maintains the electrical potential of cell membranes and neural tissues, and works as the cofactor of several enzymes which are involved in synthesis of protein, synthesis of DNA & RNA, metabolism of energy, and other processes. The function of magnesium in controlling potassium fluxes & their participation in the metabolism of the calcium are particularly significant with regard to the pathologic implications of magnesium deficiency (Classen 1984).

8.5.2 Sources of Magnesium

Magnesium may be found in a wide array of nutrient-dense foods like as nuts, seeds, legumes, whole grains, and a variety of vegetables and fruits.

8.5.3 Dietary Sources of Magnesium

Magnesium deficit in the diet that is severe enough to cause pathologic alterations is uncommon. Magnesium is extensively distributed in meals from both plants and animals, and its concentration in food is rarely significantly influenced by geochemical processes or other environmental factors. Magnesium is abundant in many foods, including soy flour, certain shellfish, beans nuts, legume seeds, and peas. These foods often contain more than 500 mg of magnesium per kilogram of fresh weight. Many highly refined flours, tubers, fruits, and mushrooms, as well as most fats and oils, supply little dietary magnesium i.e., 100 kg/mg fresh weight, despite the fact that most unprocessed cereal grains are reasonable suppliers (Paul and Southgate 1978).

8.5.4 Deficiency of Magnesium

Pathologic consequences of primary dietary magnesium shortage are uncommon in babies and much less prevalent in adults unless the relatively lower intake of magnesium is followed by persistent diarrhea or significant urine magnesium losses (Lönnerdal 1995). As demands for the magnesium increased significantly with the return of tissue development during the rehabilitation from basic malnutrition, susceptibility to the consequences of magnesium insufficiency increases (Shils 1988). The majority of the early pathologic effects of magnesium deficiency are neuromuscular or neurologic abnormalities, some of which are likely due to the element's influence on potassium transport within tissues (Gibson 1990).

8.6 Assessment of Selenium

8.6.1 Function of Selenium

Selenium has been associated with the maintenance of immune defenses, safety of biological tissues against, the control of development & growth and the oxidative stress.

8.6.2 Sources of Selenium

Protein foods from animals are the major sources of the selenium.

8.6.3 Dietary Sources of Selenium

Cooked plantains, shelled soybeans, cooked Brussels sprouts, raw broccoli, and cooked fresh spinach are some other vegetables that consist of more than 2 mg of selenium whereas 4 cooked spears of asparagus contain nearly 4 μg (Hu et al. 2021).

8.6.4 Absorption and Bioavailability of Selenium

Humans typically absorb selenium compounds quite effectively, and selenium absorption does not seem to be controlled by homeostasis. For instance, selenite form of selenium absorbs more than 80%, and absorb more than 90% in case of selenomethionine or selenate (Patterson et al. 1993).

8.6.5 Deficiency of Selenium

Selenium deficiency causes selenium sensitive illnesses in a variety of animal species, including muscular dystrophy, exudative diathesis, and hepatosis dietetica. Selenium deficiency causes irregular electrocardiograms and blood pressure abnormalities in rats and lambs. Human heart disease mortality is also reduced in places with high levels of selenium (Shamberger 1981).

8.7 *Assessment of Zinc*

8.7.1 Function of Zinc

The immune system is heavily reliant on zinc, which has a variety of effects on cellular and humoral immunity (Shankar and Prasad 1998).

8.7.2 Sources of Zinc

Poultry, low-fat cheese, pork, milk, yogurt, nuts, leafy, eggs, crustaceans, whole grains, beef kidney, lamb, liver, mollusks, heart and root vegetables are some examples of sources of zinc (Solomons 2001).

8.7.3 Availability and Dietary Sources of Zinc

Zinc concentrations in whole-grain cereals, legumes, red meat, and lentils range from 380 to 760 mol/kg (25–50 mg/kg) raw weight. Low extraction rate processed cereals, polished rice, and high or lean fat content meat have a reasonable zinc level of 150–380 mol/kg (10–25 mg/kg). Fruits, green leafy vegetables, roots and tubers, and fish are all low in zinc (10 mg/kg (150 mol/kg)) (Alloway 2008). The zinc level in separated oils and fats, sugar, and alcohol is quite low. Utilization of zinc is influenced by the diet's general make-up. Numerous dietary components have been discovered by experimental research as possible boosters or antagonists of zinc absorption (Sandström and Lönnerdal 1989).

8.7.4 Deficiency of Zinc

Sever zinc deficiency in humans causes reduced appetite, growth retardation, diarrhea, skin lesions, baldness, delayed sexual and bone development, increased susceptibility to infections caused by immune system deficiencies, and the emergence of behavioral disorders. The effect of mild and marginal zinc deficiency is very rare (Hambidge 1986).

9 Conclusion

Consuming a diversified nutrient-dense diet is essential for maintaining and promoting suboptimal health throughout the ageing process. Food nutrient quality is critical since inadequate nutrition is a major cause of many chronic and infectious illnesses in humans. Human nutrition science is primarily concerned with identifying the dietary requirements for the protection, promotion, and maintenance of wellbeing in all populations. Such information is required to evaluate the nutritional sufficiency of a diet for the growth of babies, children, and adolescents, as well as for the preservation of health in adults of both sexes and during pregnancy and breastfeeding in women. A variety of terminologies have been used in this context to specify the quantity needed, minimum requirements, suggested intakes, or allowances, and safe level of intake. Overall, following a nutritious diet and nutrients that are vital elements including lipids, proteins, carbohydrates, vitamins, minerals, water, fruit, whole grains, vegetables, nuts, chicken, fish, legumes, as well as dairy products with low amount of fat, can be good for treating a number of diseases. This book chapter has been extensively focused on the relation between the dietary nutrients and suboptimal health status. The role of various microorganisms, nutrients and biomarkers is discussed. This will surely give readers basic and deep understandings and deep insights related to dietary nutrients for the maintenance of their health.

References

Adami S, Giannini S, Bianchi G, Sinigaglia L, Di Munno O, Fiore CE, Minisola S, Rossini M. Vitamin D status and response to treatment in post-menopausal osteoporosis. Osteoporos Int. 2009;20(2):239–44.

Ahmadieh H, Arabi A. Vitamins and bone health: beyond calcium and vitamin D. Nutr Rev. 2011;69(10):584–98.

Alloway BJ. Micronutrients and crop production: an introduction. In: Micronutrient deficiencies in global crop production. Springer; 2008. p. 1–39.

Ananthakrishnan AN, Khalili H, Konijeti GG, Higuchi LM, de Silva P, Fuchs CS, Willett CW, Richter JM, Chan AT. Long-term intake of dietary fat and risk of ulcerative colitis and Crohn's disease. Gut. 2014;63(5):776–84.

Anderson GJ, Frazer DM, McLaren GD. Iron absorption and metabolism. Curr Opin Gastroenterol. 2009;25(2):129–35.

Appel LJ, Moore TJ, Obarzanek E, Vollmer WM, Svetkey LP, Sacks FM, Bray GA, Vogt TM, Cutler JA, Windhauser MM, Lin PH, Karanja N, Morton DS, MaCullough M, Swain J, Steele P, Evans MA, Niller ER, Harsha DW. A clinical trial of the effects of dietary patterns on blood pressure. N Engl J Med. 1997;336(16):1117–24.

Aronson JK, Ferner RE. Biomarkers – a general review. Curr Protoc Pharmacol. 2017;9(23):1–9.

Arora M, Baldi A. Comparative account of quality management and regulatory aspects of products with health claims: existing approaches and future challenges for probiotics and herbal products. Applied Clinical Research, Clinical Trials and Regulatory Affairs. 2015a;2(3):123–32.

Arora M, Baldi A. Regulatory categories of probiotics across the globe: a review representing existing and recommended categorization. Indian J Med Microbiol. 2015b;33(Suppl 1):S2–S10.

Arora M, Baldi A. Comparative study of regulatory framework for probiotics: current status and future recommendations. Appl Clin Res Clin Trials Regul Aff. 2017a;4(2):140–56.

Arora M, Baldi A. Selective identification and charcterization of potential probiotic strains: a review on comprehensive polyphasic approach selective identification and charcterization of potential probiotic strains. Appl Clin Res Clin Trials Regul Aff. 2017b;4(1):60–76.

Arora M, Sharma S, Baldi A. Comparative insight of regulatory guidelines for probiotics in USA, India and Malaysia: a critical review. Int J Biotechnol Wellness Industries. 2013;2(2):51–64.

Arora M, Arora M, Bansal P, Baldi A. Need and recommendations for universal guidelines on regulatory status and quality control/safety of probiotic products. Appl Clin Res Clin Trials Regul Aff. 2019a;6(3):231–49.

Arora M, Kaur N, Bansal P, Baldi A. Emergence of traditionally used foods as today's probiotics-long journey. Curr Tradit Med. 2019b;5(2):114–25.

Arora M, Baldi A, Kapila N, Bhandari S, Jeet K. Impact of probiotics and prebiotics on colon cancer: mechanistic insights and future approaches. Curr Cancer Ther Rev. 2019c;15(1):27–36.

Bailey R, Arab L. Nutritional prevention of cognitive decline. Adv Nutr. 2012;3(5):732–3.

Bakaev VV, Duntau AP. Ascorbic acid in blood serum of patients with pulmonary tuberculosis and pneumonia. Int J Tuberc Lung Dis. 2004;8(2):263–6.

Baldi A. Enforcement of unified global regulatory guidelines for probiotic products: the framework for future. J Forensic Sci & Criminal Inves. 2018a;5(1):8–11.

Baldi A. Probiotics for health benefits: the regulatory concerns and suggestive roadmap. Biomed J Sci Tech Res. 2018b;1(5):1230–1.

Beal T, White JM, Arsenault JE, Okronipa H, Hinnouho GM, Morris SS. Comprehensive Nutrient Gap Assessment (CONGA): a method for identifying the public health significance of nutrient gaps. Nutr Rev. 2021;79:4–15.

Bedford MR, Apajalahti JH. The role of feed enzymes in maintaining poultry intestinal health. J Sci Food Agric. 2022;102(5):1759–70.

Bender D. An introduction to nutrition and metabolism. CRC Press; 2002.

Benedik E. Sources of vitamin D for humans. Int J Vitam Nutr Res. 2021;92(2):118–25.

Bilsborough S, Mann N. A review of issues of dietary protein intake in humans. Int J Sport Nutr Exerc Metab. 2006;16(2):129–52.

Bin DZ, Ren XL, Xue YL, Tian Y, He BB, Xu CL, Yang B. Association of dietary intake and biomarker of α-linolenic acid with incident colorectal cancer: a dose-response meta-analysis of prospective cohort studies. Front Nutr. 2022;9:948604.

Boyd CE. Water quality: an introduction. Springer Nature; 2019.

Camaschella C. Iron deficiency. Blood. 2019;133(1):30–9.

Chandra RK. 1990 McCollum award lecture. Nutrition and immunity: lessons from the past and new insights into the future. Am J Clin Nutr. 1991;53(5):1087–101.

Chang KV, Wu WT, Chen LR, Wang HI, Wang TG, Han DS. Suboptimal tongue pressure is associated with risk of malnutrition in community-dwelling older individuals. Nutrients. 2021;13(6):1–11.

Chen J, Cheng J, Liu Y, Tang Y, Sun X, Wang T, Xiao Y, Li F, Xiang L, Jiang P, Wu S, Wu L, Luo R, Zhao X. Associations between breakfast eating habits and health-promoting lifestyle, suboptimal health status in southern China: a population based, cross sectional study. J Transl Med. 2014;12(1):1–10.

Chiodini I, Bolland MJ. Calcium supplementation in osteoporosis: useful or harmful? Eur J Endocrinol. 2018;178(4):D13–25.

Chrousos GP. Stress, chronic inflammation, and emotional and physical Well-being: concurrent effects and chronic sequelae. J Allergy Clin Immunol. 2000;106(5):S275–91.

Classen HG. Magnesium and potassium deprivation and supplementation in animals and man: aspects in view of intestinal absorption. Magnesium. 1984;3(4–6):257–64.

Conley M, Campbell KL, Hawley CM, Lioufas NM, Elder GJ, Badve SV, Pedagogos E, Milanzi E, Pascoe EM, Valks A, Toussaint ND. Relationship between dietary phosphate intake and biomarkers of bone and mineral metabolism in Australian adults with chronic kidney disease. J Ren Nutr. 2022;32(1):58–67.

D'Angelo C, Reale M, Costantini E. Microbiota and probiotics in health and HIV infection. Nutrients. 2017;9(6):615.

De Blaauw I, Deutz NE, Von Meyenfeldt MF. *In-vivo* amino acid metabolism of gut and liver during short and prolonged starvation. Am J Physiol Liver Physiol. 1996;270(2):G298–306.

Dekker E, Tanis PJ, Vleugels JLA, Kasi PM, Wallace M. Colorectal cancer. Lancet. 2019;394:1467–80.

Devaki SJ, Raveendran RL. Vitamin C: sources, functions, sensing and analysis. In: Vitamin C. IntechOpen; 2017.

Duthie GG. Lipid peroxidation. Eur J Clin Nutr. 1993;47(11):759–64.

Echouffo-Tcheugui JB, Ahima RS. Does diet quality or nutrient quantity contribute more to health? J Clin Invest. 2019;129(10):3969–70.

Eggersdorfer M, Akobundu U, Bailey RL, Shlisky J, Beaudreault AR, Bergeron G, Blancato RB, Blumberg JB, Bourassa MW, Gomes F, Jensen G, Johnson MW, Mackay D, Marshall K, Meydani SN, Tucker KL. Hidden hunger: solutions for America's aging populations. Nutrients. 2018;10(9):1–15.

Erickson RH, Kim YS. Digestion and absorption of dietary protein. Annu Rev Med. 1990;41(1):133–9.

Fallah S, Musa-Veloso K, Cao J, Venditti C, Lee HY, Hamamji S, Hu J, Appelhans K, Frankos V. Liver biomarkers in adults: evaluation of associations with reported green tea consumption and use of green tea supplements in U.S. NHANES. Regul Toxicol Pharmacol. 2022;129:105087.

Ferrie S. What is nutritional assessment? A quick guide for critical care clinicians. Aust Crit Care. 2020;33(3):295–9.

Finck H, Hart AR, Jennings A, Welch AA. Is there a role for vitamin C in preventing osteoporosis and fractures? A review of the potential underlying mechanisms and current epidemiological evidence. Nutr Res Rev. 2014;27(2):268–83.

Gaewkhiew P, Sabbah W, Bernabé E. Does tooth loss affect dietary intake and nutritional status? A systematic review of longitudinal studies. J Dent. 2017;67:1–8.
García Almeida JM, García García C, Vegas Aguilar IM, Bellido Castañeda V, Bellido Guerrero D. Morphofunctional assessment of patient's nutritional status: aglobal approach. Nutr Hosp. 2021;38(3):592–600.
Gibson RS. Principles of nutritional assessment. Oxford University Press; 1990.
Gilbert C. What is vitamin A and why do we need it? Commun Eye Heal. 2013;26(84):65.
Goldszmid RS, Trinchieri G. The price of immunity. Nat Immunol. 2012;13(10):932–8.
Goswami PK, Shukla AK, Roy A, Soni G, Singh V, Kumar V. Balanced diet for all age group. World J Pharm Res. 2017;6(3):448–58.
Griel AE, Kris-Etherton PM, Hilpert KF, Zhao G, West SG, Corwin RL. An increase in dietary n-3 fatty acids decreases a marker of bone resorption in humans. Nutr J. 2007;6(1):1–8.
Griffiths HR, Møller L, Bartosz G, Bast A, Bertoni-Freddari C, Collins A, Cooke M, Coolen S, Haenen G, Hoberg A-M, Loft S, Lunec J, Olinski R, Parry J, Pompella A, Poulsen H, Verhagen H, Astley SB. Biomarkers. Mol Aspects Med. 2002;23(1–3):101–208.
Gupta CP. Role of iron (Fe) in body. IOSR J Appl Chem. 2014;7(11):38–46.
Hambidge KM. Trace elements in human and animal nutrition. Zinc. 1986;2:13–9.
Hayes A, Duffy S, O'Grady M, Jakobsen J, Galvin K, Teahan-Dillon J, Kerry J, Kelly A, O'Doherty J, Higgins S, Seamans M, Cashman KD. Vitamin D–enhanced eggs are protective of wintertime serum 25-hydroxyvitamin D in a randomized controlled trial of adults. Am J Clin Nutr. 2016;104(3):629–37.
Haytowitz DB. Information from USDA's nutrient data bank. J Nutr. 1995;125(7):1952–5.
Hedden T, Gabrieli JDE. Insights into the ageing mind: a view from cognitive neuroscience. Nat Rev Neurosci. 2004;5(2):87–96.
Herrmann M, Farrell C-JL, Pusceddu I, Fabregat-Cabello N, Cavalier E. Assessment of vitamin D status–a changing landscape. Clin Chem Lab Med. 2017;55(1):3–26.
Hoffman JR, Falvo MJ. Protein–which is best? J Sports Sci Med. 2004;3(3):118–30.
Holick MF. Vitamin D deficiency. N Engl J Med. 2007;357(3):266–81.
Hu W, Zhao C, Hu H, Yin S. Food sources of selenium and its relationship with chronic diseases. Nutrients. 2021;13(5):1739.
Institute of Medicine (US) Standing Committee on the Scientific Evaluation of Dietary Reference Intakes. Dietary reference intakes for calcium, phosphorus, magnesium, vitamin D, and fluoride. Washington (DC): National Academies Press (US); 1997. p. 1–324.
Janghorbani M, Young VR. Stable isotope methods for bioavailability assessment of dietary minerals in humans. Adv Nutr Res. 1980;7919112:127–55.
Jeet K, Baldi A. Management of acute mountain sickness: an exploratory review on herbal drugs. Indian J Nat Prod. 2015;29(1):5–22.
Jilcott Pitts SB, Moran NE, Wu Q, Harnack L, Craft NE, Hanchard N, Bell R, Moe SG, Johnson N, Obasohan J, Carr-Manthe PL, Laska MN. Pressure-mediated reflection spectroscopy criterion validity as a biomarker of fruit and vegetable intake: a 2-site cross-sectional study of 4 racial or ethnic groups. J Nutr. 2022;152(1):107–16.
Kant S, Gupta H, Ahluwalia S. Significance of nutrition in pulmonary tuberculosis. Crit Rev Food Sci Nutr. 2015;55(7):955–63.
Karyadi E, Schultink W, Nelwan RHH, Gross R, Amin Z, Dolmans WMV, Meer JWMVD, Hautvast JGA, West CE. Poor micronutrient status of active pulmonary tuberculosis patients in Indonesia. J Nutr. 2000;130(12):2953–8.
Kong SH, Kim JH, Hong AR, Lee JH, Kim SW, Shin CS. Dietary potassium intake is beneficial to bone health in a low calcium intake population: the Korean National Health and Nutrition Examination Survey (KNHANES)(2008–2011). Osteoporos Int. 2017;28(5):1577–85.
Kumar S, Baldi A, Sharma DK. Phytosomes: a modernistic approach for novel herbal drug delivery enhancing bioavailability and revealing endless frontier of phytopharmaceuticals. J Dev Drugs. 2019;8(1):1–8.
Lentjes MAH. The balance between food and dietary supplements in the general population. Proc Nutr Soc. 2019;78(1):97–109.

Li G, Xie F, Yan S, Hu X, Jin B, Wang J, Wu J, Yin D. Subhealth: definition, criteria for diagnosis and potential prevalence in the central region of China. BMC Public Health. 2013;13(1):1–8.

Lips P. Worldwide status of vitamin D nutrition. J Steroid Biochem Mol Biol. 2010;121(1–2): 297–300.

Lönnerdal B. Magnesium nutrition of infants. Magnes Res. 1995;8(1):99–105.

Lupton JR, Blumberg JB, L'Abbe M, LeDoux M, Rice HB, von Schacky C, Yaktine A, Griffiths JC. Nutrient reference value: Non-communicable disease endpoints – a conference report. Eur J Nutr. 2016;55(1):1–10.

Lv L, Mu D, Du Y, Yan R, Jiang H. Mechanism of the immunomodulatory effect of the combination of live bifidobacterium, lactobacillus, enterococcus, and bacillus on immunocompromised rats. Front Immunol. 2021;12:2342.

Macdonald IA. Arterio–venous differences to study macronutrient metabolism: introduction and overview. Proc Nutr Soc. 1999;58(4):871–5.

Madhavi N, Kumar D, Naman S, Singh M, Singh PA, Bajwa N, Baldi A. Formulation and evaluation of novel herbal formulations incorporated with amla extract for improved stability. J Drug Deliv Ther. 2019;9(4):212–21.

Madrigal C, Soto-Méndez MJ, Hernández-Ruiz Á, Valero T, Lara Villoslada F, Leis R, Martinez de Vicotoria E, Moreno Manuel J, Ortega MR, Ruiz-Lopez Dolores M, Varela-Moreiras G, Gil A. Dietary intake, nutritional adequacy, and food sources of protein and relationships with personal and family factors in spanish children aged one to < 10 years: findings of the esnupi study. Nutrients. 2021;13(4):1062.

Majewski M, Kozlowska A, Thoene M, Lepiarczyk E, Grzegorzewski WJ. Overview of the role of vitamins and minerals on the kynurenine pathway in health and disease. J Physiol Pharmacol. 2016;67(1):3–19.

Maqbool MA, Aslam M, Akbar W, Iqbal Z. Biological importance of vitamins for human health: a review. J Agric Basic Sci. 2018;2(3):50–8.

Mateş L, Popa D-S, Rusu ME, Fizeşan I, Leucuţa D. Walnut intake interventions targeting biomarkers of metabolic syndrome and inflammation in middle-aged and older adults: a systematic review and meta-analysis of randomized controlled trials. Antioxidants. 2022;11(7):1412.

Mayeux R. Biomarkers: potential uses and limitations. NeuroRx. 2004;1(2):182–8.

McEwen BS, Wingfield JC. What's in a name? Integrating homeostasis, allostasis and stress. Horm Behav. 2010;57(2):105–11.

McNulty H, Strain JJ, Pentieva K, Ward M. C1 metabolism and CVD outcomes in older adults. Proc Nutr Soc. 2012;71(2):213–21.

Mente A, O'Donnell M, Rangarajan S, Dagenais G, Lear S, McQueen M, Diaz R, Avezum A, Jaramillo PL, Lanas F, Li W, Lu Y, Yi S, Rensheng L, Iqbal R, Mony P, Yusuf R, Yusoff K, Szuba A, Oguz A, Yusuf S. Associations of urinary sodium excretion with cardiovascular events in individuals with and without hypertension: a pooled analysis of data from four studies. Lancet. 2016;388(10043):465–75.

Mitra R, Sapolsky RM. Acute corticosterone treatment is sufficient to induce anxiety and amygdaloid dendritic hypertrophy. Proc Natl Acad Sci. 2008;105(14):5573–8.

Moustarah F, Mohiuddin SS. Dietary iron. Treasure Island: StatPearls Publishing; 2019.

Mugusi FM, Rusizoka O, Habib N, Fawzi W. Vitamin A status of patients presenting with pulmonary tuberculosis and asymptomatic HIV-infected individuals, Dar es Salaam, Tanzania. Int J Tuberc Lung Dis. 2003;7(8):804–7.

Muñoz-garach A, García-fontana B, Muñoz-torres M. Nutrients and dietary patterns related to osteoporosis. Nutrients. 2020;12(7):1–15.

Nesheim MC. What is the research base for the use of dietary supplements? Public Health Nutr. 1999;2(1):35–8.

Niki E, Abe K. Vitamin E: structure, properties and functions. Food Chem Funct Anal. 2019:1–11.

Nowson CA, Margerison C. Vitamin D intake and vitamin D status of Australians. Med J Aust. 2002;177(3):149–52.

Nowson C, O'Connell S. Protein requirements and recommendations for older people: a review. Nutrients. 2015;7(8):6874–99.

O'Donnell M, Mente A, Rangarajan S, McQueen MJ, Wang X, Liu L, Yan H, Lee SF, Mony P, Devanath A, Rosengren A, Jaramillo PL, Diaz R, Avezum A, Lanas F, Yusoff K, Iqbal R, Ilow R, Mohammadifard N, Gulec S, Yusufali AH, Lruger L, Yusuf R, Chifamba J, Kabali C, Dagenais G, Lear SA, Teo K, Yusuf S. Urinary sodium and potassium excretion, mortality, and cardiovascular events. N Engl J Med. 2014;371(7):612–23.

Oh RC, Brown DL. Vitamin B12 deficiency. Am Fam Physician. 2003;67(5):979–86.

Patterson BH, Zech LA, Swanson CA, Levander OA. Kinetic modeling of selenium in humans using stable isotope tracers: conference on trace elements in health and disease. J Trace Elem Electrolytes Health Dis. 1993;7(2):117–20.

Paul AA, Southgate DAT. McCance and Widdowson's the composition of foods. London: HMSO; 1978.

Pearlin BV, Muthuvel S, Govidasamy P, Villavan M, Alagawany M, Ragab Farag M, Dhama K, Gopi M. Role of acidifiers in livestock nutrition and health: a review. J Anim Physiol Anim Nutr (Berl). 2020;104(2):558–69.

Plotkin A, Taani MH. Factors associated with food intake, nutritional status, and function among nursing home residents with dementia. Geriatr Nurs (Minneap). 2020;41(5):559–63.

Porter K, Hoey L, Hughes CF, Ward M, McNulty H. Causes, consequences and public health implications of low B-vitamin status in ageing. Nutrients. 2016;8(11):725.

Rao RK, Samak G. Protection and restitution of gut barrier by probiotics: nutritional and clinical implications. Curr Nutr Food Sci. 2013;9(2):99–107.

Rizzoli R, Stevenson JC, Bauer JM, van Loon LJC, Walrand S, Kanis JA, Cooper C, Brandi ML, Perez MD, Reginster JY. The role of dietary protein and vitamin D in maintaining musculoskeletal health in postmenopausal women: A consensus statement from the European Society for Clinical and Economic Aspects of Osteoporosis and Osteoarthritis (ESCEO). Maturitas. 2014;79(1):122–32.

Roberfroid MB. Defining functional foods. Funct foods. 2000;9:9–27.

Rodriguez-Amaya DB. Carotenoids and food preparation: the retention of provitamin A carotenoids in prepared, processed and stored foods. Arlington: John Snow Incorporated/OMNI Project; 1997.

Rust P, Ekmekcioglu C. Impact of salt intake on the pathogenesis and treatment of hypertension. Adv Exp Med Biol. 2017;956:61–84.

Sandström B, Lönnerdal B. Promoters and antagonists of zinc absorption. In: Zinc in human biology. Springer; 1989. p. 57–78.

Semba RD. The role of vitamin A and related retinoids in immune function. Nutr Rev. 1998;56(1):S38–48.

Shamberger RJ. Selenium in the environment. Sci Total Environ. 1981;17(1):59–74.

Shankar AH, Prasad AS. Zinc and immune function: the biological basis of altered resistance to infection. Am J Clin Nutr. 1998;68(2):447S–63S.

Shils ME. Magnesium in health and disease. Annu Rev Nutr. 1988;8(1):429–60.

Shils ME, Olson JA, Shike M. Modern nutrition in health and disease. 1994.

Singh A. Role of microbial metabolites in cardiovascular and human health. In: Microbiome, immunity, digestive health and nutrition. Elsevier; 2022. p. 137–48.

Sokol RJ. Vitamin E deficiency and neurologic disease. Annu Rev Nutr. 1988;8(1):351–73.

Solomons NW. Dietary sources of zinc and factors affecting its bioavailability. Food Nutr Bull. 2001;22(2):138–54.

Spence JT. Challenges related to the composition of functional foods. J Food Compos Anal. 2006;19:S4–6.

Stephen A, Alles M, De Graaf C, Fleith M, Hadjilucas E, Isaacs E, Maffeis C, Zeinstra G, Mtthys C, Gil A. The role and requirements of digestible dietary carbohydrates in infants and toddlers. Eur J Clin Nutr. 2012;66(7):765–9.

Stringham NT, Holmes PV, Stringham JM. Supplementation with macular carotenoids reduces psychological stress, serum cortisol, and sub-optimal symptoms of physical and emotional health in young adults. Nutr Neurosci. 2018;21(4):286–96.

Sumit K, Vivek S, Sujata S, Ashish B. Herbal cosmetics: used for skin and hair. Inventi Rapid Cosmeceuticals. 2012;2012(4):1–7.

Sun D, Krishnan A, Zaman K, Lawrence R, Bhattacharya A, Fernandes G. Dietary n-3 fatty acids decrease osteoclastogenesis and loss of bone mass in ovariectomized mice. J Bone Miner Res. 2003;18(7):1206–16.

Tsegaye D, Tamiru D, Belachew T. Factors associated with dietary practice and nutritional status of pregnant women in rural communities of Illu Aba Bor Zone, Southwest Ethiopia. Nutr Diet Suppl Appl Clin Res Clin Trials Regul Aff. 2017;4(2):140–56.

Tucker KL, Qiao N, Scott T, Rosenberg I, Spiro A III. High homocysteine and low B vitamins predict cognitive decline in aging men: the veterans affairs normative aging study. Am J Clin Nutr. 2005;82(3):627–35.

Ustianowski A, Shaffer R, Collin S, Wilkinson RJ, Davidson RN. Prevalence and associations of vitamin D deficiency in foreign-born persons with tuberculosis in London. J Infect. 2005;50(5):432–7.

Veronese N, Bano G, Bertozzo G, Granziera S, Solmi M, Manzato E, Sergi G, Cohen AT, Correll CU. Vitamin K antagonists' use and fracture risk: results from a systematic review and meta-analysis. J Thromb Haemost. 2015;13(9):1665–75.

Voland L, Le Roy T, Debédat J, Clément K. Gut microbiota and vitamin status in persons with obesity: a key interplay. Obes Rev. 2022;23(2):e13377.

Wang S, Wang Y, Wan X, Guo J, Zhang Y, Tian M, Fang S, Yu B. Cobalamin intake and related biomarkers: examining associations with mortality risk among adults with type 2 diabetes in NHANES. Diabetes Care. 2022;45(2):276–84.

Watanabe F. Vitamin B12 sources and bioavailability. Exp Biol Med. 2007;232(10):1266–74.

Who J, Consultation FAOE. Diet, nutrition and the prevention of chronic diseases. World Health Organ Tech Rep Ser. 2003;916(i–viii):1–149.

Wolfe RR, Miller SL, Miller KB. Optimal protein intake in the elderly. Clin Nutr. 2008;27(5):675–84.

Wolffe AP, Matzke MA. Epigenetics: regulation through repression. Science. 1999;286(5439):481–6.

World Health Organization. Global prevalence of vitamin A deficiency in populations at risk 1995-2005: WHO Global Database on Vitamin A Deficiency. 2009.

Yassine T, Khalafalla MM, Mamdouh M, Elbialy ZI, Salah AS, Ahmedou A, Mamoon A, AM EL-S, Daon HV, Dawood MAO. The enhancement of the growth rate, intestinal health, expression of immune-related genes, and resistance against suboptimal water temperature in common carp (*Cyprinus carpio*) by dietary paraprobiotics. Aquac Reports. 2021;20:100729.

Zapata HJ, Quagliarello VJ. The microbiota and microbiome in aging: potential implications in health and age-related diseases. J Am Geriatr Soc. 2015;63(4):776–81.

Zeisel SH, da Costa K-A. Choline: an essential nutrient for public health. Nutr Rev. 2009;67(11):615–23.

Zent C, Smith P. Study of the effect of concomitant food on the bioavailability of rifampicin, isoniazid and pyrazinamide. Tuber Lung Dis. 1995;76(2):109–13.

Zhao S, Gardner K, Taylor W, Marks E, Goodson N. Vitamin D assessment in primary care: changing patterns of testing. London J Prim Care (Abingdon). 2015;7(2):15–22.

[illegible]

Chapter 12
Effect of Excess Nutraceuticals on Human Health

Likhit Akotkar and Ankit Ganeshpurkar

1 Introduction

The human body relies on the intake and metabolism of nutrients in appropriate proportion and balance to sustain normal physiological functions. However, an excessive intake of nutrients can have significant consequences on metabolic reprogramming, particularly during early life. This programming refers to the long-term effects that early nutritional experiences have on an individual's metabolic regulation and overall health. Excessive nutrient intake during this critical developmental period can alter various mechanisms that can profoundly impact the body's energy sensing, appetite regulation, insulin sensitivity of adipose tissue and the functioning of brown adipose tissue. Understanding the potential mechanisms underlying metabolic programming resulting from excess nutrition intake during early life is crucial for evaluating the long-term consequences of overnutrition and its impact on overall health. This chapter sheds light on the complex interplay between excessive nutrient intake and metabolic outcomes. World Health Organization (WHO) establishes a minimum protein requirement of 0.8 g/day for adults of an ideal weight (Mathur and Pillai 2019; Council NR 1989).

Nutrition plays a vital role in sustaining life along with promoting optimal growth and health. A balanced and healthy diet is essential to provide the necessary nutrients that the human body requires. Extensive research and millennia of eating practices have consistently highlighted the significant impact of diet on human

L. Akotkar
Department of Pharmacology, Poona College of Pharmacy, Bharati Vidyapeeth (Deemed to be University), Pune, Maharashtra, India

A. Ganeshpurkar (✉)
Department of Pharmaceutical Chemistry, Poona College of Pharmacy, Bharati Vidyapeeth (Deemed to be University), Pune, Maharashtra, India

S. M. Jafari et al. (eds.), *Nutraceuticals: A New Millennium Approach for Preventive Medicine*, Food Bioactive Ingredients,
https://doi.org/10.1007/978-3-031-76481-3_12

well-being. It is widely acknowledged that nutritional intake varies across different demographic groups reflecting the diverse dietary habits and cultural influences around the world. The reports on the Global Burden of Disease have shed light on the alarming statistics regarding the health implications of poor nutrition. Approximately one in five global deaths are primarily attributed to cancer and cardiovascular diseases and can be linked to inadequate nutritional values. Furthermore, a significant portion of the population, approximately one-third, experiences excess nutrient intake. The consequences of excessive nutrient intake extend beyond mere overconsumption; they contribute to the development of various illnesses and metabolic disorders. It results in binge eating disorder, Alzheimer's disease (AD), renal complications, infertility and Non-alcoholic fatty liver disease (NAFLD). The implications of excessive nutrient intake lead to alterations in cellular pathways and abnormal tissue remodelling. These changes can ultimately lead to the manifestation of pathophysiological conditions such as diabetes mellitus, hypercholesterolemia, obesity, fatty liver, and erectile dysfunction (Siepmann et al. 2011). The impact of excessive nutrient intake on health outcomes is multifaceted, and it depends on both the quantity and duration of nutrient consumption. Consequently, understanding the upper limits of nutrient consumption becomes crucial as it helps to distinguish safe intake levels from potentially dangerous ones. Establishing preventive measures to mitigate the negative health effects of excess nutrient intake necessitates considering various factors, including the dosage, frequency, and duration of nutrient intake, as well as the potency of the nutrients involved. By carefully monitoring and regulating nutrient intake, individuals can strive to maintain a balance that promotes health and minimizes the risks associated with excessive consumption (Penagini et al. 2015; Candela et al. 2011).

In this era of abundant food availability and sedentary lifestyles, it is imperative to address the issue of excess nutrient intake and its far-reaching consequences. By gaining a deeper understanding of the intricate relationship between nutrient intake, metabolic processes, and health outcomes, we can develop effective strategies to promote moderation and balance in our diets. Additionally, fostering awareness about the risks of excessive nutrient intake and the importance of making informed dietary choices can empower individuals to take control of their health and well-being. In this comprehensive exploration of the implications of excessive nutrient intake, we will delve into the various mechanisms through which excessive consumption can affect the body (Fig. 12.1). By examining the dose-dependent nature of nutrient intake, the potential risks associated with specific nutrients, and the development of preventive measures, we can gain valuable insights into the complexities of maintaining a healthy nutritional balance. Through a collective effort to address this issue, one can work towards a healthier future where nutrition is optimized, and the risks of excessive intake are mitigated. This chapter sheds light on the complex interplay between excessive nutrient intake and metabolic outcomes.

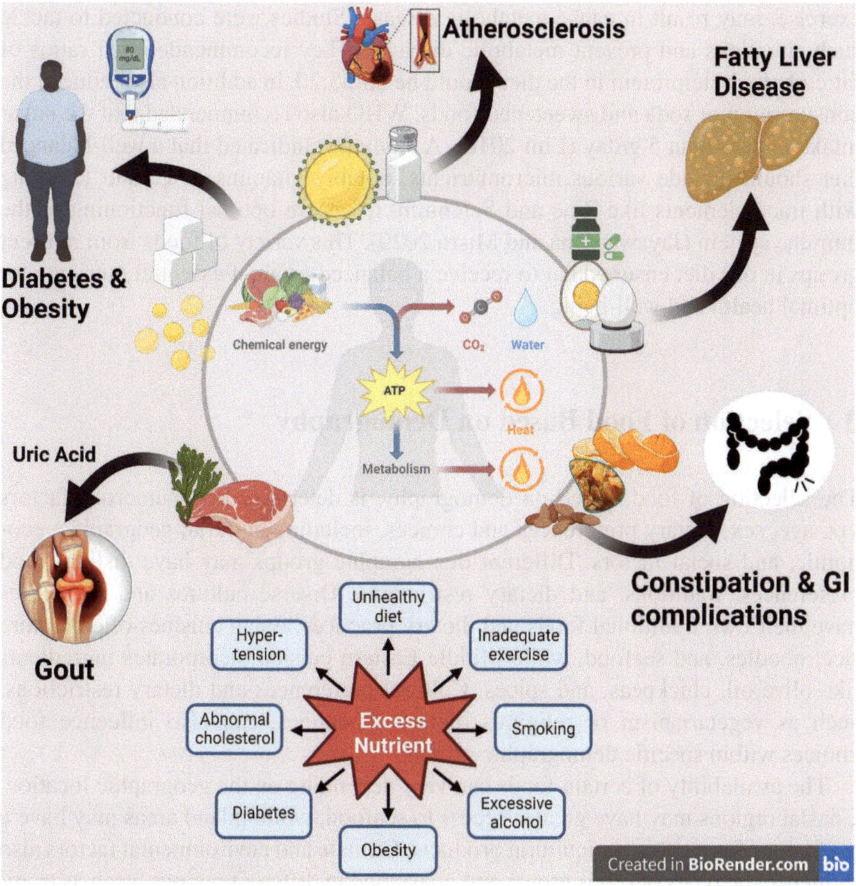

Fig. 12.1 Effect of excessive nutrient intake on Health

2 Balanced Diet and Nutrient Groups

The foundation of good health and well-being lies in maintaining a balanced diet and ensuring proper nutrition. Food is the primary supplier of the essential nutrients necessary for body growth, development and vital functions. A balanced diet involves consuming a variety of nutrients in appropriate proportions to fulfil body's nutritional requirements (Price 2005). A recent study given by Hu et al. found a correlation between rheumatoid arthritis (RA) and a balanced diet. In rheumatoid arthritis, the eating of healthy foods such as ω-2-fatty acid-rich nuts, fruits and whole grains is considered a balanced diet and unhealthy red and processed meat, sugar-sweetened beverages, trans-fats and sodium affect human health. Higher scores are associated with a healthy lifestyle and reduced risk of RA (Hu et al. 2017).

Obesity is a major health concern as a metabolic disease. Advanced lifestyle changes, health habits, unbalanced meals, and lack of body movements and exercise, may result in cause metabolic disease. Studies were conducted to tackle such situations and prevent metabolic diseases. They recommended that ratios of fat:carbohydrate:protein in the diet should be 25:55:20, in addition also reduced the consumption of soda and sweetened foods. WHO also recommended that the sugar intake be less than 5 g/day (Lim 2018). A study has indicated that a well-balanced diet should include various micronutrients, notably vitamins A, C, and D, along with trace elements like Zinc and Selenium, to ensure optimal functioning of the immune system (Jayawardena and Misra 2020). This variety of foods from nutrient groups in our diet ensures that to receive a balanced array of essential nutrients for optimal health and well-being.

3 Selection of Food Based on Demography

The selection of food, based on demography, is determined by numerous factors viz. age, sex, dietary preferences and choices, including cultural, geographic, economic, and social factors. Different demographic groups may have distinct food preferences, traditions, and dietary restrictions. Diverse cultures and ethnicities have their own traditional foods and dietary practices. Asian cuisines often feature rice, noodles, and seafood, while Middle Eastern cuisine incorporates ingredients like olive oil, chickpeas, and spices. Cultural preferences and dietary restrictions, such as vegetarianism or religious dietary guidelines, can also influence food choices within specific demographic groups.

The availability of certain foods can vary depending on the geographic location. Coastal regions may have greater access to seafood, while inland areas may have a preference for meat or agricultural products. Climate and environmental factors also influence the types of crops grown and consumed in different regions, such as tropical fruits in tropical areas or root vegetables in colder climates (Trichopoulou et al. 2007). Economic factors play a significant role in food selection. Lower-income individuals or communities may have limited fresh nutritious foods and rely more on processed or less expensive options (Wolfson et al. 2019). Conversely, higher-income individuals may have more access to a variety of high-quality and organic foods. Different age groups may have specific dietary needs and preferences. For example, infants and young children have different nutritional requirements than older adults. Adolescents may have specific preferences influenced by peer groups, while older adults may need to consider dietary modifications to address age-related health concerns. Demographic groups may have specific dietary restrictions or allergies that impact their food choices. For example, individuals with lactose intolerance may avoid dairy products, while those with celiac disease must avoid gluten-containing foods. Religious dietary practices, such as kosher or halal, also influence food selection within certain populations (Verbeke 2005).

4 Hypervitaminosis

4.1 Vitamin A and Carotenoids

Vitamin A is a fat-soluble vitamin responsible for vision, immune function and cellular growth and differentiation. The recommended dietary intake for vitamin A varies by age and gender but generally ranges from 600 to 900 μg per day for adults. Carotenoids are plant pigments that have antioxidant properties and are important for maintaining healthy skin and preventing chronic diseases. α- and β-carotene and β-cryptoxanthin are provitamin A carotenoids and can be converted to vitamin A (Weber and Grune 2012). While these nutrients are important, excess intake of carotenoids and vitamin A can have negative health effects. The excess intake of vitamin A also known as hypervitaminosis A leads to health problems, including liver damage, bone abnormalities and increased pressure on the brain. The symptoms of hypervitaminosis A include skin rash, nausea, and joint pain.

A study conducted on female rats indicated a significant fall in pregnancy rate and litter-carrying capacity on excessive intake of vitamin A. The anomaly in skull and brain development was also observed in the offspring. Further, intermittent defects such as glossia, harelip, cleft palate and eye defects were also observed (Cohlan 1953). Another independent study on rats also indicated the shortening of the mandible and maxilla, anencephaly, eye defects, cleft palate, spina bifida with meningocele and hydrocephalus caused to hypervitaminosis A (Geelen and Peters 1979). Spina bifida and cleft palate are congenital deformities caused due to excessive intake of vitamin A. It is postulated that the palatogensis is affected due to interference in cell division during shelf outgrowth, epithelial differentiation and apoptosis in shelf fusion. It also prevents hydration of the extracellular matrix leading to a hampered shelf elevation process. A high level of vitamin A also alters the action of growth factors like platelet-derived growth factor and transforming growth factor-β (Ackermans et al. 2011). The long-term excessive intake of vitamin A has also been associated with an increased risk of fractures and osteoporosis. It inhibits osteoblast activity and increases the rate of osteoclast formation. The regulation and maintenance of calcium levels by vitamin D in serum at a normal range is also hampered. It apparently shows an increased level of calcium in the serum and faster bone reabsorption, which lowers density and increases the biomarker of bones (Feskanich et al. 2002).

It is important to note that excessive intake of vitamin A is rare and usually only occurs with high-dose supplements. Carotenoids are not associated with the same risk of toxicity as vitamin A, but excessive intake of carotenoids can still have negative health effects. High doses of carotenoids can cause a yellowing of the skin called carotenemia, which is not harmful but can be mistaken for jaundice (Lascari 1981). In addition, high doses of carotenoids may increase the risk of lung cancer in smokers. This is thought to be due to the antioxidant properties of carotenoids, which can protect cells from damage caused by smoking but also protect cancer cells from being destroyed by the immune system. However, some other studies show inhibition of carcinogen-induced neoplastic transformation of lung cells (Epstein 2003).

4.2 *Vitamin B_1*

Thiamin or vitamin B_1 plays a vital role in energy metabolism and neuronal function. It is considered safe and essential for the proper body functioning. The recommended daily allowance (RDA) for adults is around 1.1–1.2 mg and 0.8–0.9 mg for males and females, respectively (Lonsdale 2006). However, excessive intake of thiamin is uncommon and usually not associated with significant health hazards. As vitamin B_1 is water-soluble, hence consumption of an excess amount is usually excreted through urine. Unlike other fat-soluble vitamins with a tendency to accumulate at toxic levels, the body does not store large amounts of thiamin. It is obtained from natural food sources and is generally well-tolerated, and rare to excessive consumption through a regular diet. The adverse effects include ataxia, impaired gut motility, lethargy, nausea and urticaria (Martel et al. 2021). In rare cases, at extremely high doses than RDA thiamin may cause adverse events. These can include allergic reactions such as itching, rash, or difficulty in breathing. These reactions are uncommon but should be taken seriously. It may also cause mild and temporary digestive disturbances such as stomach pain, nausea, and diarrhoea.

4.3 *Vitamin B_2*

Vitamin B_2, also known as riboflavin, plays several important roles viz. synthesis of flavin adenine dinucleotide and flavin mononucleotide which are responsible for energy production. Further, it is a cofactor for the anti-oxidant enzyme in the body and maintains healthy eyes and skin. The RDA for riboflavin is approximately 1.3 and 1.1 mg for adult males and females per day, respectively (Schellack et al. 2019).

Excessive intake of riboflavin is generally considered safe. However, some potential effects may occur in rare cases of extremely high doses. These effects include discolouration of urine that results in urine turning yellow and fluorescent (Rivlin and Pinto 2001). It also causes mild gastrointestinal (GIT) disturbance viz. diarrhoea, nausea, or an upset stomach. Some individuals may experience increased photosensitivity to sunlight, but this effect is more common with certain metabolic disorders like erythropoietic protoporphyria (Cardoso et al. 2012). It may also result in angular cheilitis, dysfunctional liver, photophobia, itching, burning or prickling sensation, cardiomyopathies and hypotension (Elango et al. 2015).

4.4 Vitamin B_3

Vitamin B_3 or Niacin, a water-soluble vitamin, with various important roles in the body including energy metabolism. The RDA dose of niacin is 20 mg per day (Capuzzi et al. 2000). It is responsible for the synthesis of coenzymes nicotinamide adenine dinucleotide (NAD) and nicotinamide adenine dinucleotide phosphate (NADP). It is further responsible for the repair of damaged DNA, gene expression, cell signalling and maintaining skin health (Henderson 1983). It also increases high-density lipoprotein (HDL) and decreases low-density lipoprotein (LDL) and triglycerides (Kamanna and Kashyap 2008). Excessive intake of niacin can have certain effects on the body. It may be responsible for harmless and temporary flushing at high doses due to vasodilation of blood vessels present in skin mediated through prostaglandin D_2, leading to a warm, tingling, and reddening sensation known as niacin flush (Kamanna et al. 2009). Its excess intake also causes nausea, vomiting, diarrhea, or abdominal pain, similar to other vitamin B.

4.5 Vitamin B_5

Pantothenic acid, vitamin B_5, is having various body functions including the synthesis of coenzyme A (CoA). The CoA is responsible for the metabolism of carbohydrates, fats, and proteins. Further, the acetyl CoA synthesised from pantothenic acid is responsible for the synthesis of fatty acids, cholesterol, and certain neurotransmitters. It is responsible for the production of skin cells and plays a role in wound healing (Novelli 1953). The RDA for vitamin B_5 is 1.4–4 mg in children and 5 mg in adolescents and adults. No specific side effects related to high pantothenic acid intake have been reported in the literature. However, extremely high doses may lead to gastrointestinal disturbances, such as diarrhoea (Tahiliani and Beinlich 1991).

4.6 Vitamin B_6

Vitamin B_6, pyridoxine, is naturally found in food and is present in the form of cofactor pyridoxal 5′ phosphate. It is involved in the metabolism of carbohydrates, proteins, and fats, and in bioenergetics. It is also involved in the synthesis of neurotransmitters, including dopamine, gamma-aminobutyric acid and serotonin. The production of haemoglobin, immune cells and antibodies (Bender 1994). The daily RDA for vitamin B_6 is between 0.1–1 mg for children and 1.2–1.3 mg for adolescents and adults. However, its demand increases during pregnancy and lactation up to 1.9–2 mg per day.

The excessive intake of vitamin B_6 leads to sensory neuropathy resulting in numbness, tingling, or burning sensations in the hands and feet (Institute of Medicine

(US) Standing Committee on the Scientific Evaluation of Dietary Reference Intakes and its Panel on Folate, Other B Vitamins, and Choline 1998). This condition is often characterised by impairments in sensory perception, particularly touch and vibration sensations. The other neurological symptoms include difficulty in walking, poor coordination, and impaired muscle control (Hemminger and Wills 2020).

4.7 *Vitamin B_7*

Vitamin B_7 or biotin, a water-soluble vitamin, acts as a coenzyme and is involved in the metabolism of carbohydrates, fats, and proteins and the associated energy production. It promotes healthy nails, skin, and hair, and maintains structural integrity. It is involved in gene expression, specifically in the activation of certain genes that are responsible for regulating cell growth and development. The RDA for biotin varies depending on age and life stage with infant: 5 μg, children: 8–20 μg and adults: 30 μg daily (Schellack et al. 2016). It is observed that excessive biotin intake results in allergic reactions including skin rashes, itching and difficulty in breathing. The excess intake of biotin can improve the symptoms associated with vitamin B_{12} deficiency, such as anaemia and nerve damage, without addressing the underlying deficiency itself (Powe et al. 2003). This may delay the proper diagnosis and treatment of vitamin B_{12} deficiency. High doses of biotin can interfere with certain laboratory tests, leading to inaccurate results. Biotin can bind to assay components used in these tests, which may affect the measurement of various analytes, including hormones and other biomarkers (Odhaib et al. 2019).

4.8 *Vitamin B_9*

Folic acid, also known as vitamin B_9, is a vital nutrient and is responsible for DNA synthesis and cell division. It is involved in the production of new cells, including red blood cells, and supports overall tissue growth and development. It aids in the production of the heme group of haemoglobin. Sufficient folic acid levels can subside the defects in the neural tube, such as anencephaly and spina bifida. The RDA of folate is 65–300 μg for children and 400 μg for adolescents and adults.

Excessive folic acid supplement intake may increase the risk of certain cancers. Research suggests a high-dose folic acid supplementation may promote the growth of existing precancerous or cancerous cells (Wien et al. 2012; Ebbing et al. 2009). Very high doses of folic acid can interfere with certain laboratory tests, particularly those measuring vitamin B_{12} levels. Elevated folic acid levels can mask a true vitamin B_{12} deficiency, leading to inaccurate test results and potential misdiagnosis (de Benoist 2008). Excessive folic acid intake can mask the presence of anaemia by improving the haematological parameters and haemoglobin levels, without addressing the underlying cause of the anaemia.

4.9 Vitamin B_{12}

Cyanocobalamin, a synthetic vitamin B_{12} analogue, plays a crucial role in the formation of red blood cells. It is essential for the synthesis of DNA and RNA. It is necessary for maintaining proper nervous system function. It plays a role in the synthesis of myelin, a protective covering around nerve fibres that ensures proper transmission of nerve impulses. There is no known toxicity associated with excessive intake of cyanocobalamin. High doses of cyanocobalamin can affect laboratory test results used to measure vitamin B_{12} levels. This can cause artificially high levels of serum vitamin B_{12}, potentially leading to misdiagnosis or inaccurate assessment of vitamin B_{12} status (Hvas and Nexo 2006).

4.10 Vitamin C

Vitamin C is a strong antioxidant involved in protecting the body from free radicals and helps in maintaining the health of skin, bones, and blood vessels. It is involved in collagen synthesis, a protein providing structure to various connective tissues, including skin, tendons, ligaments, and blood vessels. It supports the healing of wounds and maintains the integrity of the skin. It enhances the production and function of immune cells that protect against infections (Carr and Maggini 2017). It boosts non-heme iron absorption that is present in plant-based foods (Bendich et al. 1986). The RDA for vitamin C varies and is typically around 75–90 mg per day for adults.

Very high doses of vitamin C, typically above 2000–3000 mg per day, can cause GIT disturbances such as diarrhoea, nausea, and stomach cramps. These symptoms usually resolve once vitamin C intake is reduced. Excessive intake of vitamin C may increase the risk of kidney stones, particularly in patients with a history of pre-existing kidney conditions or kidney stones. It can be metabolized into oxalate that contributes to the formation of calcium oxalate stones (D'Costa et al. 2019). High serum vitamin C can interfere with some laboratory tests, including blood glucose and urine tests for various substances (Ten Berge et al. 2021).

4.11 Vitamin D

Vitamin D is involved in calcium and phosphate absorption from the intestines and promotes their incorporation into bones and teeth. It helps to maintain proper levels of these minerals in the blood, ensuring optimal bone health and preventing conditions like rickets and osteomalacia in children and adults, respectively. It supports bone strength and mineralization along with calcium. It modulates immune system function, including promoting the function of immune cells and reducing inflammation. It acts as a hormone that can affect gene expression regulating cellular growth, differentiation, and proliferation. The RDA for vitamin D for most adults ranges from 600–800 international units per day (Holick 2003).

Excessive intake of vitamin D, known as vitamin D toxicity or hypervitaminosis D. It leads to elevated levels of calcium in the blood resulting in nausea, vomiting, increased thirst and urination, abdominal pain, constipation, fatigue, and confusion. It can lead to the deposition of calcium in soft tissues, such as blood vessels, the heart, lungs, and kidneys. Calcium can accumulate in the kidneys and form stones, which can be painful. It may experience GIT symptoms, such as nausea, vomiting, and poor appetite (Morita et al. 1993).

4.12 Vitamin E

Vitamin E is a strong antioxidant and protects cells from oxidative damage caused by reduced oxidative stress. It maintains the integrity and stability of cell membranes and prevents lipid oxidation. It supports the immune system's function by enhancing immune cells' activity and promoting antibodies production. It helps to moisturize the skin and improve its texture by protecting against damage (Brigelius-Flohé and Traber 1999).

Hypervitaminosis E is rare and primarily associated with high-dose supplementation. It acts as a blood thinner and interferes with blood clotting which leads to an increased risk of bleeding. It can suppress the immune response to infections and impair immune function. Some individuals may experience GIT symptoms, such as diarrhoea, nausea, and stomach cramps due to hypervitaminosis E. It can interact with vitamin K and blood coagulation (Murphy 1974).

4.13 Vitamin K

Vitamin K is involved in the synthesis of clotting factors in the liver. It is involved in bone protein metabolism and contributes to bone health by promoting proper bone mineralization. It may help to prevent arterial calcification potentially reducing the risk of cardiovascular diseases. The RDA for vitamin K is around 90–120 μg per day for adults (Olson 1984). Excessive vitamin K intake can interfere with the effectiveness of anticoagulants, such as warfarin. In rare cases, extremely high doses of vitamin K may promote excessive blood clotting. Certain individuals may experience allergic reactions, such as skin rashes or itching, with excessive vitamin K intake (Vermeer 2012).

The details of dietary intake, sources, RDA and upper limits of vitamins are included in Table 12.1.

Table 12.1 Role of vitamins, their dietary sources and their daily limits

Vitamin name	Major function	Source(s)	RDA	Excess upper limit	References
Retinol, retinal, retinoic acid, (Beta carotene) (A)	Vision	Fortified milk, eggs, liver (dark green leafy and yellow/orange vegetables)	300–400 μg/day	500 μg/day	Du et al. (2018)
Thiamine (B_1)	Energy metabolism	Whole and enriched grain products, leafy greens, pork	1.2–1.5 mg/day	10 mg/day	Martel et al. (2023)
Riboflavin (B_2)	Energy metabolism	Whole and enriched grain products, milk products	1.1–1.3 mg/day	500 mg/day	Turck et al. (2017)
Niacin (B_3)	Energy metabolism	Whole and enriched grain products, protein-rich foods	20–30 mg/day	40 mg/day	Li et al. (2010)
Pantothenic acid (B_5)	Protein, fat and carbohydrate metabolism	Almost all foods, especially avocadoes, broccoli, meats	2–8 mg/day	1000 mg/day	Lee et al. (2018)
Ascorbic acid (C)	Antioxidant, collagen synthesis, immune function	Fruits and vegetables	200–400 mg/day	1000 mg/day	Lykkesfeldt (2020)
Cholecalciferol (D)	Bone growth	Sunlight, fortified milk, fatty fish, eggs, liver	*100 μg/day*	1250 *μg/day*	Balvers et al. (2015)
Tocopherol (E)	Protects cell membranes	Vegetable and seed/ nut oils, seeds and nuts, wheat germ and whole grains	800–1000 mg/day	1200 mg/day	Tucker and Townsend (2005)
Phylloquinone (K)	Blood clotting	Dark leafy greens, cabbage family, liver	120–15 *μg/day*	1000 *μg/day*	Popa et al. (2021)

5 Excess Mineral Uptake

5.1 Calcium

Calcium is crucial for the development and maintenance of healthy bones. It is essential for muscle contraction, including all types of muscle and triggers the release of calcium ions, that initiate muscle contraction. It is involved in transmitting nerve impulses throughout the body and is necessary for the blood clotting process (Peacock 2010).

Excessive intake of calcium, known as hypercalcemia leads accumulation of calcium with other substances such as oxalate or phosphate forming calculi in the urinary tract and causing pain and discomfort. Excessive calcium intake causes constipation, bloating, and abdominal discomfort. It also deposits in other soft tissues and can interfere with the absorption process of other essential minerals, such as magnesium, zinc and iron (Li et al. 2018).

5.2 Iron

Iron plays a crucial role in various physiological body functions. It is a fundamental component of haemoglobin, a protein responsible for carrying oxygen in the body tissues. It is also involved in other vital processes, including energy production, DNA synthesis, and immune function. Iron-containing enzymes, such as cytochromes, are vital components of the electron transport chain that produce energy as adenosine triphosphate (Wessling-Resnick 2017).

Hereditary hemochromatosis is a genetic condition causing iron overload. It results in repeated blood transfusions. Certain medical conditions or excessive iron supplementation cause an overload of iron. Excess iron can cause oxidative stress and free radical damage, leading to tissue injury, and organ damage, especially in the liver, heart, and pancreas. There is an increased risk of chronic liver cirrhosis, diabetes, heart disease and joint pain (Fleming and Ponka 2012).

5.3 Copper

Copper is an essential trace mineral that plays a vital role in numerous physiological functions in the body. It is required in micro quantity for proper growth, development and maintenance of overall health. Copper is a cofactor particularly important for enzymes that are involved in energy production, iron metabolism, connective tissue formation and antioxidant defence. It helps in the activation of these enzymes. Copper plays a role in the absorption, transport, and utilization of iron and is involved in the conversion of ferric to ferrous iron, i.e. its storage to a usable form.

It is involved in the synthesis and maintenance of connective tissues, such as collagen and elastin. Copper assists in the cross-linking of collagen fibres. It is involved in the synthesis of certain neurotransmitters, including norepinephrine, dopamine and serotonin. Antioxidant enzymes, such as superoxide dismutase (SOD), have copper ions which help to neutralize free radicals (Angelova et al. 2011).

Excessive levels of copper can lead to a condition called copper toxicity or copper overload. Genetic disorders like Wilson's disease, excessive copper intake, contaminated water or impaïred copper metabolism lead to copper toxicity. Abdominal pain, nausea, vomiting, liver damage, neurological problems, and anaemia are symptoms associated with copper toxicity. Long-term copper overload can cause significant organ damage, particularly affecting the liver and brain (Angelova et al. 2011).

5.4 *Zinc*

Zinc is an essential trace mineral and plays a crucial role in various physiological processes and is involved in over 300 enzymatic reactions. It is essential for the proper functioning of the immune system and aids in the activation of immune cells, and antibodies, and helps in the regulation of inflammatory responses. It is necessary for normal growth and development, as involved in cell division, protein and DNA synthesis, in rapid growth periods such as childhood, adolescence, and pregnancy. It is involved in the synthesis of collagen and wound healing. It is required for the proper functioning of taste and smell receptors and for maintaining the sensitivity of taste buds. It acts as an antioxidant and helps to protect cells against oxidative damage caused by free radicals and supports the activity of antioxidant enzymes (Jackson 1989).

Excessive levels of zinc can lead to toxicity and occur through excessive supplementation or exposure to high levels of zinc from other sources. Its symptoms are nausea, vomiting, diarrhoea, abdominal pain, and headache. In severe cases, it can lead to copper deficiency, impaired immune function and organ damage (Plum et al. 2010).

5.5 *Magnesium*

Magnesium is an essential mineral that is involved in the production and utilization of adenosine triphosphate (ATP). It is crucial for proper muscle function, including muscle contraction and relaxation. It works in conjunction with calcium to regulate the flow of ions across muscle cell membranes, enabling normal muscle contraction and preventing muscle cramps and spasms. It plays a role in nerve signal transmission and supports the regulation of neurotransmitters, such as serotonin and dopamine. It contributes to bone mineralization and works in conjunction with other minerals, such as vitamin D and calcium, to promote bone strength and density. It helps to relax the smooth muscles within blood vessels, promoting vasodilation and proper blood flow. It is involved in maintaining a regular heart rhythm and normal blood pressure (Faryadi 2012).

Excessive intake of magnesium supplements or certain medications can lead to hypermagnesemia. Symptoms of magnesium toxicity can include nausea, diarrhoea, vomiting, muscle weakness, low blood pressure, and, in severe cases, cardiac arrhythmias and impaired breathing (Rude and Singer 1981).

5.6 Manganese

Manganese, a cofactor for several enzymes, is involved in the metabolism of nutrients and neurotransmitters, and the production of collagen. It is involved in the functioning of antioxidant enzymes, such as manganese superoxide dismutase (MnSOD), and neutralizes harmful free radicals. It plays role in the formation and maintenance of bones. It supports the synthesis of glycosaminoglycans, which are essential components of cartilage and connective tissues. It is involved in proteoglycans synthesis, which is crucial for the structure and integrity of connective tissues, including tendons, ligaments, and blood vessels. It supports healing and repair of damaged tissues (Chen et al. 2018).

Excessive intake of manganese, particularly from environmental or occupational exposure, can lead to manganese toxicity. Chronic exposure to high levels of manganese has been associated with neurological symptoms of tremors, muscle rigidity, and impaired motor function similar to Parkinson's disease. Individuals with certain medical conditions, such as liver disease, may be more susceptible to manganese toxicity (Crossgrove and Zheng 2004; Keen et al. 2000).

5.7 Molybdenum

Molybdenum is a cofactor for enzymes involved in sulfur amino acid metabolism such as sulfite oxidase which converts sulfite to sulfate. It also plays a role in the metabolism of purines. It is essential for the conversion of nitrogen from dietary proteins into a usable form, including the synthesis of amino acids and nucleotides. It is indirectly involved in iron metabolism by activation of hephaestin which is involved in iron absorption (Rajagopalan 1988). Molybdenum from supplements or occupational exposure can lead to adverse effects. High levels of molybdenum can interfere with copper metabolism and utilization, leading to copper deficiency. Symptoms of molybdenum toxicity may include GIT disturbances, joint pain, anaemia, and neurological symptoms (Vyskočil and Viau 1999).

5.8 Selenium

Selenium is a crucial component of antioxidant enzymes, such as glutathione peroxidase and also works in conjugation with other antioxidants to neutralize harmful free radicals. It is necessary for the proper metabolism of thyroid hormones. It is involved in the conversion of the inactive thyroid T_4 hormone into the active T_3 form in the body. It plays a role in the production of cytokines and other immune system components (Berry and Larsen 1992). In males, selenium is involved in the production and maturation of sperm cells. In females, selenium plays a role in fertility and may support healthy pregnancy outcomes (Minich 2022). Excessive intake of selenium can lead to selenium toxicity, although it is relatively rare. Chronic excessive selenium intake, typically from supplements or contaminated food sources, can result in selenosis. Symptoms of selenium toxicity may include GIT disturbances, hair and nail changes, neurological symptoms, and in severe cases, respiratory and cardiovascular problems (Genchi et al. 2023).

The details of dietary intake, sources, RDA and upper limits of minerals are included in Table 12.2.

Table 12.2 Role of minerals, their dietary sources and their daily limits

Minerals	Uses	Source(s)	RDA	Excess upper limit	References
Magnesium	Muscle function & a healthy immune system	Nuts, spinach	360-400 mg/day	6000 mg/day	Grissinger (2014)
Calcium	Bones strength and blood clotting	Leafy greens, seeds, nuts, dates	1000–1233 mg/day	1500 mg/day	Cormick and Belizán (2019)
Zinc	Helps in the recovery of wounds	Dairy, meat	15-20 mg/day	30 mg/day	Li et al. (2022)
Phosphorus	Utilized in all cell functions	Fish, poultry, eggs	1000 mg/day	1500 mg/day	Goyal and Jialal (2019)
Copper	Absorption and metabolism of iron	Dried fruit	1–3 mg/day	800–1000 mg/day	Royer and Sharman (2020)
Iron	Red blood production	Whole grains, spinach, liver, tofu	15–20 mg/day	45 mg/day	Ghosh et al. (2019)
Sodium	Maintains water balance	Table salt, fish, nuts, seeds	2-3 g/day	5 g/day	Mente et al. (2021)
Potassium	Muscle contraction, regulates heartbeat	Nuts, seeds, fish	1.6–5.9 g/day	17 g/day	Lanham-New et al. (2012)

6 Toxic Effects Associated with Excess Nutrients Intake

Nutrients are widely accepted as a source of vital health and play a role in balancing our physiological functions as well as the strongest source of antioxidant and immune function regulation. Zinc, a crucial trace element, is widely recognized for its vital biological roles in regulating various cellular processes. These functions encompass DNA synthesis, proper growth, brain development, behavioural responses, reproductive processes, foetal development, bone formation, and wound healing. The recommended intake of zinc is not more than 8 mg/kg for females and 11 mg/kg for males considered safe in daily meals (Aggett and Comerford 1995). In humans, reaching a dose associated with acute toxicity is highly uncommon and typically occurs under extremely rare circumstances. Instances of zinc poisoning in individuals have been linked to beverages containing elevated concentrations of zinc, reaching levels as high as 2500 mg/L, resulting in estimated doses of 325–650 mg. Symptoms of zinc poisoning include nausea, vomiting, abdominal cramping, tenesmus and diarrhoea, with or without bleeding. In addition, excess zinc in pregnancy may cause teratogenicity (Chow et al. 1969; Brown et al. 1964; Maret and Sandstead 2006). In concern to the use of chromium as a nutrient, a study suggested that chromium has a genotoxic activity which alters the DNA functions (Levina and Lay 2008). Selenium, an important micronutrient for humans, has been the subject of research for approximately 60 years. It has been observed that selenium can potentially cause toxic reactions when consumed at levels significantly higher than those typically found in the human diet. The median lethal dose (LD_{50}) of selenium, which represents the amount that can be lethal to 50% of the test subjects, was found to be 12.6 mg/kg and 9.26 mg/kg body weight in female and male mice, respectively. This indicates a high health hazard potential upon acute oral exposure. However, it is worth noting that selenium was found to be safe in the AMES test, as it did not exhibit any genotoxic effects (Yang and Jia 2014).

Manganese has several vital roles in the human body. It is abundantly found in grain, tea and leafy vegetables. The adequate intake of manganese is 2.3 mg/day and 1.8 mg/day for males and females, respectively, is considered as safe. Previous studies have indicated that an excessive intake of manganese in the diet can lead to neurodegeneration. The basal ganglia, a specific region of the brain, is particularly susceptible to the toxic effects of manganese. Oxidative stress and disruptions in mitochondrial function are recognized as significant indicators of manganese toxicity (Dobson et al. 2004).

Hypervitaminosis D is accompanied by hypercalcemia when its large dose or metabolites are used without proper control. Additionally, excessive production of 1,25-dihydroxy vitamin D (1, 25(OH)2D) in lymphomas, granulomatous disorders and idiopathic infantile hypercalcemia can also lead to hypervitaminosis D with hypercalcemia. Vitamin D toxicity results in elevated levels of 25(OH)D (>150 ng/ml or >375 nmol/l), hypercalciuria, hypercalcemia and typically normal or slightly increased concentrations of 1,25(OH)2D (Marcinowska-Suchowierska et al. 2018).

Excessive consumption of protein could be a significant risk factor for the development of kidney stones. When protein is ingested, it increases the excretion of acid in the kidneys, resulting in higher acid loads. To counterbalance this acidity, the body may draw calcium from the bones, which is then excreted by the kidneys. This increase in urinary calcium caused by protein intake can contribute to calcium kidney stones. In a study that examined varying levels of protein intake, ranging from 47 to 142 g/day, a significant increase in urinary calcium was observed. This finding suggests that the calciuric response induced by animal protein consumption could potentially be a risk factor for the development of osteoporosis (Delimaris 2013).

6.1 *Obesity*

Excessive intake of vitamins can impact neurotransmitter metabolism and carbon metabolism, potentially contributing to obesity through various mechanisms. These include promoting fat synthesis, inducing insulin resistance, disrupting the metabolism of neurotransmitters, and initiating epigenetic changes. Studies have shown that niacin can increase daily food intake, while thiamin and pyridoxal are necessary for fatty acid synthesis from carbohydrates and proteins. Emerging evidence suggests that an excessive intake of vitamins, particularly the vitamin B group, might be involved in the development of obesity (Zhou and Zhou 2014). Weight gain is influenced independently by changes in diet and physical activity. Long-term weight gain is associated with specific factors related to diet and lifestyle, which have individual impacts on weight (Mozaffarian et al. 2011). According to surveys conducted by the National Nutrition Monitoring Bureau, there was a notable increase in overweight/obesity among individuals consuming energy intake exceeding 1587 calories per day. Moreover, protein and fat intake above 41–57 and 19–32 g/day, respectively, were also associated with increased obesity. The per-capita daily consumption of certain food groups, including sugars and jaggery (including condiments, biscuits, etc.), milk and its related products, fats, oils, meat and meat products, and salts/sodium, displayed a significant dose-response relationship with the prevalence of obesity (Mathur and Pillai 2019).

7 Excess Nutrients and Associated Diseases

7.1 *Alzheimer's Disease*

A well-established environmental risk factor for AD is the excessive consumption of simple sugars and saturated fatty acids. Recent evidence suggests that an unbalanced diet and lack of physical activity play a significant role in AD development and are further contributed by conditions like diabetes, hypercholesterolemia, hypertension and obesity. These modifiable factors account for nearly 40% of all risk factors associated with the development of AD. "Western diet" refers to a modern eating pattern prevalent in Western societies, characterized by the consumption of highly processed foods made from refined ingredients. This diet is rich in simple carbohydrates, salt, saturated fats, and cholesterol while lacking in whole grains, fibres and beneficial mono- and polyunsaturated fatty acids including ω-3, -6, and -9 fatty acids. Furthermore, the Western diet negatively impacts gut function and the composition of gut microbiota, indirectly leading to reduced nutrient absorption and vitamin intake from food (Christ et al. 2019).

7.2 *Non-alcoholic Fatty Liver Disease*

The precise relationship between nutrition, diet, and the onset of non-alcoholic fatty liver disease (NAFLD) remains partly comprehended. Specifically, a diet higher in fructose compared to sucrose has been associated with elevated hepatic fat deposition, inflammation and potential fibrosis (Luo and Lin 2021). Previously investigated preclinical and clinical studies suggested the association between excess nutrition and NAFLD. The development of NAFLD in overweight children may be significantly influenced by factors such as body weight and dietary patterns, notably, the consumption of sweetened beverages. The dietary consumption of fat, fructose, and glucose in children with NAFLD was found to be higher than in normal-weight healthy children. Specifically, the intake of fat, fructose, and glucose in NAFLD children was measured at 86, 61 and 53 g/day, respectively, while normal-weight healthy children had lower intakes of 78, 55 and 49 g/day, respectively (Nier et al. 2018).

7.3 *Renal Complications*

Excessive intake of vitamin C can lead to the development of oxalosis, which is characterized by the accumulation of calcium oxalate deposits in tissues. A case report highlighted a situation where women who consumed an excessive amount of vitamin C experienced acute kidney failure. The recommended daily intake of vitamin C is 75 and 90 mg/day for females and males, respectively (Sunkara et al. 2015). Another clinical report investigated monitoring over excess intake of vitamin C. Scrutiny is necessary when considering the use of vitamin C, both in individuals with normal kidney function and those with underlying renal insufficiency. It has been reported that excessive intake of vitamin C can induce acute oxalate nephropathy. Therefore, clinicians must carefully evaluate the potential risks associated with high-dose consumption of vitamin C, particularly in individuals with pre-existing kidney disease or advanced age (Lamarche et al. 2011).

7.4 Infertility

A diet that is high in calories with excessive intake of saturated fats and trans-fatty acids, along with a high glycemic index and low nutritional density, has been linked to elevated oxidative stress. This oxidative stress is considered the underlying cause of various metabolic disorders such as obesity, intestinal dysbiosis, type 2 diabetes, and insulin resistance. These metabolic disorders, in turn, contribute to fertility issues primarily due to the generation of oxidative stress. Oxidative stress is recognized as one of the primary factors leading to a decline in sperm quality and an increased risk of infertility, as well as hormonal and immunological disturbances. The presence of excess adipose tissue leads to heightened aromatase activity, which converts testosterone to estrogen, resulting in elevated estrogen levels and decreased testosterone levels in males (Skoracka et al. 2020).

7.5 Binge Eating Disorder

Binge eating disorder (BED) is an eating disorder characterized by recurrent episodes of consuming large quantities of food in a short period of time, accompanied by a loss of control and feelings of guilt, shame, or distress. Individuals with BED often eat even when they are not physically hungry and may continue eating past the point of discomfort. When it comes to excessive intake of nutrients, individuals with BED may consume a large amount of food during binge episodes, resulting in the intake of excessive calories and nutrients. The types of foods consumed can vary, but they often include high-calorie fat and sugar-rich foods. Hence, they consume an excess of macronutrients and micronutrients during binge episodes.

The excessive intake of calories during binge episodes can lead to weight gain and, in some cases, obesity. Obesity increases the risk of various health issues, including heart disease, type 2 diabetes, and certain types of cancer. It may result in imbalances in nutrient intake, as individuals with BED often consume large quantities of certain foods while neglecting others (Dingemans et al. 2002). This can lead to deficiencies or excesses of certain nutrients, affecting overall nutritional status. Binge-eating episodes can disrupt normal metabolic processes, including insulin regulation and energy expenditure. This can contribute to metabolic abnormalities, such as insulin resistance and a decreased metabolic rate. Consuming large quantities of food can lead to gastrointestinal discomfort, bloating, and other digestive issues. Excessive nutrient intake during binge episodes can exacerbate feelings of guilt, shame, and distress associated with BED. This can perpetuate a cycle of negative emotions and further contribute to disordered eating patterns. Treatment for BED typically involves a multidimensional approach, including therapy, counselling, and, in some cases, medication (Wilson et al. 2010). Selective serotonin reuptake inhibitors, such as fluoxetine, sertraline and escitalopram can be effective in reducing the frequency and severity of binge eating episodes. Vyvanse is a stimulant medication that helps to reduce binge eating episodes and may also assist with weight management. Topiramate, an anticonvulsant, has shown promise in reducing binge eating episodes and promoting weight loss in individuals with BED. Naltrexone and bupropion, used together in a specific combination, can help to reduce food cravings and regulate eating behaviour (Appolinario and McElroy 2004).

8 Preventive Measures for Excess Nutrient Intake

In light of our understanding of nutritional requirements and their potential impact on health, healthy dietary patterns can generally be described as those that prioritize the consumption of health-promoting foods. These include plant-based foods, fresh fruits and vegetables, antioxidants, soya, nuts, and sources of ω-3 fatty acids. On the other hand, healthy diets should limit the intake of saturated fats, trans fats, animal-derived proteins and refined sugars. To meet the recommended level of carbohydrates (130 g/day), a simple example would be to consume a meal consisting of a cup of rice or spaghetti with sauce, a cup of milk, a cup of orange juice, an apple, and a carrot. For individuals weighing 75 kg, a daily protein intake of 60 g is recommended, while a 55 kg woman requires 44 g of protein/day. It is important to strike a balance between meeting nutritional requirements for a healthy diet and maintaining appropriate energy intake. Food labels and individuals' beliefs about the satisfaction derived from specific foods can influence food choices and intake. Therefore, it is crucial to consider the information provided on food labels and to have accurate beliefs about the nutritional value of different foods. Furthermore, the initial palatability of foods plays a significant role in preventing a decreased intake due to repetitive consumption. Therefore, it is beneficial to ensure that foods are initially appealing to delay potential monotony effects. Offering opportunities for self-selection of foods can also help in avoiding such monotony and maintaining a varied and enjoyable diet (Ryan-Harshman and Aldoori 2006; Cena and Calder 2020).

Conflict of Interest The author(s) declares no conflict of interest.

References

Ackermans MM, Zhou H, Carels CE, Wagener FA, Von den Hoff JW. Vitamin A and clefting: putative biological mechanisms. Nutr Rev. 2011;69(10):613–24.

Aggett PJ, Comerford JG. Zinc and human health. Nutr Rev. 1995;53(9):S16.

Angelova M, Asenova S, Nedkova V, Koleva-Kolarova R. Copper in the human organism. Trakia J Sci. 2011;9(1):88–98.

Appolinario JC, McElroy SL. Pharmacological approaches in the treatment of binge eating disorder. Curr Drug Targets. 2004;5(3):301–7.

Balvers MG, Brouwer-Brolsma EM, Endenburg S, de Groot LC, Kok FJ, Gunnewiek JK. Recommended intakes of vitamin D to optimise health, associated circulating

25-hydroxyvitamin D concentrations, and dosing regimens to treat deficiency: workshop report and overview of current literature. J Nutr Sci. 2015;4:e23.
Bender DA. Novel functions of vitamin B6. Proc Nutr Soc. 1994;53(3):625–30.
Bendich A, Machlin L, Scandurra O, Burton G, Wayner D. The antioxidant role of vitamin C. Adv Free Radic Biol Med. 1986;2(2):419–44.
Berry MJ, Larsen PR. The role of selenium in thyroid hormone action. Endocr Rev. 1992;13(2):207–19.
Brigelius-Flohé R, Traber MG. Vitamin E: function and metabolism. FASEB J. 1999;13(10): 1145–55.
Brown MA, Thom JV, Orth GL, Cova P, Juarez J. Food poisoning involving zinc contamination. Arch Environ Health Int J. 1964;8(5):657–60.
Candela CG, López LB, Kohen VL. Importance of a balanced ω 6/ω 3 ratio for the maintenance of health. Nutritional recommendations. Nutr Hosp. 2011;26(2):323–9.
Capuzzi DM, Morgan JM, Brusco OA, Intenzo CM. Niacin dosing: relationship to benefits and adverse effects. Curr Atheroscler Rep. 2000;2(1):64–71.
Cardoso DR, Libardi SH, Skibsted LH. Riboflavin as a photosensitizer. Effects on human health and food quality. Food Funct. 2012;3(5):487–502.
Carr AC, Maggini S. Vitamin C and immune function. Nutrients. 2017;9(11):1211.
Cena H, Calder PC. Defining a healthy diet: evidence for the role of contemporary dietary patterns in health and disease. Nutrients. 2020;12(2)
Chen P, Bornhorst J, Aschner MA. Manganese metabolism in humans. 2018.
Chow B, Sherwin R, Hsueh AM, Blackwell B, Blackwell R. Growth and development of rats in relation to the maternal diet: a review. Beeinflussung des Stoffwechsels durch die Ernährung. 1969;11:45–56.
Christ A, Lauterbach M, Latz E. Western diet and the immune system: an inflammatory connection. Immunity. 2019;51(5):794–811.
Cohlan SQ. Excessive intake of vitamin A as a cause of congenital anomalies in the rat. Science. 1953;117(3046):535–6.
Cormick G, Belizán JM. Calcium intake and health. Nutrients. 2019;11(7):1606.
Council NR. Dietary intake and nutritional status: trends and assessment. Diet and health: implications for reducing chronic disease risk. National Academies Press (US); 1989.
Crossgrove J, Zheng W. Manganese toxicity upon overexposure. NMR Biomed. 2004;17(8):544–53.
D'Costa MR, Winkler NS, Milliner DS, Norby SM, Hickson LJ, Lieske JC. Oxalosis associated with high-dose vitamin C ingestion in a peritoneal dialysis patient. Am J Kidney Dis. 2019;74(3):417–20.
de Benoist B. Conclusions of a WHO Technical Consultation on folate and vitamin B12 deficiencies. Food Nutr Bull. 2008;29(2_suppl 1):S238–S44.
Delimaris I. Adverse effects associated with protein intake above the recommended dietary allowance for adults. ISRN Nutr. 2013;2013:126929.
Dingemans A, Bruna M, Van Furth E. Binge eating disorder: a review. Int J Obes (Lond). 2002;26(3):299–307.
Dobson AW, Erikson KM, Aschner M. Manganese neurotoxicity. Ann N Y Acad Sci. 2004;1012(1):115–28.
Du W, Wang H, Wang Z, Zhang J, Su C, Jia X, et al. Dietary vitamin a intake among Chinese adults: findings from CNTCS2015. Nutr J. 2018;17(1):60.
Ebbing M, Bønaa KH, Nygård O, Arnesen E, Ueland PM, Nordrehaug JE, et al. Cancer incidence and mortality after treatment with folic acid and vitamin B12. JAMA. 2009;302(19):2119–26.
Elango G, Venkataraman D, Rao SV, Kiran VR. Hypervitaminosis. Int J Biomed Res. 2015;6(3):151–4.
Epstein KR. The role of carotenoids on the risk of lung cancer. Semin Oncol. 2003;30(1):86–93.
Faryadi Q. The magnificent effect of magnesium to human health: a critical review. Int J Appl Sci Technol. 2012;2(3):118–26.
Feskanich D, Singh V, Willett WC, Colditz GA. Vitamin A intake and hip fractures among postmenopausal women. JAMA. 2002;287(1):47–54.

Fleming RE, Ponka P. Iron overload in human disease. N Engl J Med. 2012;366(4):348–59.
Geelen JAG, Peters PWJ. Hypervitaminosis a induced teratogenesis. CRC Crit Rev Toxicol. 1979;6(4):351–75.
Genchi G, Lauria G, Catalano A, Sinicropi MS, Carocci A. Biological activity of selenium and its impact on human health. Int J Mol Sci. 2023;24(3):2633.
Ghosh S, Sinha S, Thomas T, Sachdev HS, Kurpad AV. Revisiting dietary iron requirement and deficiency in Indian women: implications for food iron fortification and supplementation. J Nutr. 2019;149(3):366–71.
Goyal R, Jialal I. Hyperphosphatemia. 2019.
Grissinger M. Avoiding patient harm from a magnesium bolus dose. Pharm Ther. 2014;39(2):81.
Hemminger A, Wills BK. Vitamin B6 toxicity. 2020.
Henderson LM. Niacin. Annu Rev Nutr. 1983;3(1):289–307.
Holick MF. Evolution and function of vitamin D. Vitamin D analogs in cancer prevention and therapy. 2003; p. 3–28.
Hu Y, Sparks JA, Malspeis S, Costenbader KH, Hu FB, Karlson EW, et al. Long-term dietary quality and risk of developing rheumatoid arthritis in women. Ann Rheum Dis. 2017;76(8):1357–64.
Hvas A-M, Nexo E. Diagnosis and treatment of vitamin B12 deficiency--an update. Haematologica 2006;91(11):1506–1512.
Institute of Medicine (US) Standing Committee on the Scientific Evaluation of Dietary Reference Intakes and its Panel on Folate, Other B Vitamins, and Choline. Dietary reference intakes for thiamin, riboflavin, niacin, vitamin B6, folate, vitamin B12, pantothenic acid, biotin, and choline. Washington (DC): National Academies Press (US); 1998.
Jackson M. Physiology of zinc: general aspects. In: Zinc in human biology. Berlin/Heidelberg: Springer; 1989. p. 1–14.
Jayawardena R, Misra A. Balanced diet is a major casualty in COVID-19. Diabetes Metab Syndr. 2020;14(5):1085.
Kamanna VS, Kashyap ML. Mechanism of action of niacin. Am J Cardiol. 2008;101(8):S20–S6.
Kamanna V, Ganji S, Kashyap M. The mechanism and mitigation of niacin-induced flushing. Int J Clin Pract. 2009;63(9):1369–77.
Keen CL, Ensunsa JL, Clegg MS. Manganese metabolism in animals and humans including the toxicity of manganese. Met Ions Biol Syst. 2000:137–70.
Lamarche J, Nair R, Peguero A, Courville C. Vitamin C-induced oxalate nephropathy. Int J Nephrol. 2011;2011:146927.
Lanham-New SA, Lambert H, Frassetto L. Potassium. Adv Nutr. 2012;3(6):820–1.
Lascari AD. Carotenemia: a review. Clin Pediatr. 1981;20(1):25–9.
Lee JH, Ahn SY, Lee HA, Won KS, Chang HW, Oh JS, et al. Dietary intake of pantothenic acid is associated with cerebral amyloid burden in patients with cognitive impairment. Food Nutr Res. 2018:62.
Levina A, Lay PA. Chemical properties and toxicity of chromium (III) nutritional supplements. Chem Res Toxicol. 2008;21(3):563–71.
Li D, Sun WP, Zhou YM, Liu QG, Zhou SS, Luo N, et al. Chronic niacin overload may be involved in the increased prevalence of obesity in US children. World J Gastroenterol. 2010;16(19):2378–87.
Li K, Wang X-F, Li D-Y, Chen Y-C, Zhao L-J, Liu X-G, et al. The good, the bad, and the ugly of calcium supplementation: a review of calcium intake on human health. Clin Interv Aging. 2018:2443–52.
Li J, Cao D, Huang Y, Chen B, Chen Z, Wang R, et al. Zinc Intakes and health outcomes: an umbrella review. Front Nutr. 2022;9:798078.
Lim S. Eating a balanced diet: a healthy life through a balanced diet in the age of longevity. J Obes Metab Syndr. 2018;27(1):39.
Lonsdale D. A review of the biochemistry, metabolism and clinical benefits of thiamin (e) and its derivatives. Evid Based Complement Alternat Med. 2006;3(1):49–59.

Luo Y, Lin H. Inflammation initiates a vicious cycle between obesity and nonalcoholic fatty liver disease. Immun Inflamm Dis. 2021;9(1):59–73.

Lykkesfeldt J. On the effect of vitamin C intake on human health: how to (mis)interprete the clinical evidence. Redox Biol. 2020;34:101532.

Marcinowska-Suchowierska E, Kupisz-Urbańska M, Łukaszkiewicz J, Płudowski P, Jones G. Vitamin D toxicity – a clinical perspective. Front Endocrinol. 2018;9:550.

Maret W, Sandstead HH. Zinc requirements and the risks and benefits of zinc supplementation. J Trace Elem Med Biol. 2006;20(1):3–18.

Martel JL, Kerndt CC, Doshi H, Franklin DS. Vitamin B1 (thiamine). StatPearls [Internet]: StatPearls Publishing; 2021.

Martel JL, Kerndt CC, Doshi H, Franklin DS. Vitamin B1 (Thiamine). StatPearls. Treasure Island: StatPearls Publishing; 2023.

Mathur P, Pillai R. Overnutrition: Current scenario & combat strategies. Indian J Med Res. 2019;149(6):695.

Mente A, O'Donnell M, Yusuf S. Sodium intake and health: what should we recommend based on the current evidence? Nutrients. 2021;13(9)

Minich WB. Selenium metabolism and biosynthesis of selenoproteins in the human body. Biochemistry (Mosc). 2022;87(Suppl 1):S168–S77.

Morita R, Yamamoto I, Takada M, Ohnaka Y, Yuu I. Hypervitaminosis D. Nihon Rinsho. 1993;51(4):984–8.

Mozaffarian D, Hao T, Rimm EB, Willett WC, Hu FB. Changes in diet and lifestyle and long-term weight gain in women and men. N Engl J Med. 2011;364(25):2392–404.

Murphy BF. Hypervitaminosis E. JAMA. 1974;227(12):1381.

Nier A, Brandt A, Conzelmann IB, Özel Y, Bergheim I. Non-alcoholic fatty liver disease in overweight children: role of fructose intake and dietary pattern. Nutrients. 2018;10(9):1329.

Novelli GD. Metabolic functions of pantothenic acid. Physiol Rev. 1953;33(4):525–43.

Odhaib SA, Mansour AA, Haddad NS. How biotin induces misleading results in thyroid bioassays: case series. Cureus. 2019;11(5):e4727.

Olson RE. The function and metabolism of vitamin K. Annu Rev Nutr. 1984;4(1):281–337.

Peacock M. Calcium metabolism in health and disease. Clin J Am Soc Nephrol. 2010;5(Suppl 1):S23–30.

Penagini F, Mameli C, Fabiano V, Brunetti D, Dilillo D, Zuccotti GV. Dietary intakes and nutritional issues in neurologically impaired children. Nutrients. 2015;7(11):9400–15.

Plum LM, Rink L, Haase H. The essential toxin: impact of zinc on human health. Int J Environ Res Public Health. 2010;7(4):1342–65.

Popa DS, Bigman G, Rusu ME. The role of vitamin K in humans: implication in aging and age-associated diseases. Antioxidants (Basel). 2021;10(4)

Powe D, Jagger C, Kleinjan A, Carney AS, Jenkins D, Jones N. 'Entopy': localized mucosal allergic disease in the absence of systemic responses for atopy. Clin Exp Allergy. 2003;33(10):1374–9.

Price S. Understanding the importance to health of a balanced diet. Nurs Times. 2005;101(1):30–1.

Rajagopalan K. Molybdenum: an essential trace element in human nutrition. Annu Rev Nutr. 1988;8(1):401–27.

Rivlin RS, Pinto JT. Riboflavin (vitamin B2). Handbook of vitamins. 3rd ed. New York: Marcel Dekker; 2001. p. 255–73.

Royer A, Sharman T. Copper toxicity. 2020.

Rude R, Singer F. Magnesium deficiency and excess. Annu Rev Med. 1981;32(1):245–59.

Ryan-Harshman M, Aldoori W. New dietary reference intakes for macronutrients and fibre. Can Fam Physician. 2006;52(2):177–9.

Schellack G, Harirari P, Schellack N. B-complex vitamin deficiency and supplementation. SA Pharm J. 2016;83(4):14–9.

Schellack G, Harirari P, Schellack N. Vitamin B-complex deficiency, supplementation and management. SA Pharm J. 2019;86(3):23–9.

Siepmann T, Roofeh J, Kiefer FW, Edelson DG. Hypogonadism and erectile dysfunction associated with soy product consumption. Nutrition. 2011;27(7–8):859–62.

Skoracka K, Eder P, Łykowska-Szuber L, Dobrowolska A, Krela-Kaźmierczak I. Diet and nutritional factors in male (in) fertility—underestimated factors. J Clin Med. 2020;9(5):1400.

Sunkara V, Pelkowski TD, Dreyfus D, Satoskar A. Acute kidney disease due to excessive vitamin C ingestion and remote roux-en-y gastric bypass surgery superimposed on CKD. Am J Kidney Dis. 2015;66(4):721–4.

Tahiliani AG, Beinlich CJ. Pantothenic acid in health and disease. Vitam Horm. 1991;46:165–228.

Ten Berge D, Muller W, Beishuizen A, Cornet AD, Slingerland R, Krabbe J. Significant interference on specific point-of-care glucose measurements due to high dose of intravenous vitamin C therapy in critically ill patients. Clin Chem Lab Med. 2021;59(5):e197–e9.

Trichopoulou A, Soukara S, Vasilopoulou E. Traditional foods: a science and society perspective. Trends Food Sci Technol. 2007;18(8):420–7.

Tucker JM, Townsend DM. Alpha-tocopherol: roles in prevention and therapy of human disease. Biomed Pharmacother. 2005;59(7):380–7.

Turck D, Bresson JL, Burlingame B, Dean T, Fairweather-Tait S, Heinonen M, et al. Dietary reference values for riboflavin. EFSA J. 2017;15(8):e04919.

Verbeke W. Consumer acceptance of functional foods: socio-demographic, cognitive and attitudinal determinants. Food Qual Prefer. 2005;16(1):45–57.

Vermeer CV. Vitamin K: the effect on health beyond coagulation–an overview. Food Nutr Res. 2012;56(1):5329.

Vyskočil A, Viau C. Assessment of molybdenum toxicity in humans. J Appl Toxicol. 1999;19(3):185–92.

Weber D, Grune T. The contribution of β-carotene to vitamin A supply of humans. Mol Nutr Food Res. 2012;56(2):251–8.

Wessling-Resnick M. Iron: basic nutritional aspects. Molecular, genetic, and nutritional aspects of major and trace minerals. Elsevier; 2017. p. 161–73.

Wien TN, Pike E, Wisløff T, Staff A, Smeland S, Klemp M. Cancer risk with folic acid supplements: a systematic review and meta-analysis. BMJ Open. 2012;2(1):e000653.

Wilson GT, Wilfley DE, Agras WS, Bryson SW. Psychological treatments of binge eating disorder. Arch Gen Psychiatry. 2010;67(1):94–101.

Wolfson JA, Ramsing R, Richardson CR, Palmer A. Barriers to healthy food access: associations with household income and cooking behavior. Prev Med Rep. 2019;13:298–305.

Yang H, Jia X. Safety evaluation of se-methylselenocysteine as nutritional selenium supplement: acute toxicity, genotoxicity and subchronic toxicity. Regul Toxicol Pharmacol. 2014;70(3):720–7.

Zhou S-S, Zhou Y. Excess vitamin intake: an unrecognized risk factor for obesity. World J Diabetes. 2014;5(1):1.

Index

S. M. Jafari et al. (eds.), *Nutraceuticals: A New Millennium Approach for Preventive Medicine*, Food Bioactive Ingredients,
https://doi.org/10.1007/978-3-031-76481-3

MIX
Papier aus verantwortungsvollen Quellen
Paper from responsible sources
FSC® C105338

If you have any concerns about our products,
you can contact us on
ProductSafety@springernature.com

In case Publisher is established outside the EU,
the EU authorized representative is:
Springer Nature Customer Service Center GmbH
Europaplatz 3, 69115 Heidelberg, Germany

Printed by Libri Plureos GmbH
in Hamburg, Germany